ISNM
International Series of Numerical Mathematics
Vol. 132

New Developments in Approximation Theory

2nd International Dortmund Meeting (IDoMAT) '98, Germany,
February 23–27, 1998

Edited by

M.W. Müller
M.D. Buhmann
D.H. Mache
M. Felten

Springer Basel AG

Editors:

Manfred W. Müller
Martin D. Buhmann
Detlef H. Mache
Michael Felten
Lehrstuhl VIII für Mathematik
Universität Dortmund
Vogelpothsweg 87
44221 Dortmund
Germany

current address of Detlef H. Mache:

Ludwig-Maximilians-Universität München
Numerische Analysis
Theresienstrasse 39
80333 München
Germany

1991 Mathematics Subject Classification 65Dxx

A CIP catalogue record for this book is available from the Library of Congress, Washington D.C., USA

Deutsche Bibliothek Cataloging-in-Publication Data
New developments in approximation theory / 2nd International
Dortmund Meeting (IDoMAT) '98, Germany, February 23–27,
1998 / ed. by M. W. Müller ... – Springer Basel AG, 1999
   (International series of numerical mathematics ; Vol. 132)
   ISBN 978-3-0348-9733-4   ISBN 978-3-0348-8696-3 (eBook)
   DOI 10.1007/978-3-0348-8696-3

© 1999 Springer Basel AG
Originally published by Birkhäuser Verlag in 1999
Softcover reprint of the hardcover 1st edition 1999
Printed on acid-free paper produced of chlorine-free pulp. TCF ∞
Cover design: Heinz Hiltbrunner, Basel

ISBN 978-3-0348-9733-4

9 8 7 6 5 4 3 2 1

Preface

This book contains the refereed papers which were presented at the second International Dortmund Meeting on Approximation Theory (IDoMAT 98) at Haus Bommerholz, the conference center of Dortmund University, during the week of February 23-27,1998. At this conference 52 researchers and specialists from Bulgaria, China, France, Great Britain, Hungary, Israel, Italy, Roumania, South Africa and Germany participated and described new developments in the fields of univariate and multivariate approximation theory. The papers cover topics such as radial basis functions, bivariate spline interpolation, multilevel interpolation, multivariate triangular Bernstein bases, Padé approximation, comonotone polynomial approximation, weighted and unweighted polynomial approximation, adaptive approximation, approximation operators of binomial type, quasi interpolants, generalized convexity and Peano kernel techniques. This research has applications in areas such as computer aided geometric design, as applied in engineering and medical technology (e.g. computerised tomography).

Again this international conference was wholly organized by the Dortmund Lehrstuhl VIII for Approximation Theory. The organizers attached great importance to inviting not only well-known researchers but also young talented mathematicians. IDoMAT 98 gave an excellent opportunity for talks and discussions between researchers from different fields of Approximation Theory. In this way the conference was characterized by a warm and cordial atmosphere. The success of IDoMAT 98 was above all due to everyone of the participants.

Our thanks go to the referees for their prompt cooperation as well as their accurate work, so that we can present in this volume an interesting impression of the good quality of our meeting.

Finally we would like to thank Deutsche Forschungsgemeinschaft for the financial support and Birkhäuser-Verlag for agreeing to publish the proceedings in the ISNM series.

The editors

March 1999

Contents

International Series of Numerical Mathematics
Vol. 132, © 1999 Birkhäuser Verlag Basel/Switzerland

Feller Semigroups, Bernstein type Operators and Generalized Convexity Associated with Positive Projections

Dedicated to Professor Giuseppe Mastroianni on the occasion of his 60th birthday

Francesco Altomare* – Ioan Rasa

Abstract

We study the majorizing approximation properties of both Bernstein type operators and the corresponding Feller semigroups associated with a positive projection acting on the space of all continuous functions defined on a convex compact set.

The relationship between these properties and some generalized form of convexity is investigated as well.

1 Introduction

Starting from a positive linear projection acting on the space of all continuous functions defined on a convex compact subset of a locally convex space, it is possible to construct several positive linear approximation methods such as Lototsky-Schnabl operators and Bernstein-Schnabl operators.

Under suitable assumptions, these operators determine a Feller semigroup whose generator $(A, D(A))$ can be described in a core of its domain. Indeed, in the finite dimensional case, $(A, D(A))$ is the closure of a degenerate second-order elliptic differential operator.

A detailed analysis of these operators, of the associated Feller semigroups and of their corresponding Markov processes can be found in the monograph [4].

During the last years, for both theoretical and practical questions, some attention was also devoted to investigating the majorizing approximation properties

---

*The contribution of the first author is due to work done under the auspices of the G.N.A.F.A (C.N.R) and partially supported by Ministero dell' Università R.S.T. (Quote 60% and 40%). The work was carried out in Oct. 1997 while the second author was Visiting Professor at the University of Bari.

of Bernstein-Lototsky-Schnabl operators and of their corresponding Feller semi-group, ([15], [16]); see also [2]).

These properties have interesting applications in the study of the qualitative properties of the solutions of the initial-boundary problems associated with the operator $(A, D(A))$ and of the corresponding probability transition function.

Moreover, it turns out that these properties are strongly related to a generalized form of convexity, whose study seems to be of independent interest.

In this paper we survey some known results on this topic and in addition, we present some new advances together with some open problems.

2   Differential operators associated with positive projections

Let K be a convex compact subset of $\mathbf{R}^p$, $p \geq 1$, having non-empty interior. We shall denote by $\mathcal{C}(K)$ the Banach lattice of all real-valued continuous functions on K endowed with the sup-norm and the natural order.

The symbol $\mathcal{C}^2(K)$ stands for the subspace of all functions $f \in \mathcal{C}(K)$ which are two times continuously differentiable in the interior $\operatorname{int} K$ of K and whose partial derivatives of order ≤ 2 can be continuously extended to K.

For every $u \in \mathcal{C}^2(K)$ and $i, j = 1, \ldots, p$ we shall continue to denote by $\frac{\partial u}{\partial x_i}$ and $\frac{\partial^2 u}{\partial x_i \partial x_j}$ the continuous extensions to K of the partial derivatives $\frac{\partial u}{\partial x_i}$ and $\frac{\partial^2 u}{\partial x_i \partial x_j}$ defined on $\operatorname{int} K$.

For every $j = 1, \ldots, p$ we shall denote by $\operatorname{pr}_j \in \mathcal{C}(K)$ the continuous function on K defined by

$$\operatorname{pr}_j(x) = x_j \qquad \text{for each } x = (x_i)_{1 \leq i \leq p} \in K. \tag{1}$$

As a starting point for the construction of the differential operators in question we fix a positive linear projection $T: \mathcal{C}(K) \to \mathcal{C}(K)$, i.e., T is a positive linear operator such that $T \circ T = T$.

We also assume that T is not trivial, i.e., T is not the identity operator on $\mathcal{C}(K)$, and moreover, that

$$T(\mathbf{1}) = \mathbf{1} \tag{2}$$

and

$$T(\operatorname{pr}_j) = \operatorname{pr}_j \qquad (j = 1, \ldots, p), \tag{3}$$

(here $\mathbf{1}$ denotes the function having constant value 1).

Furthermore, set

$$H_T := T(\mathcal{C}(K)) = \{h \in \mathcal{C}(K) \mid Th = h\} \tag{4}$$

and assume that for every $z \in K$, $\alpha \in [0, 1]$ and $h \in H_T$ we have

$$h_{z,\alpha} \in H_T, \tag{5}$$

where
$$h_{z,\alpha}(x) := h(\alpha x + (1-\alpha)z) \qquad (x \in K). \tag{6}$$

If we denote by $\partial_e K$ and $\partial_{H_T} K$ the set of extreme points of K and the Choquet boundary of K with respect to H_T (see, e.g., [4, sect. 2.6]) we have

$$\partial_e K \subset \partial_{H_T} K \subset \partial K \tag{7}$$

(see [4, sect. 6.1]).

Moreover, we know ([4, Prop. 3.3.1 and Prop. 3.3.2]) that

$$\begin{aligned}\partial_{H_T} K &= \{x \in K \mid T(f)(x) = f(x) \text{ for every } f \in \mathcal{C}(K)\} \\ &= \{x \in K \mid T(e)(x) = e(x)\}, \end{aligned} \tag{8}$$

where

$$e(x) := \|x\|^2 = \sum_{i=1}^{p} x_i^2 \qquad (x = (x_i)_{1 \le i \le p} \in K). \tag{9}$$

In particular, from (8) it follows that for every $f, g \in \mathcal{C}(K)$ we have

$$T(f) = T(g) \qquad \text{provided } f = g \text{ on } \partial_{H_T} K. \tag{10}$$

We are now in the position to introduce the differential operator $W_T : \mathcal{C}^2(K) \to \mathcal{C}(K)$ which is defined by

$$W_T(u)(x) := \frac{1}{2} \sum_{i,j=1}^{p} a_{ij}(x) \frac{\partial^2 u(x)}{\partial x_i \partial x_j} \tag{11}$$

for every $u \in \mathcal{C}^2(K)$ and $x \in K$, where for each $i, j = 1, \ldots, p$

$$a_{ij} := T(\mathrm{pr}_i \mathrm{pr}_j) - \mathrm{pr}_i \mathrm{pr}_j. \tag{12}$$

The operator W_T is elliptic and degenerates on $\partial_{H_T} K$ (in particular, on $\partial_e K$), i.e.

$$\sum_{i,j=1}^{p} a_{ij}(x)\, \xi_i \xi_j \ge 0 \qquad (x \in K, \ (\xi_1, \ldots, \xi_p) \in \mathbf{R}^p), \tag{13}$$

and
$$W_T(u)(x) = 0 \qquad \text{for each } x \in \partial_{H_T} K. \tag{14}$$

The operator W_T will be called *the elliptic second order differential operator associated with the projection T*.

Here we indicate some (fundamental) examples (for more details see [4, Ch.6], [16], [17]).

Example 2.1 *1) Let K be a simplex of $\mathbf{R}^p$ with vertices $v_0, \ldots, v_p$ $(p \geq 1)$. If $x \in K$, then $x = \sum_{h=0}^{p} b_h(x)v_h$, where $b_h(x) \geq 0$ $(h = 0, \ldots, p)$ and $\sum_{h=0}^{p} b_h(x) = 1$.*

Consider the positive linear projection defined by

$$T(f)(x) := \sum_{h=0}^{p} f(v_h)b_h(x) \tag{15}$$

for every $f \in C(K)$ and $x \in K$.

In fact, $T(f)$ is the unique continuous affine function on K which coincides with f on $\{v_0, \ldots, v_p\} = \partial_e K$.

The projection T is called the canonical projection associated with K.

In this case H_T is the space $A(K)$ of all continuous affine functions on K. Moreover

$$\partial_{H_T} K = \partial_e K \tag{16}$$

and the coefficients a_{ij} of the differential operator W_T are given by

$$a_{ij}(x) = \sum_{h,k=0}^{p} b_h(x)b_k(x)pr_i(v_h)(pr_j(v_h) - pr_j(v_k)). \tag{17}$$

In particular, if K is the standard simplex of $\mathbf{R}^p$

$$K_p := \\ \{(x_1, \ldots, x_p) \in \mathbf{R}^p \mid x_1 + \cdots + x_p \leq 1, \; x_i \geq 0 \; (1 \leq i \leq p)\}, \tag{18}$$

(thus $v_0 = (0, \ldots, 0)$, $v_1 = (1, 0, \ldots, 0)$, $\ldots$, $v_p = (0, \ldots, 0, 1)$) then

$$W_T(u)(x) = \frac{1}{2} \sum_{i=1}^{p} x_i(1 - x_i) \frac{\partial^2 u(x)}{\partial x_i^2} - \sum_{1 \leq i < j \leq p} x_i x_j \frac{\partial^2 u(x)}{\partial x_i \partial x_j} \tag{19}$$

($u \in C^2(K_p)$, $x \in K_p$), where, when $p = 1$ the second sum is by convention $= 0$.

2) Assume that $K = \prod_{i=1}^{p}[a_i, b_i]$, where $a_i, b_i \in R$, $a_i < b_i$ $(1 \leq i \leq p)$. For every $f \in C(K)$ denote by $S(f)$ the unique continuous multiaffine function on K which coincides with f on $\partial_e K$. The operator S is a positive linear projection which verifies (2), (3), and (5).

In this case H_S is the space of all continuous multiaffine functions on K and

$$\partial_{H_S} K = \partial_e K. \tag{20}$$

In order to compute the coefficients a_{ij} of the differential operator W_S we point out that for every $f \in \mathcal{C}(K)$ and $x = (x_1, \ldots, x_p) \in K$ we have

$$S(f)(x) = \frac{1}{(b_1 - a_1) \cdots (b_p - a_p)} \times$$

$$\times \sum_{h_1, \ldots, h_p = 0}^{1} f(a_1 + \delta_{h_1 1}(b_1 - a_1), \ldots, a_p + \delta_{h_p 1}(b_p - a_p)) \times \tag{21}$$

$$\times (x_1 - a_1)^{h_1}(b_1 - x_1)^{1 - h_1} \cdots (x_p - a_p)^{h_p}(b_p - x_p)^{1 - h_p},$$

(where δ_{ij} stands for the Kronecker symbol of indices i and j).
From (21) we obtain that

$$a_{ij} = \begin{cases} 0, & \text{if } i \neq j, \\ x_i(a_i + b_i - x_i) - a_i b_i, & \text{if } i = j, \end{cases} \tag{22}$$

and hence

$$W_S(u)(x) = \frac{1}{2} \sum_{i=1}^{p} (x_i(a_i + b_i - x_i) - a_i b_i) \frac{\partial^2 u(x)}{\partial x_i^2}. \tag{23}$$

In the case when K is the hypercube $[0,1]^p$ of $\mathbf{R}^p$, if we denote by S_p the corresponding positive projection, then we obtain

$$W_{S_p}(u)(x) = \frac{1}{2} \sum_{i=1}^{p} x_i(1 - x_i) \frac{\partial^2 u(x)}{\partial x_i^2}. \tag{24}$$

3) Let K be an arbitrary convex compact subset of $\mathbf{R}^p$ having non-empty interior and consider a symmetric matrix $(c_{ij})_{1 \leq i,j \leq p}$ of Hölder continuous functions on int K with exponent $\beta \in]0, 1[$.
Consider the differential operator

$$L(u)(x) = \sum_{i,j=1}^{p} c_{ij}(x) \frac{\partial^2 u(x)}{\partial x_i \partial x_j} \quad (u \in \mathcal{C}^2(\text{int } K), \ x \in K) \tag{25}$$

and assume that L is strictly elliptic, i.e., for every $x \in$ int K the matrix $(c_{ij}(x))$ is positive definite and, if $\sigma(x)$ denotes its smallest eigenvalue, then we have $\sigma(x) \geq \sigma_0 > 0$ for some $\sigma_0 \in \mathbf{R}$.
Denote by $T_L : \mathcal{C}(K) \to \mathcal{C}(K)$ the Dirichlet operator associated with L. Thus, for every $f \in \mathcal{C}(K)$, $T_L(f)$ denotes the unique solution of the Dirichlet problem

$$\begin{cases} L(u) = 0 & \text{on int } K, \ u \in \mathcal{C}(K) \cap \mathcal{C}^2(\text{int } K), \\ u = f & \text{on } \partial K. \end{cases} \tag{26}$$

The operator T_L is a positive projection, the subspace H_{T_L} coincides with the space of all $u \in \mathcal{C}(K) \cap \mathcal{C}^2(\text{int } K)$ which are L-harmonic on int K (i.e., $L(u) = 0$ on int K) and

$$\partial_{H_{T_L}} K = \partial K. \tag{27}$$

In the particular case when ∂K is an ellipsoid defined by a quadratic form

$$Q(x) := \sum_{i,j=1}^{p} q_{ij} x_i x_j \ (x \in \mathbf{R}^p), \ with \ center \ 0, \ i.e.,$$

$$K := \{ x \in \mathbf{R}^p \mid Q(x) \le 1 \}, \tag{28}$$

and if, moreover, the functions c_{ij} $(1 \le i,j \le p)$ are constant and satisfy the relation

$$\sum_{i,j=1}^{p} q_{ij} c_{ij} = 1 \tag{29}$$

then for every $u \in \mathcal{C}^2(K)$ and $x \in K$, we have

$$W_{T_L}(u)(x) = \frac{1 - Q(x)}{2} L(u)(x). \tag{30}$$

For further examples see [5], where the reader can find a complete description of those convex compact subsets of $\mathbf{R}^2$ for which there exist positive projections (all of which are classified as well).

3   Feller semigroups generated by multiplicative perturbations of W_T

First, we introduce the positive linear operators in terms of which we shall represent the Feller semigroups quoted in the title of this section.

We shall keep the same notation as in Section 2.

Fix $\lambda \in \mathcal{C}(K)$, $0 \le \lambda \le 1$, and for each $x \in K$ denote by $v_{x,\lambda}$ the probability Radon measure on K defined by

$$v_{x,\lambda}(f) := \lambda(x)T(f)(x) + (1 - \lambda(x))f(x) \qquad (f \in \mathcal{C}(K)). \tag{31}$$

For a given $n \in \mathbf{N}$, $n \ge 1$, the n-th Lototsky–Schnabl operator associated with the projection T and the function λ is the positive linear operator $L_{n,\lambda} : \mathcal{C}(K) \to \mathcal{C}(K)$ defined by

$$L_{n,\lambda}(f)(x) := \int_K \!\! \cdots \!\! \int_K f(\frac{x_1 + \cdots + x_n}{n}) \, dv_{x,\lambda}(x_1) \cdots dv_{x,\lambda}(x_n) \tag{32}$$

$(f \in \mathcal{C}(K)$, $x \in K)$.

In the case $\lambda = \mathbf{1}$, then $L_{n,\lambda}$ is simply denoted by B_n and is called the n-th Bernstein–Schnabl operator associated with T.

By knowing the analytic expression of T (as for instance, in Examples 2.1), it is possible to describe more explicitly the operators $L_{n,\lambda}$ ([4, Sect. 6.1]).

Furthermore, among the several interesting properties of these operators we quote the following ones:

$$L_{n,\lambda}(h) = h \qquad (h \in H_T,\ n \geq 1); \tag{33}$$

$$L_{n,\lambda}(f)(x) = f(x) \quad (f \in \mathcal{C}(K),\ x \in \partial_{H_T} K,\ n \geq 1); \tag{34}$$

$$\lim_{n \to \infty} L_{n,\lambda}(f) = f \quad uniformly\ on\ K \quad (f \in \mathcal{C}(K)); \tag{35}$$

$$\lim_{n \to \infty} n\,(L_{n,\lambda}(u) - u) = \lambda W_T(u)\ uniformly\ on\ K \quad (u \in \mathcal{C}^2(K)). \tag{36}$$

Some estimates of the rate of convergence of the operators $L_{n,\lambda}$ are also available. More precisely, for every $n \geq 1$ we have

$$|L_{n,\lambda}(f)(x) - f(x)| \leq (1 + \lambda(x))\,\omega(f, \sqrt{\tfrac{1}{n}}), \quad (f \in \mathcal{C}(K),\ x \in K) \tag{37}$$

$$\|L_{n,\lambda}(f) - f\| \leq M\sqrt{\tfrac{1}{n}\|\lambda(T(e) - e)\|}, \quad (f \in Lip_M 1). \tag{38}$$

Here, $\omega(f, \cdot)$ denotes the ordinary modulus of continuity, $Lip_M 1$ is the class of all functions $f \in \mathcal{C}(K)$ satisfying the property $|f(x) - f(y)| \leq M\|x - y\|,\ (x, y \in K)$ and the function e is defined by (9).

When $K = [0, 1]$ and T is defined by (15), then we have the following further estimates:

$$|L_{n,\lambda}(f)(x) - f(x)| \leq K\omega_2\left(f, \sqrt{\frac{x(1-x)\lambda(x)}{n}}\right), \tag{39}$$

$(f \in \mathcal{C}([0, 1]),\ x \in [0, 1])$

$$\|L_{n,\lambda}(f) - f\| \leq C\omega_\varphi^2\left(f, \sqrt{\tfrac{1}{n}}\right) \quad (f \in \mathcal{C}([0, 1])), \tag{40}$$

where the constant K is independent of f, x and n, the constant C is independent of f and n, $\omega_2(f, \cdot)$ denotes the second modulus of smoothness, and $\omega_\varphi^2(f, \cdot)$ is the second Ditzian-Totik modulus of smoothness with step-weight function $\varphi(x) := \sqrt{x(1-x)},\ (x \in [0, 1])$.

We recall that a Feller semigroup on $\mathcal{C}(K)$ is a C_0–semigroup $(T(t))_{t \geq 0}$ of positive linear operators on $\mathcal{C}(K)$ satisfying $T(t)\mathbf{1} = \mathbf{1}\ (t \geq 0)$ (for more details on C_0–semigroups see, e.g., [14]).

Given a closed operator $A: D(A) \to \mathcal{C}(K)$ defined on a dense subspace $D(A)$ of $\mathcal{C}(K)$, a *core* for A is a subspace D_0 of $D(A)$ which is dense in $D(A)$ for the graph norm

$$\|u\|_A = \|u\| + \|Au\| \qquad (u \in D(A)). \tag{41}$$

This also means that A is determined by its restriction to D_0.

In the sequel, for each $m \geq 1$, we shall denote by $A_m(K)$ the subspace of the restrictions to K of all polynomials of degree $\leq m$.

We shall denote by $A(K)$ $(= A_1(K))$ the space of all continuous affine functions on K. Moreover, if $k \geq 1$, then the symbol $L_{n,\lambda}^k$ stands for the power (or the iterate) of $L_{n,\lambda}$ of order k.

We are in the position to state one of the main results of the theory.

Theorem 3.1 *([3]) Suppose that the function* $\lambda \in \mathcal{C}(K)$, $0 \leq \lambda \leq 1$, *is strictly positive, i.e.,* $\lambda(x) > 0$ *for each* $x \in K$. *Moreover, assume that*

$$T(A_2(K)) \subset A(K) \tag{42}$$

or, alternatively,

$$T(A_m(K)) \subset A_m(K), \qquad \text{for every } m \geq 1. \tag{43}$$

Then the following statements hold true:

(i) *the operator* $(\lambda W_T, \mathcal{C}^2(K))$ *is closable and its closure* $(A_\lambda, D(A_\lambda))$ *generates a Feller semigroup* $(T_\lambda(t))_{t \geq 0}$ *on* $\mathcal{C}(K)$.

(ii) *for each* $t \geq 0$, *for every sequence* $(k(n))_{n \geq 1}$ *of positive integers satisfying* $\lim\limits_{n \to \infty} \dfrac{k(n)}{n} = t$ *(in particular for* $k(n) = [nt]$, *the integer part of* nt*) and for every* $f \in \mathcal{C}(K)$ *we have*

$$T_\lambda(t)f = \lim_{n \to \infty} L_{n,\lambda}^{k(n)}(f) \qquad \text{uniformly on } K. \tag{44}$$

(iii) $\mathcal{C}^2(K)$ *is a core for* A_λ.

(iv) *If* $(A, D(A))$ *denotes the operator* $(A_1, D(A_1))$, *then we have*

$$D(A_\lambda) = D(A) \quad \text{and} \quad A_\lambda = \lambda A. \tag{45}$$

Finally note that from (36) it also follows that

$$A_\lambda u = \lim_{n \to \infty} n(L_{n,\lambda}(u) - u) \qquad (u \in \mathcal{C}^2(K)). \tag{46}$$

By the representation formula (44) and by means of a careful analysis of the operators $L_{n,\lambda}$ we hope to obtain qualitative information about the semigroup $T_\lambda(t))_{t \geq 0}$ and, consequently, about the solutions of the Cauchy problems

$$\begin{cases} \dfrac{\partial u(x,t)}{\partial t} = A_\lambda(u(\cdot,t))(x), & (x \in K, \ t > 0) \\[2mm] u(x,0) = u_0(x), & u_0 \in D(A_\lambda), \end{cases} \tag{47}$$

which, as it is well-known, are given by

$$u(x,t) = T_\lambda(t)(u_0)(x) \qquad (x \in K, \ t > 0). \tag{48}$$

Note also that the boundary conditions for problem (47) are incorporated into the domain $D(A_\lambda)$. They include the so-called Wentcel's boundary conditions

$$A_\lambda u = 0 \quad \text{on } \partial_{H_T} K \quad (u \in D(A_\lambda)) \tag{49}$$

which follow from (44) and (34).

Typical problems that have been investigated via formula (44) both in the general context of this section as well as in particular ones are:

(1) Asymptotic behaviour of $(T_\lambda(t))_{t \geq 0}$; in fact it has been shown that

$$\lim_{t \to \infty} T_\lambda(t)f = T(f) \quad \text{for each} \quad f \in \mathcal{C}(K) \tag{50}$$

(see [3]).

(2) Investigation of those closed subsets M of $\mathcal{C}(K)$ satisfying

$$L_{n,\lambda}(M) \subset M \qquad \text{for every } n \geq 1.$$

In this case we also obtain that $T_\lambda(t)(M) \subset M$ ($t \geq 0$) (regularity results for problem (47)).

Examples include spaces of Hölder continuous functions ([15], [16]), cones of (axially) convex functions ([15], [20]), polyhedral convex functions ([20]), subharmonic functions ([6], [17], [20]).

(3) Determination of the Favard class of the semigroup $(T_\lambda(t))_{t \geq 0}$, i.e., the subspace of all $f \in \mathcal{C}(K)$ such that $\sup\limits_{t > 0} \|\frac{1}{t}(T_\lambda(t)f - f)\| < +\infty$.

Often, this can be done by knowing the saturation class of the sequence $(L_{n,\lambda})_{n \geq 1}$ which is defined as the subspace of all $f \in \mathcal{C}(K)$ such that

$$\sup_{n \geq 1} n \|L_{n,\lambda}(f) - f\| < +\infty.$$

(see [6], [11], [12], [13], [17], [20]).

Note that from (50) it follows that the projection T can be recovered from the semigroup $(T_\lambda(t))_{t \geq 0}$ for some $\lambda \in \mathcal{C}(K)$.

In fact, T can be also recovered from the generator $(A, D(A))$ in the sense that for every $f \in \mathcal{C}(K)$, $T(f)$ is the unique solution of the problem

$$\begin{cases} Au = 0, & u \in D(A), \\ u = f, & \text{on } \partial_{H_T} K, \end{cases} \tag{51}$$

([15, Th. 5.5]).

We finally mention the following new identity, which in fact contains the converse of (49):

$$\partial_{H_T} K \;\; = \{x \in K \mid A_\lambda u(x) = 0 \text{ for every } u \in D(A_\lambda)\}$$
$$= \{x \in K \mid Au(x) = 0 \text{ for every } u \in D(A)\}. \tag{52}$$

Indeed, if we set $\operatorname{Ker} T := \{f \in C(K) \mid Tf = 0\}$, then by virtue of (50) and Theorem 5.1 of [10] we have $\operatorname{Ker} T = \overline{A(D(A))}$. So, on account of (8), for every $x \in K$ we get

$$(x \in \partial_{H_T} K) \iff \Big((Tf - f)(x) = 0 \text{ for every } f \in C(K) \Big) \iff$$

$$\iff (g(x) = 0 \text{ for every } g \in \operatorname{Ker} T)$$
$$\iff (g(x) = 0 \text{ for every } g \in \overline{A(D(A))})$$
$$\iff (Au(x) = 0 \text{ for every } u \in D(A)).$$

Another field where qualitative properties of $(T_\lambda(t))_{t \geq 0}$ can be usefully applied is that of Markov processes.

Indeed, from general results (see, e.g., [21, Th. 9.2.6]) it follows that $(T_\lambda(t))_{t \geq 0}$ is the transition semigroup of a right-continuous normal Markov process

$$(\Omega, \mathcal{U}, (P^x)_{x \in K}, (Z_t)_{t \geq 0}) \tag{53}$$

with state space K, whose paths have left-hand limits on $[0, \zeta[$ almost surely, where $\zeta : \Omega \to \mathbf{R}$ is the lifetime of the process defined by

$$\zeta(\omega) := \inf\{t \geq 0 \mid Z_t(\omega) \in \partial K\}, \quad \omega \in \Omega. \tag{54}$$

Intuitively, we may think of a particle which moves in K after a random experiment $\omega \in \Omega$. Then, for every $t \geq 0$ and $x \in K$ and for every Borel set B of K, $Z_t(\omega)$ expresses the position (in K) of the particle at the time t and $P_t(x, B) := P^x\{Z_t \in B\}$ is the probability that a particle starting at position x is in B at time t.

Finally, the Borel measure $P_t(x, \cdot)$ is exactly the measure which corresponds to the Radon measure

$$\mu_{x,t}(f) := T_\lambda(t)(f)(x) \qquad (f \in C(K)), \tag{55}$$

via the Riesz representation theorem.

In other words, for every $f \in C(K)$ we have

$$T_\lambda(t)(f)(x) = \int_\Omega f \circ Z_t \, dP^x =: E_x(f(Z_t)). \tag{56}$$

From (44) it follows that the measure $P_t(x, \cdot)$ can be represented as a weak limit of a sequence of suitable measures (see [4, pp. 443-444, 462-465]). Moreover, formula (50) implies that

$$\lim_{t \to +\infty} P_t(x, \cdot) = \tilde{\mu}_x \quad \text{weakly}, \tag{57}$$

where $\tilde{\mu}_x$ denotes the Borel measure on K corresponding to $\mu_x := v_{x,1}$ (see (31)).

Thus, for every $f \in C(K)$

$$\mu_x(f) = T(f)(x). \tag{58}$$

By using (44) and (45) it is possible to obtain some information about the (vector) expected value and the variance of Z_t (with respect to P^x) and the asymptotic behaviour of $P_t(x, \cdot)$ as $t \to \infty$ (see [5] for details).

Remark 3.2 *The results we presented here can be extended to the case where K is a convex compact subset of a (not necessarily finite dimensional) locally convex space.*

In this case, the subset $\{1, pr_1, \ldots, pr_p\}$ must be replaced by the subspace $A(K)$ of all real-valued affine continuous functions on K, the subspace $C^2(K)$ by the subspace

$$A_\infty(K) := \bigcup_{m=1}^{\infty} A_m(K), \tag{59}$$

where for every $m \geq 1$ $A_m(K)$ denotes the subspace generated by

$$\left\{ \prod_{i=1}^{m} h_i \mid h_i \in A(K), \ i = 1, \ldots, m \right\},$$

and, finally, the operator W_T by the operator $V_T : A_\infty(K) \to C(K)$ defined by

$$V_T\left(\prod_{i=1}^{m} h_i\right) := \begin{cases} 0, & m = 1 \\ T(h_1 h_2) - h_1 h_2, & m = 2 \\ \displaystyle\sum_{1 \leq i < j \leq m} (T(h_i h_j) - h_i h_j) \prod_{\substack{r=1 \\ r \neq i,j}}^{m} h_r, & m \geq 3 \end{cases} \tag{60}$$

In this framework, Theorem 3.1 holds true at least under the assumption (42).

An important (infinite dimensional) case where the theory can be applied, occurs when K is a Bauer simplex (see [4, §1.5]). In this case there is a unique positive projection $T : C(K) \to C(K)$ satisfying the above assumptions and which generalizes the projection (15) (see [4, Cor 1.5.9]).

The results we shall discuss and develop in the next sections will be obtained both in the finite and infinite dimensional case.

We shall be mainly interested in investigating the *semigroup majorizing approximation property*

$$f \leq T_\lambda(t)(f), \quad (f \in C(K), \ t \geq 0). \tag{61}$$

We shall present a survey of the known results on this topic together with some new results and some open problems.

4 The semigroup majorizing approximation property

As we explained above, the main aim of this paper is to discuss property (61).

A possible reason to investigate it rests with the remark that, if $u_0 \in D(A)$ satisfies it, then for the solution $u(x,t)$ of the Cauchy problem (47) we have

$$u_0(x) \leq u(x,t) \quad (t \geq 0, \ x \in K). \tag{62}$$

Moreover, for an arbitrary $f \in C(K)$ satisfying (61) we also have

$$f(x) \leq E_x\left(f(Z_t)\right) \quad (t \geq 0, \ x \in K) \tag{63}$$

which is a useful information about the expected value of $f(Z_t)$.

In any case property (61) implies the following maximum principle. We shall keep the same notation as in Section 3. Thus we fix a convex compact subset K of a (finite or infinite dimensional) locally convex space and we consider a positive linear projection $T : C(K) \to C(K)$ satisfying the assumptions

$$T(h) = h \text{ for every } h \in A(K) \tag{64}$$

together with (5) and (42) or, alternatively, (43) in the finite dimensional case. Finally we fix a strictly positive function $\lambda \in C(K)$ such that $0 < \lambda \leq 1$.

Proposition 4.1 *Assume that a function $f \in C(K)$ verifies property (61). Then $f \leq T(f)$ and*

$$\max_K f = \max_{\partial_{H_T} K} f.$$

Proof. From (50) it follows that $f \leq Tf$. Moreover, since $Tf \in H_T$, there exists $x \in \partial_{H_T} K$ such that $\max_K Tf = Tf(x)$.

Therefore, for every $y \in K$ we obtain

$$f(y) \leq Tf(y) \leq Tf(x) = f(x).$$

■

Note that, taking (44) into account, property (61) will be true if f verifies the following *discrete majorizing approximation property*:

$$f \leq L_{n,\lambda}(f) \quad \text{for every } n \geq 1. \tag{65}$$

Moreover, if $f \in D(A)$, then

$$T_\lambda(t)f - f = \int_0^t T_\lambda(s)A_\lambda f \, ds$$

and

$$\lambda A f = A_\lambda f = \lim_{t \to 0^+} \frac{T_\lambda(t)f - f}{t}.$$

So, property (61) is equivalent to the property

$$Af \geq 0 \quad (f \in D(A)). \tag{66}$$

In the sequel, the functions $f \in D(A)$ satisfying (66) will be simply called *generalized A-subharmonic functions*.

Finally, note that if $f \in C(K)$ satisfies (61), then there exists a sequence $(f_n)_{n \geq 1}$ in $D(A)$ such that $Af_n \geq 0$ for every $n \geq 1$ and $f = \lim_{n \to \infty} f_n$ uniformly on K.

Indeed , for every $t \geq 0$

$$\int_0^t T_\lambda(s)f \, ds \in D(A_\lambda) = D(A)$$

and

$$\lambda A(\int_0^t T_\lambda(s)f \, ds = A_\lambda(\int_0^t T_\lambda(s)f \, ds) = T_\lambda(t)f - f \geq 0;$$

furthermore

$$f = \lim_{t \to 0^+} \frac{1}{t} \int_0^t T_\lambda(s)f \, ds \text{ uniformly on } K.$$

In the next sections we shall discuss properties (65) and (66) separately.

5 The discrete majorizing approximation property and T-convex functions

Although we shall study property (65) in connection with (61), this property seems to be of independent interest. In particular, from it we can deduce a maximum principle which is stronger than the one of Proposition 4.1.

Theorem 5.1 *Let $f \in C(K)$ be a function satisfying (65) and set $M = \max\{f(x)|$ $x \in K\}$. Suppose that λ is not identically 0 on $\{x \in K| \ f(x) = M\}$. Then for each $x_0 \in K$ satisfying $f(x_0) = M$ and $\lambda(x_0) > 0$ we have*

$$f(x) = M \ \text{for every} \ x \in \overline{co}(Supp \ \mu_{x_0})$$

(see(58)). In particular $\max_K f = \max_{\partial_{H_T} K} f$ and, finally,

$$L_{n,\lambda}(f)(x_0) = f(x_0) \quad (n \geq 1).$$

This result was obtained jointly by the present authors (1993) and a proof of it can be found in [4, Th. 6.1.17 and Cor.6.1.18].

It unifies several maximum principles obtained by different and more complicated methods (see, for instance, Chang and Zhang ([7]) for the standard simplex of $\mathbf{R}^2$, Sauer ([19]) for the standard simplex of $\mathbf{R}^p$, Dahmen and Micchelli ([9]) for a finite product of simplices).

Before going further we point out that, because of (35), our property (65) will be satisfied if we require that

$$L_{n+1,\lambda}(f) \leq L_{n,\lambda}(f) \ \text{for every} \ n \geq 1. \tag{67}$$

Also this property has its own interest but we do not discuss it here. For more details we refer to [4, Sect. 6.1].

In order to find a suitable class of functions which satisfy (65) we recall the following definition.

Definition 5.2 *([15]) A function $f \in C(K)$ is said to be T-convex if*

$$f_{z,\alpha} \leq T(f_{z,\alpha}) \ \text{for every} \ z \in K \ \text{and} \ \alpha \in [0,1], \tag{68}$$

where

$$f_{z,\alpha}(x) := f(\alpha x + (1 - \alpha)z) \quad (x \in K). \tag{69}$$

If f is T-convex, we obtain in particular (for $\alpha = 1$) that $f \leq T(f)$.

If $f \in C(K)$ is convex, then by virtue of [4, (1.5.4)] and (2) and (3) we get $f \leq T(f)$. On the other hand every $f_{z,\alpha}$ is convex too and hence $f_{z,\alpha} \leq T(f_{z,\alpha})$.

In other words, *every convex function $f \in C(K)$ is T-convex as well*. In general, the converse is not true.

Among other things, T-convex functions satisfy property (65) as the next result shows.

Theorem 5.3 *([15]) Let $f \in C(K)$ be a T-convex function and $\eta, \lambda \in C(K)$ satisfying $0 \leq \eta \leq \lambda \leq 1$. Then for every $n \geq 1$ we have*

$$L_{n,\eta}(f) \leq L_{n,\lambda}(f)$$

and, in particular,

$$f \leq L_{n,\lambda} \leq B_n(f) \leq T(f). \tag{70}$$

Here, we present some fundamental examples and results for T-convex functions.

Theorem 5.4 *1) ([15]; see also [8], [18]) Let K be a Bauer simplex (in particular a simplex of $\mathbf{R}^p$) and consider the canonical projection T on $C(K)$ (see (15) and Remark 3.2). Then for a given $f \in C(K)$ the following statements are equivalent:*

i) f is T-convex;

ii) f is axially convex, i.e. f is convex on each segment parallel to a segment joining two extreme points of K.

iii) $B_n(f)$ is axially convex for every $n \geq 1$.

Moreover, if i) holds true, then

$$B_{n+1}(f) \leq B_n(f) \quad \text{for every } n \geq 1.$$

Thus, on $K = [0,1]$ the T-convex functions are exactly the convex functions. However, in this case we have a more precise result.

2) ([2],[1]) Let $f \in C([0,1])$. Then the following statements are equivalent:

i) f is convex;

ii) $L_{n+1,\lambda}(f) \leq L_{n,\lambda}(f)$ for every $n \geq 1$ and for any strictly positive $\lambda \in C([0,1])$, $\lambda \leq 1$;

iii) $f \leq L_{n,\lambda}(f)$ for every $n \geq 1$ and for any strictly positive $\lambda \in C([0,1])$, $\lambda \leq 1$;

iv) $f \leq T_\lambda(t)f$ for every $t \geq 0$ and for any strictly positive $\lambda \in C([0,1])$, $\lambda \leq 1$;

v) $L_{n,\lambda}(f)$ is convex for every $n \geq 1$ and for any constant $\lambda \in]0,1]$.

vi) $T_\lambda(t)f$ is convex for every $t \geq 0$ and for any strictly positive $\lambda \in C([0,1])$, $\lambda \leq 1$.

3)([16]) Assume that $K = [0,1]^p$ and consider the projection S_p defined by (21). Then a function $f \in C(K)$ is T-convex if and only if is convex with respect to each variables.

4)([17]; see also [6]) Assume that K is an ellipsoid of $\mathbf{R}^p$ and consider the projection T_L associated with a strictly elliptic differential operator L of the form (25). Then for a given $f \in C(K)$ the following statements are equivalent:

i) f is T-convex;

ii) f is L-subharmonic;

iii) $f \leq L_{n,\lambda}(f)$ for every $n \geq 1$ and for any strictly positive $\lambda \in C(K)$, $\lambda \leq 1$;

iv) $f \leq T_\lambda(t)f$ for every $t \geq 0$ and for any strictly positive $\lambda \in C(K)$, $\lambda \leq 1$;

v) $T_\lambda(t)f$ is L-subharmonic for every $t \geq 0$ and for any strictly positive $\lambda \in C(K)$, $\lambda \leq 1$.

In the next result we present a new characterization of T-convexity for sufficiently smooth functions.

Theorem 5.5 *Assume that* $f = \prod_{i=1}^{m} h_i \in A_\infty(K)$ *for some* $m \geq 2$ *or, respectively, that* $f \in C^2(K)$, *if* $K \subset \mathbf{R}^p$ *and* $\operatorname{int} K \neq \emptyset$. *Then the following statements are equivalent:*

   i) f *is* T-*convex;*

   ii) $\sum_{1 \leq i < j \leq m} (T(h_i h_j) - h_i h_j)(x) \prod_{\substack{r=1 \\ r \neq i,j}}^{m} h_r(z) \geq 0$ *(where for* $m = 2$ *the empty product is, by convention,* $= 1$) *or, in the respective case,*

$$\sum_{i,j=1}^{p} a_{ij}(x) \frac{\partial^2 f(z)}{\partial x_i \partial x_j} \geq 0$$

for every $x, z \in K$ *(see* (12))*.*

   iii) $A(f_{z,\alpha}) \geq 0$ *for every* $z \in K$ *and* $\alpha \in [0,1]$.

Proof. i)$\Rightarrow$ii). For every $x, z \in K$ set

$$
V_z f(x) := \begin{cases} \sum_{1 \leq i < j \leq m} (T(h_i h_j) - h_i h_j)(x) \prod_{\substack{r=1 \\ r \neq i,j}}^{m} h_r(z) & \text{if } f = \prod_{i=1}^{m} h_i, \\[2ex] \dfrac{1}{2} \sum_{i,j=1}^{p} a_{ij}(x) \dfrac{\partial^2 f(z)}{\partial x_i \partial x_j} & \text{if } f \in C^2(K) \end{cases} \tag{71}
$$

Then, by using a proof similar to that of Theorem 1.3 of [16] it is possible to show that

$$V_z f(x) = \lim_{\alpha \to 0^+} \alpha^{-2} (T(f_{z,\alpha}) - f_{z,\alpha})(x) \tag{72}$$

and so statement ii) follows.

   ii) $\Rightarrow$ iii) Because of Theorem 3.1 and formulas (11), (60) and (71) we have that

$$A(f_{z,\alpha})(x) = \alpha^2 V_{\alpha x + (1-\alpha)z} f(x) \quad (x, z \in K, \ 0 \leq \alpha \leq 1)$$

and so the result follows.

   iii) $\Rightarrow$ i) According to the discussion of Section 4, for every $z \in K$ and $\alpha \in [0,1]$ we have

$$f_{z,\alpha} \leq T_\lambda(t)(f_{z,\alpha}) \quad (t \geq 0)$$

and hence by Proposition 4.1, $f_{z,\alpha} \leq T(f_{z,\alpha})$. ∎

In the next corollary we shall consider the particular case where $f = \prod_{i=1}^{m} h_i$, each h_i being the restriction on K of a continuous linear functional. In this case, or in the case when $f \in \mathcal{C}^2(K)$, $K \subset \mathbf{R}^p$, int $K \neq \emptyset$, there exist the Gateaux derivatives

$$f'(z,x) := \lim_{t \to 0} \frac{1}{t} \left(f(z+tx) - f(z) \right) \tag{73}$$

$$f''(z,x) := \lim_{t \to 0} \frac{1}{t} \left(f'(z+tx,x) - f'(z,x) \right) \quad (z,x \in K) \tag{74}$$

Corollary 5.6 *Assume that* $f = \prod_{i=1}^{m} h_i$ *for some* $m \geq 2$, *where each* h_i *is the restriction on* K *of a continuous linear functional or, respectively,* $f \in \mathcal{C}^2(K)$, *if* $K \subset \mathbf{R}^p$ *and* int $K \neq \emptyset$.

Then f *is* T-*convex if and only if, for every* $z \in K$,

$$T\left(f''(z, \cdot) \right) \geq f''(z, \cdot).$$

Proof. By using the properties of the functionals h_i, respectively Taylor expansions, it can be proved that

$$V_z(f) = \frac{1}{2} \left[T(f''(z, \cdot)) - f''(z, \cdot) \right], \tag{75}$$

where $V_z(f)$ is defined by (71), and so the result follows from Theorem 5.5. ■

We conclude the section with the following conjecture partially suggested by Theorem 5.4.

Conjecture 5.7 *T-convexity is, in fact, equivalent to property* (65).

6 Generalized A-subharmonicity and weak T-convexity

In this section we shall discuss property (66). First, we note that (66) is equivalent to the following two properties

$$s\,R(s,A)f \leq t\,R(t,A)f \quad \text{for every } 0 < t < s, \tag{76}$$

or

$$t\,R(t,A)f \geq f \quad \text{for every } t > 0 \tag{77}$$

where $R(t,A)$ denotes the inverse of $tI - A$ for each $t > 0$, I being the identity operator on $\mathcal{C}(K)$.

Indeed, each $R(t,A)$ $(t > 0)$ is a positive operator on $\mathcal{C}(K)$ because $(A, D(A))$ generates a positive C_0-semigroup.

On the other hand, by using (50) it is easy to show that

$$\lim_{a\to+\infty} \frac{1}{a} \int_0^a T_\lambda(t)f \, dt = T(f) \quad (f \in C(K)) \tag{78}$$

and hence (see, e.g., [10, Theorem 5.1])

$$\lim_{t\to 0^+} t \, R(t, A_\lambda)f = T(f) \quad (f \in C(K)) \tag{79}$$

We also know that (see, e.g., [14, Lemma 3.2])

$$\lim_{t\to+\infty} t \, R(t, A_\lambda)f = f \quad (f \in C(K)) . \tag{80}$$

Now, assume that $f \in D(A)$ satisfies (66). Then, for each $0 < t < s$, by using the resolvent identity we get

$$R(t, A)Af - R(s, A)Af = (s - t) \, R(t, A) \, R(s, A) \, Af \geq 0$$

Hence $tR(t, A)f - f \geq sR(s, A)f - f$ and so (76) follows.

From (76), as $s \to +\infty$ we obtain (77), taking (80) into account (with $\lambda = 1$).

Finally, since $t \, R(t, A)f = R(t, A)Af + f$ $(t > 0)$, from (77) it follows that $R(t, A)Af \geq 0$ and so

$$Af = \lim_{t\to+\infty} tR(t, A)Af \geq 0.$$

After these preliminaries, we introduce the following

Definition 6.1 *A function $f \in C(K)$ is said to be weakly T-convex if*

$$f(z) \leq T(f_{z,\alpha})(z) \quad \text{for each } z \in K, \ \alpha \in [0, 1].$$

For instance, if $K = [0, 1]$ and T is the canonical projection (15), then $f \in C([0, 1])$ is weakly T-convex if

$$f(z) \leq z f(\alpha + (1 - \alpha)z) + (1 - z) f((1 - \alpha)z) \quad (z, \alpha \in [0, 1]) . \tag{81}$$

If K is the standard simplex of $\mathbf{R}^2$ and T is the canonical projection (15), then $f \in C(K)$ is weakly T-convex if

$$f(x, y) \leq (1 - x - y) f(x - \alpha x, y - \alpha y)+ \tag{82}$$

$$+x f(x + \alpha(1 - x), y - \alpha y) + y f(x - \alpha x, y + \alpha(1 - y))$$

for every $(x, y) \in K$ and $\alpha \in [0, 1]$.

Clearly, every T-convex (and, hence, convex) function is weakly T-convex.

The converse is not true in general. For instance, for the standard simplex K of $\mathbf{R}^2$, the function $f(x,y) := -(x+x^2)y$ $((x,y) \in K)$ is weakly T-convex but not T-convex.

A possible interest in weakly T-convex functions rests with the following result.

Proposition 6.2 *Assume that $f \in A_\infty(K)$ or, respectively, $f \in C^2(K)$ if $K \subset \mathbf{R}^p$ and $\operatorname{int} K \neq \emptyset$. If f is weakly T-convex, then $Af \geq 0$.*

Proof. If $f \in A(K)$, then the result is trivial because $Af = 0$.

Assume that $f \in A_m(K)$ for some $m \geq 2$, or $f \in C^2(K)$ in the finite dimensional case. Then, combining (11), (60), (71) and (72), for every $z \in K$ we get

$$Af(z) = V_z f(z) = \lim_{\alpha \to 0^+} \alpha^{-2}\left(T(f_{z,\alpha})(z) - f_{z,\alpha}(z)\right) =$$

$$= \lim_{\alpha \to 0^+} \alpha^{-2}\left(T(f_{z,\alpha})(z) - f(z)\right) \geq 0.$$

∎

We are rather convinced that the converse of Proposition 6.2 holds true as well, but we have not been able to show it and we leave it as a conjecture.

Conjecture 6.3 *If $f \in D(A)$ and $Af \geq 0$, then f is weakly T-convex.*

Note that if this conjecture is true (or, at least, in that context where it is true), then every $f \in C(K)$ satisfying (61) would be weakly T-convex. This follows from the discussion after formula (66).

Note that the conjecture is true for $K = [0,1]$. In this case,

$$D(A) = \{f \in C([0,1]) \cap C^2(]0,1[) | \lim_{x \to 0^+} x(1-x)f''(x) =$$

$$= \lim_{x \to 1^-} x(1-x)f''(x) = 0\}$$

and $Af(x) = \frac{x(1-x)}{2} f''(x)$ for each $x \in]0,1[$.

So, if $f \in D(A)$ and $Af \geq 0$, then f is convex and hence weakly T-convex.

Another case where (a weaker version of) the conjecture is true occurs when K is an ellipsoid of $\mathbf{R}^p$ (see Ex. 2.1, 3 and [17]) or a trapezium, a triangle or a conical subset of $\mathbf{R}^2$ (see [5, Examples 4.1, 1) and 3)]).

This follows from the following result.

Proposition 6.4 *Suppose that* $K \subset \mathbf{R}^p$, $p \geq 1$, *and* int $K \neq \emptyset$. *Assume that there exist* $F \in \mathcal{C}(K)$ *and* $c_{ij} \in \mathbf{R}$, $1 \leq i,j \leq p$ *such that*

$$a_{ij}(x) = 2c_{ij}F(x) \quad (x \in K, \ 1 \leq i,j \leq p)$$

(see (12)).

If $f \in C^2(K)$ *and* $Af \geq 0$, *then* f *is* T-convex *(and hence, weakly* T-convex*).*

Proof. Consider the function $e(x) := ||x||^2$ $(x \in K)$. Then

$$0 \leq T(e)(x) - e(x) = 2F(x) \sum_{i=1}^{p} c_{ii}.$$

So, the function F cannot change sign and, without loss of generality, we can assume that $F \geq 0$; moreover, $F > 0$ on int K (see (8)).

Now, consider $f \in C^2(K)$ satisfying $Af \geq 0$. Then, for every $x, z \in K$ we have (see (11) and (71))

$$F(z)\,V_z f(x) = F(x)\,Af(z) \geq 0.$$

It follows that $V_z f \geq 0$, $z \in$ int K; by continuity we deduce $V_z f \geq 0$, $z \in K$. So f is T-convex by virtue of Theorem 5.5. ∎

We end the paper with the following result which shows that in some cases weakly T-convex functions are actually T-convex. This happens, for instance, if K is an ellipsoid of $\mathbf{R}^p$. To state this final result we need some preliminaries.

If $z \in K$ and $\alpha \in [0,1]$, we shall denote by $\varphi_{z,\alpha} : K \to K$ the mapping defined by

$$\varphi_{z,\alpha}(x) := \alpha x + (1-\alpha)z \quad (x \in K) \tag{83}$$

If $\alpha > 0$, $\varphi_{z,\alpha}$ is injective. Moreover $f_{z,\alpha} = f \circ \varphi_{z,\alpha}$ for every $f \in \mathcal{C}(K)$ (see (6)). We also set

$$K_{z,\alpha} := \varphi_{z,\alpha}(K) = \{\alpha x + (1-\alpha)z \mid x \in K\}. \tag{84}$$

Lemma 6.5 *Suppose that* card $K > 1$. *Let* $\alpha, \beta \in\,]0,1]$ *and* $z, u \in K$ *and assume that* $K_{u,\beta} \subset K_{z,\alpha}$.

Then $\beta \leq \alpha$ *and*

$$\varphi_{u,\beta} = \varphi_{z,\alpha} \circ \varphi_{w,\frac{\beta}{\alpha}}, \tag{85}$$

where w *is an arbitrary point in* K *if* $\beta = \alpha$ *or, if* $\beta < \alpha$, w *is the unique point in* K *such that* $(1-\beta)u = (\alpha - \beta)w + (1-\alpha)z$.

Proof. Firstly, assume that $u \neq z$ and set

$$m := \max \left\{ t \in \mathbf{R} \mid z + t(u - z) \in K_{u,\beta} \right\},$$

$$n := \max \left\{ t \in \mathbf{R} \mid z + t(u - z) \in K_{z,\alpha} \right\},$$

and

$$k := \max \left\{ t \in \mathbf{R} \mid z + t(u - z) \in K \right\}.$$

Then

$$1 \leq m \leq n \leq k, \; n = \alpha k \quad \text{and} \quad m - 1 = \beta(k - 1) \tag{86}$$

so that $\beta \leq \alpha$. Note that, if $\beta = \alpha$, then $\beta = \alpha = 1$.

Assume, now, that $u = z$ and set $r := \frac{\beta}{\alpha}$ and $X := K - z$. Then $r > 0$, $rX \subset X$ and $0 \in X$.

Since X does not reduce to $\{0\}$, we can choose $a \in \partial_e X$, $a \neq 0$. Since $ra \in X$ and $a = \frac{1}{r}(ra) + (1 - \frac{1}{r})0$, then necessarily $r \leq 1$ and hence $\beta \leq \alpha$.

So, in any case we conclude that $\beta \leq \alpha$; moreover, if $u \neq z$ and $\beta = \alpha$, then necessarily $\beta = \alpha = 1$.

We proceed now to discuss separately the cases $\beta = \alpha$ and $\beta < \alpha$. If $\beta = \alpha$, then for each $w \in K$ $\varphi_{w, \frac{\beta}{\alpha}}$ is the identity on K. So, the result follows either if $u \neq z$ (in this case $\beta = \alpha = 1$) or if $u = z$.

If $\beta < \alpha$, set $w := z + \frac{1-\beta}{\alpha-\beta}(u - z)$. If $u = z$, then $w = z$ and (85) can be verified by a simple computation. If $u \neq z$, from (86) it follows that $0 \leq \frac{1-\beta}{\alpha-\beta} \leq k$. Since $z \in K$ and $z + k(u - z) \in K$, we conclude that $w \in K$. After that, it is easy to check (85) and this finishes the proof. ∎

We can now state and prove our final result.

Theorem 6.6 *Assume that $K \subset \mathbf{R}^p$, $p \geq 1$ and $\mathrm{int}\, K \neq \emptyset$. Suppose that $\mathrm{Supp}\, \mu_x = \partial K$ for each $x \in \mathrm{int}\, K$ (see (58)). Then every weakly T-convex function $f \in \mathcal{C}(K)$ is T-convex as well.*

Proof. Let $x \in \mathrm{int}\, K$. Then we know that $\mathrm{Supp}\, \mu_x \subset \partial_{H_T} K \subset \partial K$ ([4, Remark 3 to Th. 3.3.3]) and hence $\partial_{H_T} K = \partial K$. So, by virtue of (8), we obtain that

$$T(f) = f \text{ on } \partial K \text{ for every } f \in \mathcal{C}(K). \tag{87}$$

Fix now a weakly T-convex function $f \in \mathcal{C}(K)$ and assume that f is not T-convex. Then there exist $x, z \in K$, $x \neq z$ and $0 < \alpha < 1$ such that

$$T(f_{z,\alpha})(x) < f_{z,\alpha}(x). \tag{88}$$

Consider the continuous function $T(f_{z,\alpha}) \circ \varphi_{z,\alpha}^{-1}$ defined on the compact $K_{z,\alpha}$ (see (83) and (84)) and extend it to a function $h \in \mathcal{C}(K)$.

Set $g := f - h$; then for every $v \in \partial K_{z,\alpha}$, there exists $u \in \partial K$ such that $v = \varphi_{z,\alpha}(u)$ and hence, by (87),

$$h(v) = Tf_{z,\alpha}(u) = f_{z,\alpha}(u) = f(\varphi_{z,\alpha}(u)) = f(v).$$

Thus we have

$$g = 0 \text{ on } \partial K_{z,\alpha} \tag{89}$$

On the other hand, $h_{z,\alpha} = T(f_{z,\alpha}) \in H_T$ and hence $T(h_{z,\alpha}) = h_{z,\alpha}$. Accordingly, by using (88), we get

$$T(g_{z,\alpha})(x) < g_{z,\alpha}(x). \tag{90}$$

But for every $p \in \partial K = \partial_{H_T} K$, since $\varphi_{z,\alpha}(p) \in \partial K_{z,\alpha}$, we have

$$Tg_{z,\alpha}(p) = g_{z,\alpha}(p) = g(\varphi_{z,\alpha}(p)) = 0$$

so that, by (10), $Tg_{z,\alpha}(x) = 0$ and hence, by (90), $g_{z,\alpha}(x) > 0$.

In conclusion

$$M := \max\{g(y) \mid y \in K_{z,\alpha}\} \geq g_{z,\alpha}(x) > 0. \tag{91}$$

Let $u \in K_{z,\alpha}$ such that $g(u) = M$. Because of (89), necessarily $u \in \text{int } K_{z,\alpha} \subset \text{int } K$ so that $\text{Supp}\,\mu_u = \partial K$.

Set $\beta := \max\{t \in \mathbf{R} \mid tK + (1-t)u \in K_{z,\alpha}\}$ and note that $\beta > 0$ because $u \in \text{int } K_{z,\alpha}$.

Moreover, since $K_{u,\beta} \in K_{z,\alpha}$, from Lemma 6.5 it follows that $\beta \leq \alpha$ and there exists $w \in K$ such that $h_{u,\beta} = h \circ \varphi_{u,\beta} = h \circ \varphi_{z,\alpha} \circ \varphi_{w,\frac{\beta}{\alpha}} = Tf_{z,\alpha} \circ \varphi_{w,\frac{\beta}{\alpha}} \in H_T$ (because of (91)). This implies that $T(h_{u,\beta}) = h_{u,\beta}$, hence

$$Tg_{u,\beta}(u) - g(u) = Tf_{u,\beta}(u) - f(u) \geq 0$$

and so

$$g(u) \leq Tg_{u,\beta}(u).$$

Furthermore, again from (85), we infer that $g_{u,\beta} \leq M$, so that $T(g_{u,\beta}) \leq M$ and $g(u) \leq Tg_{u,\beta}(u) \leq M = g(u)$. Accordingly,

$$\int g_{u,\beta}\, d\mu_u = T(g_{u,\beta})(u) = M,$$

so that $g_{u,\beta} = M$ on $\text{Supp}\,\mu_u = \partial K$.

On the other hand, from the definition of β it also follows that there exists $v \in \partial K_{z,\alpha}$ that can be written as $v = \beta p + (1-\beta)u$ for some $p \in \partial K$. Then (89) implies that $g_{u,\beta}(p) = g(v) = 0$ and this contradicts (91). ∎

ACKNOWLEDGMENTS

We wish to thank M. Ivan and D. Leviatan for their critical reading of the manuscript.

References

[1] J. A. Adell - J. de la Cal - I. Raşa, *Lototsky-Schnabl operators on the unit interval,* preprint (1997)

[2] F. Altomare, *Lototsky–Schnabl operators on the unit interval and degenerate diffusion equations,* in: Progress in Functional Analysis, K.D. Bierstedt, J. Bonet, J. Horváth and M. Maestre (Eds.), North-Holland Mathematics Studies **170**, North-Holland Amsterdam, 1992, 259–277.

[3] F. Altomare, *Lototsky–Schnabl operators on compact convex sets and their associated semigroups,* Mh. Math. **114** (1992), 1–13.

[4] F. Altomare – M. Campiti, *Korovkin-type Approximation Theory and its Applications,* de Gruyter Series Studies in Mathematics **17**, W. de Gruyter & Co., Berlin–New York, 1994.

[5] F. Altomare - I. Raşa, *Towards a characterization of a class of differential operators associative with positive projections,* Atti Sem. Mat. Fis. Univ. Modena, Suppl. **46** (1998), 3–38.

[6] M. Blümlinger, *Approximation durch Bernsteinoperatoren auf kompakten konvexen Mengen,* Ostereich. Akad. Wiss. Math.-Natur.kl Sitzungsber. II **196** (1987), no. 4–7, 181–215.

[7] G.-Z. Chang - J.-Z. Zhang, *Converse theorems of convexity for Bernstein polynomials over triangles,* J. Approx. Theory **61** (1990), no. 3,265-278.

[8] W. Dahmen, *Convexity and Bernstein-Bézier polynomials,* in: Curves and Surfaces, P. J. Laurent, A. Le Méhauté and L. L. Schumaker (Eds.) 107-134, Academic Press, Boston, 1991.

[9] W. Dahmen - C. A. Micchelli, *Convexity and Bernstein polynomials on k-simploids,* Acta Math. Appl. Sinica **6** (1990), no. 1, 50-66.

[10] E. B. Davies, *One-Parameter Semigroups,* Academic Press, London - New York - San Francisco, 1980.

[11] T. Nishishiraho, *Saturation of positive linear operators,* Tôhoku Math. J. (2) **28** (1976), no. 2, 239–243.

[12] T. Nishishiraho, *Saturation of bounded linear operators,* Tôhoku Math. J. **30** (1978) 69–81.

[13] T. Nishishiraho, *The convergence and saturation of iterations of positive linear operators,* Math. Z. **194** (1987), 397–404.

[14] A. Pazy, *Semigroups of Linear Operators and Applications to Partial Differential Equations,* Springer–Verlag, Berlin, 1983.

[15] I. Raşa, *Altomare projections and Lototsky–Schnabl operators*, Proc. 2nd International Conference in Functional Analysis and Approximation Theory, 1992, Suppl. Rend. Circ. Mat. Palermo **33** (1993), 439–451.

[16] I. Raşa, *On some properties of Altomare projections*, Conf. Sem. Mat. Univ. Bari **253** (1993), 17 p.

[17] M. Romito, *Lototsky–Schnabl operators associated with a strictly elliptic differential operator and their corresponding Feller semigroup*, to appear in Mh. Math., 1998.

[18] T. Sauer, *Multivariate Bernstein polynomials and convexity*, Computer Aided Geometric Design **8** (1991), 465-478.

[19] T. Sauer, *On the maximum principle of Bernstein polynomials on a simplex*, J. Approx. Theory **71** (1992), no. 1, 121-122.

[20] T. Sauer, *The genuine Bernstein–Durrmeyer operators on a simplex*, Results Math. **26** (1994), no. 1–2, 99–130.

[21] K. Taira, *Diffusion Processes and Partial Differential Equations*, Academic Press, Boston - San Diego - London - Tokyo, 1988.

Università degli Studi di Bari
Dipartimento di Matematica
Via E. Orabona 4
70125 Bari, Italia
Email address: altomare@pascal.dm.uniba.it

Catedra de Matematică
Universitatea Tehnică din Cluj-Napoca
Str. C. Daicoviciu 15
3400 Cluj–Napoca, Romania
Email address: Ioan.Rasa@math.utcluj.ro

International Series of Numerical Mathematics
Vol. 132, © 1999 Birkhäuser Verlag Basel/Switzerland

Gregory's Rational Cubic Splines in Interpolation Subject to Derivative Obstacles

Marion Bastian-Walther and Jochen W. Schmidt

Abstract

This paper is concerned with the interpolation of gridded data subject to lower and upper bounds on the first order derivatives. We apply Gregory's rational cubic C^1 splines and corresponding tensor products. The occurring rationality parameters can always be determined such that the considered interpolation problems turn out to be solvable.

1 Introduction

In univariate interpolation, let be given a grid

$$\Delta = \{x_0 < x_1 < \cdots < x_n\}$$

and data values $z_i \in \mathbf{R}^1$, $i = 0, \ldots, n$. The problem is to find, say, a C^1 function s with

$$s(x_i) = z_i, \ i = 0, \ldots, n. \tag{1}$$

In addition, the interpolant s should satisfy the restrictions

$$L_i \leq s'(x) \leq U_i \text{ for } x \in [x_{i-1}, x_i], \ i = 1, \ldots, n \tag{2}$$

where the obstacles L_i, U_i are prescribed. However, they should be compatible with the data set. In the present context the requirement

$$L_i \leq \tau_{i-1} \leq U_i, \ L_i < \tau_i < U_i, \quad i = 1, \ldots, n \tag{3}$$

is adequate. The τ_i are slopes, i.e.,

$$\tau_i = (z_i - z_{i-1})/h_i, \ i = 1, \ldots, n \tag{4}$$

with the step sizes $h_i = x_i - x_{i-1}$. Further we set $\tau_0 = \tau_1$, for example. For $L_i = 0$, $U_i = +\infty$, $i = 1, \ldots, n$, the problem (1), (2) reduces to monotone interpolation.

There are some models leading to the constrained interpolation problem (1), (2). We mention the following occurring in the motion of a material point. Here, depending on the time x the way $s = s(x)$ or, more clearly, the velocity $s' = s'(x)$

are to be determined such that the point reaches the positions z_i at the times x_i, $i = 0, \ldots, n$. We are led to the same model in controlling the flow of a production, e.g., in controlling the velocity of a bottle machine. Here z_i denotes the required output at the time x_i. In most problems of this type the velocity s' has to be nonnegative and less than a maximal value $U_{\max}$. Then, in the strip condition (2) we have to set $L_i = 0$, $U_i = U_{\max}$. Obviously, these problems are not solvable if $U_{\max}$ is too small. However, the compatibility (3) ensures the existence of spline solutions.

Recently in paper [9], univariate interpolation subject to the derivative obstacles (2) was treated from a numerical point of view. This was done by applying quadratic C^1 and quartic C^2 splines on refined grids. The desired strip interpolants can be constructed by appropriate placing of the knots added in refining the grid. Further classes of polynomial splines suitable in handling the present interpolation problem can be found in the review paper [10]. Most of these univariate algorithms are extended to the interpolation of bivariate data sets given on a rectangular array; see [11]. There the univariate results are utilized by applying tensor product techniques. Polynomial splines are also successfully used in univariate least squares smoothing subject to derivative obstacles like (2); see [1, 8, 12, 13].

Gregory's rational cubic splines [5] are known to be very suitable in univariate monotone and convex interpolation; see the paper [4] and the monograph [14], too. This spline class allows satisfactory numerical algorithms also in constructing solutions of the present restricted interpolation problem (1), (2). We are in the position to offer bounds for the rationality parameters occurring in Gregory's splines such that above these bounds the restrictions (2) are met. Further we described an extension of the univariate method to the restricted interpolation of gridded bivariate data. Tensor products of Gregory's splines are suitable.

Finally, we refer readers interested in interpolations subject to other types of two-sided restrictions to the review papers [3, 10]. Here, for example, results on range restricted interpolation can be found; see, in addition, paper [2].

2   Univariate Interpolation in a Derivative Strip

We begin with treating the univariate problem (1), (2). The derived results are of interest for themselves, and we need them for handling the corresponding bivariate problem on gridded data.

2.1   Gregory's Rational Cubic Splines

These splines are defined by

$$s(x) = z_{i-1}v + z_i u + h_i uv \frac{(p_{i-1} - \tau_i)v + (\tau_i - p_i)u}{1 + \alpha_i uv} \tag{5}$$

for $x \in [x_{i-1}, x_i]$, $i = 1, \ldots, n$. By u and v we denote the barycentric coordinates with respect to the considered subinterval, i.e.,

$$u = (x - x_{i-1})/h_i \, , \quad v = (x_i - x)/h_i \, , \tag{6}$$

implying $u + v = 1$. The values $\alpha_i \geq 0$ are the so-called rationality or tension parameters; in general, they should be as small as possible. The splines (5) interpolate, and they always belong to $C^1[x_0, x_n]$. The parameters p_i are the first-order derivatives at the data sites. In other words, the spline (5) is uniquely defined by the Hermite conditions

$$z_i = s(x_i) \, , \quad p_i = s'(x_i) \, , \quad i = 0, \ldots, n \, . \tag{7}$$

We denote the space of Gregory's C^1 splines (5) by $S_G^1(\Delta)$ assuming to this end the tension parameters α_i to be arbitrarily fixed.

For the obstacles L and U given by

$$L'(x) = L_i \, , \quad U'(x) = U_i \text{ for } x \in [x_{i-1}, x_i] \, , \quad i = 0, \ldots, n \, ; \tag{8}$$

we do not require their continuity at the data sites. Therefore we have to introduce the space $S_G^{-1}(\Delta)$ of piecewise rational cubic splines. These splines s are defined by

$$
\begin{aligned}
s(x) = {} & s(x_{i-1} + 0)v + s(x_i - 0)u + \\
& + \frac{uv}{1 + \alpha_i uv} \Big\{ \big(h_i s'(x_{i-1} + 0) - s(x_i - 0) + s(x_{i-1} + 0) \big) v + \\
& + \big(s(x_i - 0) - s(x_{i-1} + 0) - h_i s'(x_i - 0) \big) u \Big\}
\end{aligned}
\tag{9}
$$

for $x \in [x_{i-1}, x_i]$, $i = 1, \ldots, n$. Of course, it follows $S_G^1(\Delta) \subset S_G^{-1}(\Delta)$. Furthermore $L, U \in S_G^{-1}(\Delta)$ hold true, and we obtain

$$L(x) = h_i L_i (u - v)/2 \, , \quad U(x) = h_i U_i (u - v)/2 \text{ for } x \in [x_{i-1}, x_i] \, , \quad i = 0, \ldots, n \, . \tag{10}$$

Further, piecewise constant functions on Δ are from $S_G^{-1}(\Delta)$. However, obstacles with continuous piecewise linear derivatives, e.g.,

$$L'(x) = l_{i-1} v + l_i u \text{ for } x \in [x_{i-1}, x_i] \, , \quad i = 0, \ldots, n \, ,$$

do not belong to $S_G^{-1}(\Delta)$. For this reason the present considerations are restricted to the simpler constraints (2).

2.2 Positive Interpolation

Because of $1 = u^2 + 2uv + v^2$ we find immediately from (5) that

$$(1 + \alpha_i uv)s(x) =$$

$$z_{i-1} v^3 + \big((3 + \alpha_i) z_{i-1} + h_i p_{i-1} \big) uv^2 + \big((3 + \alpha_i) z_i - h_i p_i \big) u^2 v + z_i u^3 \, .$$

Hence, conditions sufficient for $s \in S_G^1(\Delta)$ to be nonnegative on $[x_{i-1}, x_i]$ read

$$z_{i-1} \geq 0, \ (3 + \alpha_i)z_{i-1} + h_i p_{i-1} \geq 0,$$
$$(3 + \alpha_i)z_i - h_i p_i \geq 0, \ z_i \geq 0, \ i = 1, \dots, n. \tag{11}$$

Since compatibility now means

$$z_i \geq 0, \ i = 0, \dots, n,$$

system (11) is solved for example by $p_i = 0$, $i = 0, \dots, n$. Thus, positive interpolation is always successful using Gregory's splines (5); as it is already known even $\alpha_i = 0$, $i = 1, \dots, n$, is allowed.

Next, we derive nonnegativity conditions for the splines (9) from $S_G^{-1}(\Delta)$. In view of the analogue to (11) we are led to introduce the functionals

$$\lambda_{i,1}(s) = s(x_{i-1} + 0), \ \lambda_{i,2}(s) = s(x_i - 0),$$
$$\lambda_{i,3}(s) = (3 + \alpha_i)s(x_{i-1} + 0) + h_i s'(x_{i-1} + 0), \tag{12}$$
$$\lambda_{i,4}(s) = (3 + \alpha_i)s(x_i - 0) - h_i s'(x_i - 0).$$

Thus the conditions

$$\lambda_{i,k}(s) \geq 0 \text{ for } k = 1, 2, 3, 4, i = 1, \dots, n \tag{13}$$

imply the splines $s \in S_G^{-1}(\Delta)$ to be nonnegative on $[x_0, x_n]$.

Other nonnegativity conditions not comparable with (13) are derived in paper [2].

2.3 Interpolation Subject to Derivative Obstacles

Concerning univariate problems, in this chapter we treat the main object of the present paper. After some computations we obtain for Gregory's splines (5)

$$s'(x)(1 + \alpha_i uv)^2 = p_{i-1}v^4 + p_i u^4 + 2((3 + \alpha_i)\tau_i - p_i)uv^3 +$$
$$+ ((12 + 6\alpha_i + \alpha_i^2)\tau_i - (3 + \alpha_i)(p_{i-1} + p_i))u^2 v^2 + 2((3 + \alpha_i)\tau_i - p_{i-1})u^3 v;$$

see [14], too. Thus, by considering the coefficients we get the following conditions being sufficient for the monotonicity of $s \in S_G^1(\Delta)$:

$$p_{i-1} \geq 0, \ p_i \geq 0,$$
$$(3 + \alpha_i)\tau_i - p_{i-1} \geq 0, \ (3 + \alpha_i)\tau_i - p_i \geq 0, \tag{14}$$
$$(12 + 6\alpha_i + \alpha_i^2)\tau_i - (3 + \alpha_i)(p_{i-1} + p_i) \geq 0, \ i = 1, \dots, n.$$

The last three of these inequalities (14) can be replaced by

$$(3 + \alpha_i)\tau_i - p_{i-1} - p_i \geq 0; \tag{15}$$

indeed, adding $p_{i-1} \geq 0$ and/or $p_i \geq 0$ to (15) we obtain the third and fourth inequalities of (14) as well as $\tau_i \geq 0$. The last inequality (14) is a positive combination of $\tau_i \geq 0$ and (15) with the coefficients 3 and $(3+\alpha_i)$. Therefore, in $S_G^{-1}(\Delta)$ we should define

$$\mu_{i,1}(s) = s'(x_{i-1}+0)\,, \ \mu_{i,2}(s) = s'(x_i-0)\,,$$
$$\mu_{i,3}(s) = (3+\alpha_i)(s(x_i-0) - s(x_{i-1}+0))/h_i - s'(x_{i-1}+0) - s'(x_i-0)\,.$$
$$(16)$$

Then the conditions

$$\mu_{i,k}(s) \geq 0 \text{ for } k = 1,2,3\,, \ i = 1,\dots,n \tag{17}$$

ensure for $s \in S_G^{-1}(\Delta)$ that

$$s'(x) \geq 0 \text{ for } x \in [x_{i-1}, x_i]\,, \ i = 1,\dots,n\,. \tag{18}$$

Let the bounds L and U be given by (10). Then

$$\mu_{i,k}(L) \leq \mu_{i,k}(s) \leq \mu_{i,k}(U) \text{ for } k = 1,2,3\,, \ i = 1,\dots,n \tag{19}$$

i.e.,

$$L_i \leq p_{i-1} \leq U_i\,, \ L_i \leq p_i \leq U_i\,,$$
$$(1+\alpha_i)L_i \leq (3+\alpha_i)\tau_i - p_{i-1} - p_i \leq (1+\alpha_i)U_i\,, i = 1,\dots,n \tag{20}$$

guarantee the requirements (2). Now, the compatibility (3) is assumed. If we set

$$p_i = \tau_i\,, \ i = 0,\dots,n\,, \tag{21}$$

all inequalities (20) are satisfied, with exception of the third. This inequality

$$(1+\alpha_i)(L_i - \tau_i) \leq \tau_i - \tau_{i-1} \leq (1+\alpha_i)(U_i - \tau_i)$$

holds true if

$$\alpha_i \geq -1 + \max\left\{\frac{\tau_i - \tau_{i-1}}{L_i - \tau_i}\,, \ \frac{\tau_i - \tau_{i-1}}{U_i - \tau_i}\right\}\,, \ i = 1,\dots,n\,. \tag{22}$$

Thus we have proved

Theorem 2.1 *An interpolant* $s \in S_G^1(\Delta)$, *i.e., a spline (5) with the derivatives (21) satisfies the restrictions (2) on the first-order derivatives if the requirements (22) on the nonnegative tension parameters* $\alpha_1,\dots,\alpha_n$ *are valid.*

We point out that monotone and even co-monotone interpolation are considerably easier to handle than the general problem (1), (2). The requirement (2) reduces to

$$\varepsilon_i s'(x) \geq 0 \text{ for } x \in [x_{i-1}, x_i]\,, \ i = 1,\dots,n\,, \tag{23}$$

where $\varepsilon_i = +1$ for $\tau_i \geq 0$, and $\varepsilon_i = -1$ for $\tau_i \leq 0$. Hence, in (2) we set $L_i = 0$, $U_i = +\infty$ for $\tau_i \geq 0$, and $L_i = -\infty$, $U_i = 0$ for $\tau_i \leq 0$. For $\tau_i = 0$ one should decide on one of these two possibilities. In view of (20), a sufficient condition for (23) reads

$$\varepsilon_i p_{i-1} \geq 0, \ \varepsilon_i p_i \geq 0, \ \varepsilon_i \left((3 + \alpha_i)\tau_i - p_{i-1} - p_i\right) \geq 0, \ i = 1, \ldots, n. \qquad (24)$$

Thus, because of $\varepsilon_i \tau_i \geq 0$, $i = 1, \ldots, n$, this system is solved by $p_0 = \cdots = p_n = 0$, and even $\alpha_1 = \cdots = \alpha_n = 0$ is allowed. In other words, co-monotone interpolation is always successful in the space of cubic C^1 splines.

2.4   Optimal Splines by Minimizing the Holladay Functional

We remark that the constrained interpolation problem (1), (2) is not uniquely solvable. The described spline solution is only one from the set of solutions. It is usual to select a restricted spline interpolant by minimizing a choice functional. Examples are the Holladay functional as well as modifications and approximations. Here we propose to minimize the Simpson approximation of the Holladay functional

$$\int_{x_0}^{x_n} s''(x)^2 \, dx \approx \sum_{i=1}^{n} \frac{h_i}{6} \left\{ s''(x_{i-1})^2 + 4s''(\tfrac{1}{2}(x_{i-1} + x_i))^2 + s''(x_i)^2 \right\}$$

$$= \sum_{i=1}^{n} \frac{2}{3h_i} \left\{ (-p_i - (2 + \alpha_i)p_{i-1} + (3 + \alpha_i)\tau_i)^2 + \frac{256(p_i - p_{i-1})^2}{(4 + \alpha_i)^4} + \right. \qquad (25)$$

$$\left. + ((2 + \alpha_i)p_i + p_{i-1} - (3 + \alpha_i)\tau_i)^2 \right\}$$

subject to the constraints (20) or (24). The nonnegative tension parameters are not included into the minimization procedure; they are fixed as small as possible according to (22), and in co-monotone interpolation they are set equal to zero.

3   Tensor Product Interpolation in Derivative Strips

3.1   Interpolation with Tensor Products of Gregory's Splines

On a rectangular grid

$$\Delta^x \times \Delta^y = \{x_0 < x_1 < \cdots < x_n\} \times \{y_0 < y_1 < \cdots < y_m\} \qquad (26)$$

let the data set

$$(x_i, y_j, z_{i,j}), \ i = 0, \ldots, n, \ j = 0, \ldots, m, \qquad (27)$$

be given. We are interested in a C^1 function s which interpolates, i.e.,

$$s(x_i, y_j) = z_{i,j}, \ i = 0, \ldots, n, \ j = 0, \ldots, m, \qquad (28)$$

and which satisfies, with given obstacles $L_{i,j}^x$, $U_{i,j}^x$, $L_{i,j}^y$, $U_{i,j}^y$, two-sided restrictions on the first-order derivatives,

$$L_{i,j}^x \leq \frac{\partial}{\partial x} s(x,y) \leq U_{i,j}^x, \; L_{i,j}^y \leq \frac{\partial}{\partial y} s(x,y) \leq U_{i,j}^y \tag{29}$$

for $(x,y) \in [x_{i-1}, x_i] \times [y_{j-1}, y_j]$, $i = 1, \ldots, n$, $j = 1, \ldots, m$.

In view of the mentioned interpolation property in $S_G^1(\Delta)$, it follows for tensor product interpolation in $S_G^1(\Delta^x) \otimes S_G^1(\Delta^y)$ that a spline s from this space is uniquely determined by the requirements

$$s(x_i, y_j) = z_{i,j},$$

$$\frac{\partial}{\partial x} s(x_i, y_j) = p_{i,j},$$

$$\frac{\partial}{\partial y} s(x_i, y_j) = q_{i,j}, \tag{30}$$

$$\frac{\partial^2}{\partial x \, \partial y} s(x_i, y_j) = r_{i,j}, \; i = 0, \ldots, n, \; j = 0, \ldots, m.$$

While the function values $z_{i,j}$ are assumed to be given, the partial derivatives $p_{i,j}$, $q_{i,j}$, $r_{i,j}$ are used here as parameters in order to satisfy the restrictions (29). In addition, the tension parameters $\alpha_i^x \geq 0$, $\alpha_j^y \geq 0$ are available for this purpose.

3.2 Discretizing the Constraints Using the Nonnegativity Lemma

Our notations are as follows. We denote the functionals (12), (16) in $S_G^{-1}(\Delta^x)$ and $S_G^{-1}(\Delta^y)$ by $\lambda_{i,k}^x$, $\mu_{i,k}^x$ and $\lambda_{j,l}^y$, $\mu_{j,l}^y$, respectively. Analogously, the step sizes, tension parameters, barycentric coordinates and so on, are supplemented by the superscripts x and y in $S_G^{-1}(\Delta^x)$ and $S_G^{-1}(\Delta^y)$, respectively. We mention explicitly that the superscripts x and y are labels and not variables.

Applying the nonnegativity lemma for tensor products [6, 7] to the functionals introduced by (12), (16) we obtain

Proposition 3.1 For $s \in S_G^{-1}(\Delta^x) \otimes S_G^{-1}(\Delta^y)$, the conditions

$$(\mu_{i,k}^x \otimes \lambda_{j,l}^y)(s) \geq 0 \text{ for } k = 1,2,3, \; l = 1,2,3,4, \; i = 1, \ldots, n, \; j = 1, \ldots, m, \tag{31}$$

imply

$$\frac{\partial}{\partial x} s(x,y) \geq 0 \text{ for } (x,y) \in [x_{i-1}, x_i] \times [y_{j-1}, y_j], \; i = 1, \ldots, n, \; j = 1, \ldots, m, \tag{32}$$

and

$$(\lambda_{i,k}^x \otimes \mu_{j,l}^y)(s) \geq 0 \text{ for } k = 1,2,3,4, \; l = 1,2,3, \; i = 1, \ldots, n, \; j = 1, \ldots, m, \tag{33}$$

guarantee

$$\frac{\partial}{\partial y}s(x,y) \geq 0 \ \text{for} \ (x,y) \in [x_{i-1}, x_i] \times [y_{j-1}, y_j], \ i = 1, \ldots, n, \ j = 1, \ldots, m. \quad (34)$$

This lemma allows to discretize straightforwardly the restrictions (29). Let L^x, U^x, L^y, $U^y \in S_G^{-1}(\Delta^x) \otimes S_G^{-1}(\Delta^y)$ be functions with

$$\frac{\partial}{\partial x}L^x = L_{i,j}^x, \ \frac{\partial}{\partial x}U^x = U_{i,j}^x, \ \frac{\partial}{\partial y}L^y = L_{i,j}^y, \ \frac{\partial}{\partial y}U^y = U_{i,j}^y$$

on $[x_{i-1}, x_i] \times [y_{j-1}, y_j]$. If we apply Proposition 3.1 to the differences $s - L^x$, $U^x - s$, $s - L^y$, $U^y - s$, we get the desired dicretizations. We are led to the piecewise constant obstacles (29) by setting

$$L^x(x,y) = h_i^x L_{i,j}^x (u^x - v^x)/2, \ U^x(x,y) = h_i^x U_{i,j}^x (u^x - v^x)/2,$$
$$L^y(x,y) = h_j^y L_{i,j}^y (u^y - v^y)/2, \ U^y(x,y) = h_j^y U_{i,j}^y (u^y - v^y)/2 \quad (35)$$

for $(x,y) \in [x_{i-1}, x_i] \times [y_{j-1}, y_j]$; compare with (10). Then it is easily seen that

$$(\mu_{i,k}^x \otimes \lambda_{j,l}^y)(L^x) = \begin{cases} L_{i,j}^x & \text{for } k = 1,2, \ l = 1,2, \\ (1 + \alpha_i^x)L_{i,j}^x & \text{for } k = 3, \ l = 1,2, \\ (3 + \alpha_j^y)L_{i,j}^x & \text{for } k = 1,2, \ l = 3,4, \\ (1 + \alpha_i^x)(3 + \alpha_j^y)L_{i,j}^x & \text{for } k = 3, \ l = 3,4, \end{cases} \quad (36)$$

and that

$$(\lambda_{i,k}^x \otimes \mu_{j,l}^y)(L^y) = \begin{cases} L_{i,j}^y & \text{for } k = 1,2, \ l = 1,2, \\ (1 + \alpha_j^y)L_{i,j}^y & \text{for } k = 1,2, \ l = 3, \\ (3 + \alpha_i^x)L_{i,j}^y & \text{for } k = 3,4, \ l = 1,2, \\ (1 + \alpha_j^y)(3 + \alpha_i^x)L_{i,j}^y & \text{for } k = 3,4, \ l = 3. \end{cases} \quad (37)$$

Analogous relations hold for U^x and U^y, respectively. Next, we compute $(\mu_{i,k}^x \otimes \lambda_{j,l}^y)(s)$ and $(\lambda_{i,k}^x \otimes \mu_{j,l}^y)(s)$ for $s \in S_G^1(\Delta^x) \otimes S_G^1(\Delta^y)$. To this end, we introduce the abbreviations

$$\tau_{i,j}^x = (z_{i,j} - z_{i-1,j})/h_i^x, \ \tau_{i,j}^y = (z_{i,j} - z_{i,j-1})/h_j^y, \quad (38)$$

for the partial slopes, and

$$\tau_{i,j}^{xy} = (z_{i,j} - z_{i-1,j} - z_{i,j-1} + z_{i-1,j-1})/(h_i^x h_j^y) \quad (39)$$

for the usual twist approximation. Further we set

$$[k] = \begin{cases} 0 & \text{for } k \text{ even} \\ 1 & \text{for } k \text{ odd}. \end{cases} \quad (40)$$

Then, with the notations (30), we obtain straightforwardly

$$(\mu_{i,k}^x \otimes \lambda_{j,l}^y)(s) = \begin{cases} p_{i-[k],j-[l]} \text{ for } k=1,2, \ l=1,2, \\[2mm] (3+\alpha_i^x)\tau_{i,j-[l]}^x - p_{i,j-[l]} - p_{i-1,j-[l]} \text{ for } k=3, \ l=1,2, \\[2mm] (3+\alpha_j^y)p_{i-[k],j-[l]} + (-1)^{l+1}h_j^y r_{i-[k],j-[l]} \\[2mm] \qquad\qquad\qquad \text{ for } k=1,2, \ l=3,4, \\[2mm] (3+\alpha_i^x)(3+\alpha_j^y)\tau_{i,j-[l]}^x - (3+\alpha_j^y)(p_{i,j-[l]}+p_{i-1,j-[l]})+ \\[2mm] + (-1)^{l+1}(3+\alpha_i^x)h_j^y \dfrac{q_{i,j-[l]}-q_{i-1,j-[l]}}{h_i^x} - \\[2mm] - (-1)^{l+1}h_j^y(r_{i,j-[l]}+r_{i-1,j-[l]}) \text{ for } k=3, \ l=3,4, \end{cases} \tag{41}$$

as well as

$$(\lambda_{i,k}^x \otimes \mu_{j,l}^y)(s) = \begin{cases} q_{i-[k],j-[l]} \text{ for } k=1,2, \ l=1,2, \\[2mm] (3+\alpha_j^y)\tau_{i-[k],j}^y - q_{i-[k],j} - q_{i-[k],j-1} \text{ for } k=1,2, \ l=3, \\[2mm] (3+\alpha_i^x)q_{i-[k],j-[l]} + (-1)^{k+1}h_i^x r_{i-[k],j-[l]} \\[2mm] \qquad\qquad\qquad \text{ for } k=3,4, \ l=1,2, \\[2mm] (3+\alpha_j^y)(3+\alpha_i^x)\tau_{i-[k],j}^y - (3+\alpha_i^x)(q_{i-[k],j}+q_{i-[k],j-1})+ \\[2mm] + (-1)^{k+1}(3+\alpha_j^y)h_i^x \dfrac{p_{i-[k],j}-p_{i-[k],j-1}}{h_j^y} - \\[2mm] - (-1)^{k+1}h_i^x(r_{i-[k],j}+r_{i-[k],j-1}) \text{ for } k=3,4, \ l=3. \end{cases} \tag{42}$$

Now we are in the position to formulate our basic conditions on the existence of restricted interpolants.

Proposition 3.2 *An interpolating spline $s \in S_G^1(\Delta^x) \otimes S_G^1(\Delta^y)$ of Gregory's type satisfies the restrictions (29) on the first-order derivatives if*

$$L_{i,j}^x \leq p_{i-k,j-l} \leq U_{i,j}^x,$$

$$(1+\alpha_i^x)L_{i,j}^x \leq (3+\alpha_i^x)\tau_{i,j-l}^x - p_{i,j-l} - p_{i-1,j-l} \leq (1+\alpha_i^x)U_{i,j}^x,$$

$$(3+\alpha_j^y)L_{i,j}^x \leq (3+\alpha_j^y)p_{i-k,j-l} + (-1)^{l+1}h_j^y r_{i-k,j-l} \leq (3+\alpha_j^y)U_{i,j}^x,$$

$$(1+\alpha_i^x)(3+\alpha_j^y)L_{i,j}^x \leq (3+\alpha_i^x)(3+\alpha_j^y)\tau_{i,j-l}^x -$$

$$- (3+\alpha_j^y)(p_{i,j-l}+p_{i-1,j-l}) + (-1)^{l+1}(3+\alpha_i^x)h_j^y \frac{q_{i,j-l}-q_{i-1,j-l}}{h_i^x} -$$

$$- (-1)^{l+1}h_j^y(r_{i,j-l}+r_{i-1,j-l}) \leq (1+\alpha_i^x)(3+\alpha_j^y)U_{i,j}^x, \tag{43}$$

and if

$$L_{i,j}^y \leq q_{i-k,j-l} \leq U_{i,j}^y,$$

$$(1+\alpha_j^y)L_{i,j}^y \leq (3+\alpha_j^y)\tau_{i-k,j}^y - q_{i-k,j} - q_{i-k,j-1} \leq (1+\alpha_j^y)U_{i,j}^y,$$

$$(3+\alpha_i^x)L_{i,j}^y \leq (3+\alpha_i^x)q_{i-k,j-l} + (-1)^{k+1}h_i^x r_{i-k,j-l} \leq (3+\alpha_i^x)U_{i,j}^y,$$

$$(1+\alpha_j^y)(3+\alpha_i^x)L_{i,j}^y \leq (3+\alpha_j^y)(3+\alpha_i^x)\tau_{i-k,j}^y -$$

$$- (3+\alpha_i^x)(q_{i-k,j} + q_{i-k,j-1}) + (-1)^{k+1}(3+\alpha_j^y)h_i^x \frac{p_{i-k,j} - p_{i-k,j-1}}{h_j^y} -$$

$$- (-1)^{k+1}h_i^x(r_{i-k,j} + r_{i-k,j-1}) \leq (1+\alpha_j^y)(3+\alpha_i^x)U_{i,j}^y,$$

$$(44)$$

for $k = 0,1$, $l = 0,1$, $i = 1,\ldots,n$, $j = 1,\ldots,m$.

3.3   Computing Bivariate Restricted Interpolants

In view of Proposition 3.2, we are interested in the solvability of the systems (43), (44) of nonlinear inequalities. This property can be achieved by choosing the tension parameters $\alpha_i^x \geq 0$, $\alpha_j^y \geq 0$ suitably. The first-order derivatives should be

$$p_{i,j} = \tau_{i,j}^x, \quad q_{i,j} = \tau_{i,j}^y, \quad i = 0,\ldots,n, \quad j = 0,\ldots,m, \qquad (45)$$

while the twist parameters $r_{i,j}$ may be arbitrarily. For simplicity we set

$$r_{i,j} = 0, \quad i = 0,\ldots,n, \quad j = 0,\ldots,m. \qquad (46)$$

Then the sets of inequalities (43) and (44) reduce to

$$L_{i,j}^x \leq \tau_{i-k,j-l}^x \leq U_{i,j}^x,$$

$$(1+\alpha_i^x)L_{i,j}^x \leq (1+\alpha_i^x)\tau_{i,j-l}^x + \tau_{i,j-l}^x - \tau_{i-1,j-l}^x \leq (1+\alpha_i^x)U_{i,j}^x,$$

$$(3+\alpha_j^y)L_{i,j}^x \leq (3+\alpha_j^y)\tau_{i-k,j-l}^x \leq (3+\alpha_j^y)U_{i,j}^x,$$

$$(1+\alpha_i^x)(3+\alpha_j^y)L_{i,j}^x \leq (1+\alpha_i^x)(3+\alpha_j^y)\tau_{i,j-l}^x + (3+\alpha_j^y)(\tau_{i,j-l}^x - \tau_{i-1,j-l}^x) +$$

$$+ (-1)^{l+1}(3+\alpha_i^x)h_j^y \tau_{i,j-l}^{xy} \leq (1+\alpha_i^x)(3+\alpha_j^y)U_{i,j}^x,$$

$$(47)$$

and to

$$L_{i,j}^y \leq \tau_{i-k,j-l}^y \leq U_{i,j}^y \,,$$

$$(1+\alpha_j^y)L_{i,j}^y \leq (1+\alpha_j^y)\tau_{i-k,j}^y + \tau_{i-k,j}^y - \tau_{i-k,j-1}^y \leq (1+\alpha_j^y)U_{i,j}^y \,,$$

$$(3+\alpha_i^x)L_{i,j}^y \leq (3+\alpha_i^x)\tau_{i-k,j-l}^y \leq (3+\alpha_i^x)U_{i,j}^y \,,$$

$$(1+\alpha_j^y)(3+\alpha_i^x)L_{i,j}^y \leq (1+\alpha_j^y)(3+\alpha_i^x)\tau_{i-k,j}^y + (3+\alpha_i^x)(\tau_{i-k,j}^y - \tau_{i-k,j-1}^y) +$$

$$+ (-1)^{k+1}(3+\alpha_j^y)h_i^x\tau_{i-k,j}^{xy} \leq (1+\alpha_j^y)(3+\alpha_i^x)U_{i,j}^y \,.$$

$$(48)$$

Now, we assume the obstacles to be compatible with the derivatives. Here this requirement means

$$L_{i,j}^x \leq \tau_{i-1,j-1}^x, \tau_{i-1,j}^x \leq U_{i,j}^x \,, \quad L_{i,j}^x < \tau_{i,j-1}^x, \tau_{i,j}^x < U_{i,j}^x \,,$$

$$L_{i,j}^y \leq \tau_{i-1,j-1}^y, \tau_{i,j-1}^y \leq U_{i,j}^y \,, \quad L_{i,j}^y < \tau_{i-1,j}^y, \tau_{i,j}^y < U_{i,j}^y \,, \tag{49}$$

$$i = 1, \ldots, n \,, \; j = 1, \ldots, m \,.$$

Hence, the first and third inequalities in (47), (48), respectively, are always satisfied. The remaining ones can be written as

$$(1+\alpha_i^x)V_{ijl}^x \leq A_{ijl}^x \leq (1+\alpha_i^x)W_{ijl}^x \,,$$

$$(1+\alpha_i^x)(3+\alpha_j^y)V_{ijl}^x \leq (3+\alpha_j^y)A_{ijl}^x + (3+\alpha_i^x)B_{ijl}^x \leq (1+\alpha_i^x)(3+\alpha_j^y)W_{ijl}^x \,, \tag{50}$$

$$(1+\alpha_j^y)V_{ijk}^y \leq A_{ijk}^y \leq (1+\alpha_j^y)W_{ijk}^y \,,$$

$$(1+\alpha_j^y)(3+\alpha_i^x)V_{ijk}^y \leq (3+\alpha_i^x)A_{ijk}^y + (3+\alpha_j^y)B_{ijk}^y \leq (1+\alpha_j^y)(3+\alpha_i^x)W_{ijk}^y \,, \tag{51}$$

for $k = 0, 1$, $l = 0, 1$, $i = 1, \ldots, n$, $j = 1, \ldots, m$. Here the following abbreviations are used,

$$V_{ijl}^x = L_{i,j}^x - \tau_{i,j-l}^x \,, \quad W_{ijl}^x = U_{i,j}^x - \tau_{i,j-l}^x \,,$$

$$V_{ijk}^y = L_{i,j}^y - \tau_{i-k,j}^y \,, \quad W_{ijk}^y = U_{i,j}^y - \tau_{i-k,j}^y \,,$$

$$A_{ijl}^x = \tau_{i,j-l}^x - \tau_{i-1,j-l}^x \,, \quad B_{ijl}^x = (-1)^{l+1}h_j^y\tau_{i,j-l}^{xy} \,, \tag{52}$$

$$A_{ijk}^y = \tau_{i-k,j}^y - \tau_{i-k,j-1}^y \,, \quad B_{ijk}^y = (-1)^{k+1}h_i^x\tau_{i-k,j}^{xy} \,.$$

The inequalities (50), (51) have the structure

$$1 + \alpha - A \geq 0 \,,$$

$$(1+\alpha)(3+\beta) - A(3+\beta) - B(3+\alpha) \geq 0 \,. \tag{53}$$

For example, if the right hand inequalities (50) are considered we have to set

$$A = A_{ijl}^x/W_{ijl}^x \,, \; B = B_{ijl}^x/W_{ijl}^x \,, \; \alpha = \alpha_i^x \,, \; \beta = \alpha_j^y \,. \tag{54}$$

Further systems (53) occurring in (50) and (51) are obtained taking

$$A = A^x_{ijl}/V^x_{ijl}, \ B = B^x_{ijl}/V^x_{ijl}, \ \alpha = \alpha^x_i, \ \beta = \alpha^y_j, \tag{55}$$

or

$$A = A^y_{ijk}/W^y_{ijk}, \ B = B^y_{ijk}/W^y_{ijk}, \ \alpha = \alpha^y_j, \ \beta = \alpha^x_i, \tag{56}$$

or

$$A = A^y_{ijk}/V^y_{ijk}, \ B = B^y_{ijk}/V^y_{ijk}, \ \alpha = \alpha^y_j, \ \beta = \alpha^x_i. \tag{57}$$

Proposition 3.3 *The requirements $\alpha \geq 0$, $\beta \geq 0$ and*

$$1 + \alpha \geq 2 + 2A$$
$$3 + \beta \geq 2B \tag{58}$$

being linear in α, β imply the quadratic system of inequalities (53).

Proof. In view of (58) we obtain the first inequality (53),

$$1 + \alpha \geq (1 + \alpha)/2 \geq 1 + A \geq A,$$

and $3 + \beta \geq (3 + \beta)/2 \geq B$. The second inequality (53) is equivalent to

$$(1 + \alpha - A)(3 + \beta - B) \geq (2 + A)B. \tag{59}$$

Hence, the left hand side of (59) is nonnegative. Therefore, the condition (59) holds true in both cases $2 + A \geq 0$, $B \leq 0$, and $2 + A \leq 0$, $B \geq 0$. Further, for $2 + A \geq 0$, $B \geq 0$ the inequality (59) is valid if $1 + \alpha - A \geq 2 + A$, $3 + \beta - B \geq B$, i.e., if the conditions (58) are assumed. In the fourth case $2 + A \leq 0$, $B \leq 0$, it suffices to require $1 + \alpha - A \geq -2 - A$, $3 + \beta - B \geq -B$ being valid in view of $\alpha \geq 0$, $\beta \geq 0$. ∎

Now we apply Proposition 3.3 to the mentioned systems (53), (54)–(57). The result is

Theorem 3.4 *Let the requirement (49) of compatibility be valid. Then an interpolant $s \in S^1_G(\Delta^x) \otimes S^1_G(\Delta^y)$ with the derivatives (45) and (46) satisfies the strip conditions (29) on the first-order derivatives if the nonnegative tension parameters are chosen as follows. For $i \in \{1, \ldots, n\}$ let*

$$\alpha^x_i \geq -1 + 2\left(1 + \max\left\{\frac{A^x_{ijl}}{V^x_{ijl}}, \frac{A^x_{ijl}}{W^x_{ijl}}\right\}\right),$$

$$\alpha^x_i \geq -3 + 2\max\left\{\frac{B^y_{ijk}}{V^y_{ijk}}, \frac{B^y_{ijk}}{W^y_{ijk}}\right\}, \ k = 0, 1, \ l = 0, 1, \ j = 1, \ldots, m, \tag{60}$$

and for $j \in \{1, \ldots, m\}$ let

$$\alpha_j^y \geq -1 + 2 \left(1 + \max \left\{ \frac{A_{ijk}^y}{V_{ijk}^y}, \frac{A_{ijk}^y}{W_{ijk}^y} \right\} \right),$$

$$\alpha_j^y \geq -3 + 2 \max \left\{ \frac{B_{ijl}^x}{V_{ijl}^x}, \frac{B_{ijl}^x}{W_{ijl}^x} \right\}, \quad k = 0, 1, \ l = 0, 1, \ i = 1, \ldots, n. \tag{61}$$

The abbreviations used here are defined in (38), (39), and (52).

Further, we briefly consider bivariate co-monotone interpolation. Now the restrictions (29) are

$$\varepsilon_{i,j}^x \frac{\partial}{\partial x} s(x, y) \geq 0, \ \varepsilon_{i,j}^y \frac{\partial}{\partial y} s(x, y) \geq 0 \text{ for } (x, y) \in [x_{i-1}, x_i] \times [y_{j-1}, y_j], \tag{62}$$

$i = 1, \ldots, n, \ j = 1, \ldots, m$, with the signs

$$\varepsilon_{i,j}^x = \begin{cases} +1 & \text{if } \tau_{i,j-l}^x \geq 0 \text{ for } l = 0, 1, \\ -1 & \text{if } \tau_{i,j-l}^x \leq 0 \text{ for } l = 0, 1, \\ 0 & \text{if } \tau_{i,j-1}^x > 0, \ \tau_{i,j}^x < 0 \text{ or } \tau_{i,j-1}^x < 0, \ \tau_{i,j}^x > 0, \end{cases} \tag{63}$$

$$\varepsilon_{i,j}^y = \begin{cases} +1 & \text{if } \tau_{i-k,j}^y \geq 0 \text{ for } k = 0, 1, \\ -1 & \text{if } \tau_{i-k,j}^y \leq 0 \text{ for } k = 0, 1, \\ 0 & \text{if } \tau_{i-1,j}^y > 0, \ \tau_{i,j}^y < 0 \text{ or } \tau_{i-1,j}^y < 0, \ \tau_{i,j}^y > 0. \end{cases} \tag{64}$$

For $z_{i-1,j-1} = z_{i,j-1} = z_{i-1,j} = z_{i,j}$, one should decide on one of the possibilities $+1$ or -1 in (63) as well as in (64). Since $\varepsilon_{i,j}^x \tau_{i,j-l}^x \geq 0$, $\varepsilon_{i,j}^y \tau_{i-k,j}^y \geq 0$ for $k = 0, 1$, $l = 0, 1$, $i = 1, \ldots, n$, $j = 1, \ldots, m$, the system (43), (44) adapted to the co-monotonicity, namely

$$\varepsilon_{i,j}^x p_{i-k,j-l} \geq 0, \ \varepsilon_{i,j}^x \left((3 + \alpha_i^x) \tau_{i,j-l}^x - p_{i,j-l} - p_{i-1,j-l} \right) \geq 0,$$

$$\varepsilon_{i,j}^x \left((3 + \alpha_j^y) p_{i-k,j-l} + (-1)^{l+1} h_j^y r_{i-k,j-l} \right) \geq 0,$$

$$\varepsilon_{i,j}^x \left((3 + \alpha_i^x)(3 + \alpha_j^y) \tau_{i,j-l}^x - (3 + \alpha_j^y)(p_{i,j-l} + p_{i-1,j-l}) + \right.$$

$$\left. + (-1)^{l+1}(3 + \alpha_i^x) h_j^y \frac{q_{i,j-l} - q_{i-1,j-l}}{h_i^x} - (-1)^{l+1} h_j^y (r_{i,j-l} + r_{i-1,j-l}) \right) \geq 0, \tag{65}$$

$$\varepsilon_{i,j}^y q_{i-k,j-l} \geq 0, \ \varepsilon_{i,j}^y \left((3 + \alpha_j^y) \tau_{i-k,j}^y - q_{i-k,j} - q_{i-k,j-1} \right) \geq 0,$$

$$\varepsilon_{i,j}^y \left((3 + \alpha_i^x) q_{i-k,j-l} + (-1)^{k+1} h_i^x r_{i-k,j-l} \right) \geq 0,$$

$$\varepsilon_{i,j}^y \left((3 + \alpha_j^y)(3 + \alpha_i^x) \tau_{i-k,j}^y - (3 + \alpha_i^x)(q_{i-k,j} + q_{i-k,j-1}) + \right.$$

$$\left. + (-1)^{k+1}(3 + \alpha_j^y) h_i^x \frac{p_{i-k,j} - p_{i-k,j-1}}{h_j^y} - (-1)^{k+1} h_i^x (r_{i-k,j} + r_{i-k,j-1}) \right) \geq 0,$$

$$k = 0, 1, \ l = 0, 1, \ i = 1, \ldots, n, \ j = 1, \ldots, m,$$

$$(66)$$

has the solution $p_{i,j} = q_{i,j} = r_{i,j} = 0$, $i = 0, \ldots, n$, $j = 0, \ldots, m$, and, moreover, the choices $\alpha_i^x = 0$, $i = 1, \ldots, n$, and $\alpha_j^y = 0$, $j = 1, \ldots, m$, are feasible. Thus, bivariate co-monotone interpolation is always possible in the class of bicubic C^1 splines.

In general, the bivariate constrained interpolant obtained by Theorem 3.4, and especially the described co-monotone interpolant should be improved by applying an optimization procedure; compare with Section 2.4 for the corresponding univariate approach. Suitable proposals for a bivariate choice functional can be found in [3, 7, 11], and in several other papers.

3.4   Comments

Interpolation subject to strip conditions on the first-order derivatives is of some practical interest. A rather general example is given. We have proposed algorithms to solve this problem for univariate as well as for gridded bivariate data. The point is to use Gregory's rational cubic splines. This spline class allows algorithms which are always successful provided the data and the bounds are compatible. The described procedures are easy to implement.

Acknowledgment. Both authors are supported by the Deutsche Forschungs-gemeinschaft under grant Schw 469/3–1.

References

[1] L.-E. Andersson and T. Elfving, *Best constrained approximation in Hilbert space and interpolation by cubic splines subject to obstacles*, SIAM J. Sci. Comput., **16** (1995), 1209–1232.

[2] M. Bastian-Walther and J. W. Schmidt, *Range restricted interpolation using Gregory's rational cubic splines*, J. Comp. Appl. Math., to appear.

[3] M. Bastian-Walther and J. W. Schmidt, *Shape preserving interpolation by tensor product splines on refined grids*, In H. Nowacki and P. D. Kaklis,

editors, *Creating Fair and Shape-Preserving Curves and Surfaces*, Teubner Verlag, Stuttgart.Leipzig, 1998, 201–217.

[4] R. Delbourgo and J. A. Gregory, *Shape preserving piecewise rational interpolation*, SIAM J. Sci. Statist. Comput., **6** (1985), 967–976.

[5] J. A. Gregory, *Shape preserving spline interpolation*, Computer-aided design, **18** (1986), 53–57.

[6] B. Mulansky, *Tensor products of convex cones*, In G. Nürnberger, J. W. Schmidt, and G. Walz, editors, *Multivariate Approximation and Splines*, volume 125 of Internat. Series Numer. Math., Birkhäuser, Basel, 1997, 167–176.

[7] B. Mulansky and J. W. Schmidt, *Nonnegative interpolation by biquadratic splines on refined rectangular grids*, In P. J. Laurent, A. Le Méhauté, and L. L. Schumaker, editors, *Wavelets, Images and Surface Fitting*, A K Peters, Wellesley, 1994, 379–386.

[8] J. W. Schmidt, *Convex smoothing by splines and dualization*, In Bl. Sendov, R. Lazarov, and I. Dimov, editors, *Numerical Methods and Applications*, Publishing House Bulgar. Acad. Sciences., Sofia, 1989, 425–436.

[9] J. W. Schmidt, *Interpolation in a derivative strip*, Computing, **58** (1997), 377–389.

[10] J. W. Schmidt and W. Heß, *Numerical methods in strip interpolations applying C^1, C^2, and C^3 splines on refined grids*, Mitt. Math. Ges. Hamburg, **16** (1997), 107–135.

[11] J. W. Schmidt and M. Walther, *Gridded data interpolation with restrictions on the first order derivatives*, In G. Nürnberger, J. W. Schmidt, and G. Walz, editors, *Multivariate Approximation and Splines*, volume 125 of Internat. Series Numer. Math., Birkhäuser, Basel, 1997, 291–307.

[12] T. Schütze and H. Schwetlick, *Constrained approximation by splines with free knots*, BIT, **37** (1997), 105–137.

[13] H. Schwetlick and V. Kunert, *Spline smoothing under constraints on derivatives*, BIT, **33** (1993), 512–528.

[14] H. Späth, *One Dimensional Spline Interpolation Algorithms*, A K Peters, Wellesley, 1995.

Technische Universität
Institut für Numerische Mathematik
D-01062 Dresden, Germany
Email addresses: walther@math.tu-dresden.de   jschmidt@math.tu-dresden.de

International Series of Numerical Mathematics
Vol. 132, © 1999 Birkhäuser Verlag Basel/Switzerland

Interpolation by Splines on Triangulations

Oleg Davydov, Günther Nürnberger and Frank Zeilfelder

Abstract

We review recently developed methods of constructing Lagrange and Hermite interpolation sets for bivariate splines on triangulations of general type. Approximation order and numerical performance of our methods are also discussed.

1 Introduction

Let Δ be a regular triangulation of a simply connected polygonal domain Ω in $\mathbb{R}^2$. Given integer r and q with $q \geq r + 1$, we denote by $S_q^r(\Delta) = \{s \in C^r(\Omega) : s|_T \in \Pi_q \text{ for all } T \in \Delta\}$ the space of **bivariate splines of degree q and smoothness r** (with respect to Δ). Here $\Pi_q = \operatorname{span}\{x^\alpha y^\beta : \alpha, \beta \geq 0, \ \alpha + \beta \leq q\}$ denotes the space of **bivariate polynomials of total degree q**. We investigate the following problem. Construct **Lagrange interpolation set** $\{z_1, \ldots, z_d\}$ in Ω, where $d = \dim S_q^r(\Delta)$, such that for each function $f \in C(T)$, a unique spline $s \in S_q^r(\Delta)$ exists that satisfies the **Lagrange interpolation conditions** $s(z_\nu) = f(z_\nu)$, $\nu = 1, \ldots, d$. If we consider not only function values of f but also partial derivatives, then we speak of **Hermite interpolation conditions**.

In the literature, point sets that admit unique Lagrange and Hermite interpolation by spaces $S_q^r(\Delta)$ of splines of degree q and smoothness r were constructed for crosscut partitions Δ, in particular for Δ^1 and Δ^2-partitions [1, 4, 15, 21, 22, 23, 27, 28]. Results on the approximation order of these interpolation methods were given in [4, 11, 15, 20, 21, 24, 27, 28]. A Hermite interpolation scheme for $S_q^1(\Delta)$, $q \geq 5$, where Δ is an arbitrary triangulation, can be obtained by using a nodal basis of this space constructed in [17] (see also [9]). For $q = 4$ it was shown in [2] that a spline in $S_4^1(\Delta)$ exists which coincides with a given function at the vertices of Δ. Under certain restrictions on the triangulation, analogous results were obtained in [5, 18] for function and gradient values at the vertices. (Note that the dimension of $S_4^1(\Delta)$ is about six times the number of vertices of Δ.)

In this paper we review several new methods of interpolation by bivariate splines on triangulations.

In Section 2 we describe an inductive method [10] for constructing Lagrange and Hermite interpolation points for $S_q^1(\Delta)$, $q \geq 5$, where Δ is an arbitrary triangulation. Here, in each step, one vertex is added to the subtriangulation considered before. For $q = 4$ this method works under certain assumptions on Δ or a slight modification of it.

Section 3 is devoted to an algorithm [12] for constructing point sets that admit unique Lagrange and Hermite interpolation by the space $S_3^1(\Delta)$ of splines of degree 3 defined on a general class of triangulations Δ. Note that for $S_3^1(\Delta)$ even the dimension of the space is not known for arbitrary triangulations, in contrast to the case $q \geq 4$, where dimension formulas are available (cf. [17] for $q \geq 5$ and [2] for $q = 4$). We consider triangulations Δ that consist of nested polygons whose vertices are connected by line segments. In particular, the dimension of $S_3^1(\Delta)$ is determined for triangulations of this type.

In Section 4 we describe an algorithm [25, 26], which, for given points in the plane, constructs a triangulation Δ and, subsequently, Lagrange and Hermite interpolation sets for $S_q^r(\Delta)$, with $r = 1, 2$. Moreover, this method is applied to given quadrangulations with diagonals.

In Section 5 we discuss a Hermite type interpolation scheme [13] for $S_q^r(\Delta)$, $q \geq 3r + 2$, which possesses optimal approximation order $\mathcal{O}(h^{q+1})$. Furthermore, the fundamental functions of the scheme form a stable (for the triangulations that do not contain near-degenerate edges) and locally linearly independent basis for a superspline subspace of $S_q^r(\Delta)$.

Finally, in Section 6 some numerical examples are presented.

2 Interpolation by C^1 Splines of Degree $q \geq 4$

The choice of interpolation points depends on the following properties of edges, vertices and subtriangulations of Δ.

Definition 2.1. (*i*) *An interior edge e with vertex v of the triangulation Δ is called* **degenerate** *at v if the edges with vertex v adjacent to e lie on a line.* (*ii*) *An interior vertex v of Δ is called* **singular** *if v is a vertex of exactly four edges and these edges lie on two lines.* (*iii*) *An interior vertex v of Δ on the boundary of a given subtriangulation Δ' of Δ is called* **semi-singular of type 1** *w.r.t. Δ' if exactly one edge with endpoint v is not contained in Δ' and this edge is degenerate at v.* (*iv*) *An interior vertex v of Δ on the boundary of a given subtriangulation Δ' of Δ is called* **semi-singular of type 2** *w.r.t. Δ' if exactly two edges with endpoint v are not contained in Δ' and these edges are degenerate at v.* (*v*) *A vertex v of Δ is called* **semi-singular** *w.r.t. Δ' if v satisfies (*iii*) or (*iv*).*

Definition 2.2. *We say that $\Delta' \subset \Delta$ is a* **tame subtriangulation** *if the following conditions (T1)–(T3) hold.*

(T1) *$\Omega_{\Delta'} := \bigcup_{T \in \Delta'} T$ is simply connected.*

(T2) *For any two triangles $T', T'' \in \Delta'$ there exists a sequence $\{T_1, \ldots, T_\mu\} \subset \Delta'$ such that T_i and T_{i+1} have a common edge, $i = 1, \ldots, \mu - 1$, $T_1 = T'$, $T_\mu = T''$.*

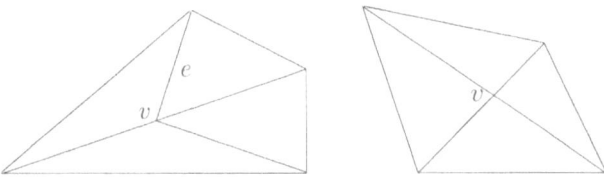

Fig. 2.1. Degenerate edge, respectively singular vertex.

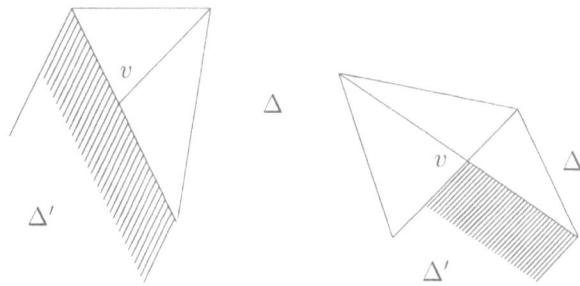

Fig. 2.2. Semi-singular vertex.

(T3) *If two vertices $v_1, v_2 \in \Omega_{\Delta'}$ are connected by an edge e of the triangulation Δ, then $e \subset \Omega_{\Delta'}$.*

Interpolation by C^1 Quartic Splines. We construct a chain of subsets Ω_i of Ω such that $\emptyset = \Omega_0 \subset \Omega_1 \subset \Omega_2 \subset \cdots \subset \Omega_m = \Omega$, and correspond to each Ω_i a set of points $\mathcal{L}_i \subset \Omega_i \setminus \Omega_{i-1}$, $i = 1, \ldots, m$.

For $i = 1$, we take $\Delta_1 = \{T_1\}$, where T_1 is an arbitrarily chosen "starting" triangle in Δ, and set $\Omega_1 := \Omega_{\Delta_1} = T_1$. We choose $\mathcal{L}_1$ to be an arbitrary set of 15 points lying on T_1 and admissible for Lagrange interpolation from Π_4. (For example, we choose five parallel line segments l_ν in T_1 and ν different points on each l_ν, $\nu = 1, 2, 3, 4, 5$.)

Proceeding by induction, we take $i \geq 2$ and suppose that Δ_{i-1} has already been defined and is a tame subtriangulation of Δ, with $\Omega_{i-1} := \Omega_{\Delta_{i-1}}$ being a proper subset of Ω. In order to construct Δ_i, we choose a vertex $v_i \in \Omega \setminus \Omega_{i-1}$ such that v_i is connected to vertices $v_{i,0}, v_{i,1}, \ldots, v_{i,\mu_i} \in \Omega_{i-1}$, where $\mu_i \geq 1$, and the subtriangulation $\Delta_i := \Delta_{i-1} \cup \{T_{i,1}, \ldots, T_{i,\mu_i}\}$, where $T_{i,j} := \langle v_i, v_{i,j-1}, v_{i,j} \rangle$, is tame. (Existence of at least one v_i with this property is shown in [10].) Thus, we set $\Omega_i := \Omega_{\Delta_i}$ (see Figure 2.3).

In order to describe $\mathcal{L}_i$, we need additional notation. Denote by $\hat{e}_{i,j}$ the edge attached to v_i and $v_{i,j}$, $j = 0, \ldots, \mu_i$, $i = 2, \ldots, m$ (see Figure 2.4). For each $i \in \{2, \ldots, m\}$ we define $J_i \subset \{0, \ldots, \mu_i\}$ as follows: 1) for $j \in \{1, \ldots, \mu_i - 1\}$, we have $j \in J_i$ if and only if $\hat{e}_{i,j}$ is nondegenerate at $v_{i,j}$; 2) for $j \in \{0, \mu_i\}$, we have

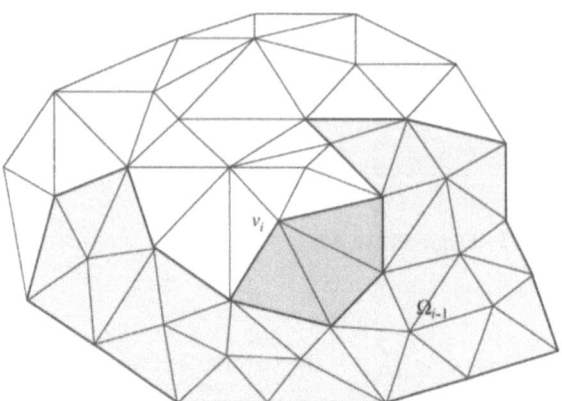

Fig. 2.3. Construction of subtriangulation Δ_i.

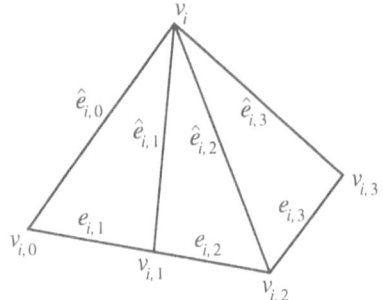

Fig. 2.4. $\Omega_i \setminus \Omega_{i-1}$.

$j \in J_i$ if and only if $v_{i,j}$ is semisingular w.r.t. Δ_i, and $\hat{e}_{i,j}$ is nondegenerate at $v_{i,j}$. Moreover, for every $i \in \{2, \ldots, m\}$ we set $\theta_i := 1$ if v_i is semisingular w.r.t. Δ_i but nonsingular, and $\theta_i := 0$ otherwise.

We consider three cases.

Case 1. Suppose that $\theta_i = 0$. Then $\mathcal{L}_i \subset \Omega_i \setminus \Omega_{i-1}$ consists of

- the vertex v_i,

- any point $w_{i,j}$ in the interior of the edge $\hat{e}_{i,j}$, for each $j \in \{0, \ldots, \mu_i\} \setminus J_i$,

- two points w_i', w_i'' in the interiors of two noncollinear edges $\hat{e}_{i,j'}$ and $\hat{e}_{i,j''}$ respectively, for some $j', j'' \in \{0, \ldots, \mu_i\}$, and

- any point z_i in the interior of a triangle $T_{i,j'''}$, for some $j''' \in \{1, \ldots, \mu_i\}$.

Case 2. Suppose that $\theta_i = 1$ and there exists $j^* \in \{0, \mu_i\} \setminus J_i$, such that $\hat{e}_{i,j^*}$ is nondegenerate at v_i. Then $\mathcal{L}_i \subset \Omega_i \setminus \Omega_{i-1}$ consists of

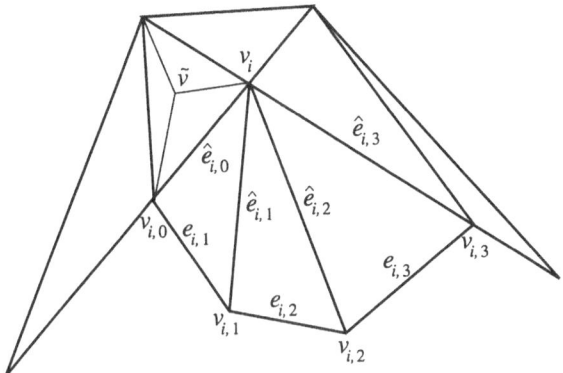

Fig. 2.5. Clough-Tocher split of a triangle in Case 3.

- the vertex v_i,

- any point $w_{i,j}$ in the interior of the edge $\hat{e}_{i,j}$, for each $j \in \{0, \ldots, \mu_i\} \setminus (J_i \cup \{j^*\})$,

- two points w_i', w_i'' in the interiors of two noncollinear edges $\hat{e}_{i,j'}$ and $\hat{e}_{i,j''}$ respectively, for some $j', j'' \in \{0, \ldots, \mu_i\} \setminus \{j^*\}$, and

- any point z_i in the interior of a triangle $T_{i,j'''}$, for some $j''' \notin \{j^*, j^*+1\}$.

Case 3. Suppose that $\theta_i = 1$ and $\hat{e}_{i,j}$ is degenerate at v_i for every $j \in \{0, \mu_i\} \setminus J_i$. Then we need to slightly modify the triangulation by performing a Clough-Tocher split of the triangle $\tilde{T}$ that lies outside Ω_i and shares the edge $\hat{e}_{i,0}$ with $T_{i,1}$. Therefore, we add a new vertex $\tilde{v}$ in the interior of $\tilde{T}$ and connect $\tilde{v}$ with three edges to each of the vertices of $\tilde{T}$ (see Figure 2.5). After this modification vertex v_i is no longer semisingular w.r.t. Δ_i, hence $\theta_i = 0$, and we choose $\mathcal{L}_i \subset \Omega_i \setminus \Omega_{i-1}$ according to the rule described in Case 1. Furthermore, we choose $v_{i+1} := \tilde{v}$. It is easy to see that Δ_{i+1} defined by adding to Δ_i the triangle with vertices v_i, $v_{i,0}$ and $\tilde{v}$, is a tame subtriangulation of Δ. Moreover, we have $\theta_{i+1} = 0$. Thus, we choose $\mathcal{L}_{i+1} \subset \Omega_{i+1} \setminus \Omega_i$ according to Case 1. We denote the resulting modified triangulation by Δ^*.

Theorem 2.3. [10] *The set of points $\mathcal{L} := \bigcup_{i=1}^m \mathcal{L}_i$ described above is a Lagrange interpolation set for $S_4^1(\Delta^*)$. In particular, $\Delta^* = \Delta$ if Case 3 does not occur.*

Remark 2.4. (i) We note that Case 3 is an exceptional case. Among other things, its occurrence requires that one vertex of Δ should be connected with five vertices lying on a line (if v_i is semisingular of type I) or two vertices of Δ should be connected with four vertices lying on a line (if v_i is semisingular of type II: see Figure 2.5). Therefore, no modification of Δ is needed if each vertex is connected

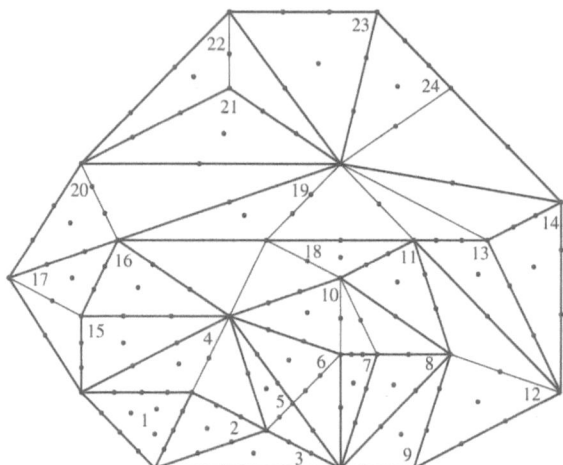

Fig. 2.6. Location of Lagrange interpolation points for $S_4^1(\Delta)$.

with at most three vertices lying on a line. In particular, this last property is satisfied for any triangulation obtained from an arbitrary convex quadrangulation by inserting one or two diagonals of each quadrilateral. (ii) We also note that our method works without modifying Δ if the total number of edges attached to v_i is odd. Then $\mathcal{L}_i$ is defined in Case 3 as in Case 1, with the point z_i being removed.

Remark 2.5. Lagrange interpolation of f at some points of the above scheme can be replaced by interpolation of appropriate first or second partial derivatives of f provided that such derivatives exist. Namely, interpolation of f at w_i', w_i'' can be replaced by the conditions $D_x s(v_i) = D_x f(v_i)$, $D_y s(v_i) = D_y f(v_i)$, interpolation of f at $w_{i,j}$ can be replaced by $D_{\hat{e}_{i,j}}^2 s(v_i) = D_{\hat{e}_{i,j}}^2 f(v_i)$, and interpolation of f at z_i can be replaced by $D_{\hat{e}_{i,j'''-1}} D_{\hat{e}_{i,j'''}} s(v_i) = D_{\hat{e}_{i,j'''-1}} D_{\hat{e}_{i,j'''}} f(v_i)$. (Here and below we use the notations $D_x f := \frac{\partial f}{\partial x}$, $D_y f := \frac{\partial f}{\partial y}$ and $D_\sigma f := \sigma_x D_x f + \sigma_y D_y f$, where $\sigma = (\sigma_x, \sigma_y)$ is a unit vector in the plane. For simplicity, we write $D_e f := D_\sigma f$, where σ is the unit vector parallel to the edge e.) Particularly, our Hermite interpolation scheme includes the function and gradient values at all vertices of the triangulation.

Remark 2.6. The computation of the interpolating spline $s \in S_4^1(\Delta)$ according to our scheme is easy to perform step by step, by constructing $s|_{\Omega_i \setminus \Omega_{i-1}}$ after $s|_{\Omega_{i-1} \setminus \Omega_{i-2}}$. This can always be done by solving small systems of linear equations. Moreover, for Δ^1 and Δ^2 triangulations our method leads to the interpolation schemes developed by Nürnberger [20] and Nürnberger & Walz [24], respectively. These schemes possess (nearly) optimal approximation order. (For certain classes of triangulations, quasi-interpolation methods for $S_4^1(\Delta)$ were developed in [5, 6].)

Interpolation by C^1 Splines of Degree $q \geq 5$. We construct a chain of subsets Ω_i of Ω as above and correspond to each Ω_i a set of points $\mathcal{L}_i^{(q)} \subset \Omega_i \setminus \Omega_{i-1}$, $i = 1, \ldots, m$, as follows. (In this case no modification of the given triangulation Δ is necessary.) Namely, we choose $\mathcal{L}_1^{(q)}$ to be an arbitrary set of $d_q := \binom{q+2}{2}$ points lying on T_1 and admissible for Lagrange interpolation from Π_q. In order to define $\mathcal{L}_i^{(q)}$ we consider two cases.

Case 1. Suppose that $\theta_i = 0$. Then $\mathcal{L}_i^{(q)} \subset \Omega_i \setminus \Omega_{i-1}$ consists of

- the vertex v_i,

- any $q - 3$ distinct points $w_{i,j}^{(1)}, \ldots, w_{i,j}^{(q-3)}$ in the interior of the edge $\hat{e}_{i,j}$, for each $j \in \{0, \ldots, \mu_i\} \setminus J_i$,

- any $q - 4$ distinct points $w_{i,j}^{(1)}, \ldots, w_{i,j}^{(q-4)}$ in the interior of the edge $\hat{e}_{i,j}$, for each $j \in J_i$,

- two points w_i', w_i'' in the interiors of two noncollinear edges $\hat{e}_{i,j'}$ and $\hat{e}_{i,j''}$ respectively, for some $j', j'' \in \{0, \ldots, \mu_i\}$,

- any d_{q-4} distinct points $z_{i,j'''}^{(1)}, \ldots, z_{i,j'''}^{(d_q-4)}$ lying in the interior of a triangle $T_{i,j'''}$, for some $j''' \in \{1, \ldots, \mu_i\}$, and admissible for Lagrange interpolation from Π_{q-4}, and

- any d_{q-5} distinct points $z_{i,j}^{(1)}, \ldots, z_{i,j}^{(d_q-5)}$ lying in the interior of $T_{i,j}$ and admissible for Lagrange interpolation from Π_{q-5}, for each $j \in \{1, \ldots, \mu_i\} \setminus \{j'''\}$.

Case 2. Suppose that $\theta_i = 1$. (Hence, there exists $j^* \in \{0, \mu_i\}$, such that $\hat{e}_{i,j^*}$ is nondegenerate at v_i.) Then $\mathcal{L}_i^{(q)} \subset \Omega_i \setminus \Omega_{i-1}$ consists of

- the vertex v_i,

- any $q - 3$ distinct points $w_{i,j}^{(1)}, \ldots, w_{i,j}^{(q-3)}$ in the interior of the edge $\hat{e}_{i,j}$, for each $j \in \{0, \ldots, \mu_i\} \setminus (J_i \cup \{j^*\})$,

- any $q - 4$ distinct points $w_{i,j}^{(1)}, \ldots, w_{i,j}^{(q-4)}$ in the interior of the edge $\hat{e}_{i,j}$, for each $j \in J_i \setminus \{j^*\}$,

- any $q - \kappa$ distinct points $w_{i,j^*}^{(1)}, \ldots, w_{i,j^*}^{(q-\kappa)}$ in the interior of the edge $\hat{e}_{i,j^*}$, where $\kappa = 5$ if $j^* \in J_i$, and $\kappa = 4$ if $j^* \notin J_i$,

- two points w_i', w_i'' in the interiors of two noncollinear edges $\hat{e}_{i,j'}$ and $\hat{e}_{i,j''}$ respectively, for some $j', j'' \in \{0, \ldots, \mu_i\} \setminus \{j^*\}$,

- any d_{q-4} distinct points $z_{i,j'''}^{(1)}, \ldots, z_{i,j'''}^{(d_q-4)}$ lying in the interior of a triangle $T_{i,j'''}$, for some $j''' \in \{1, \ldots, \mu_i\} \setminus \{j^*, j^* + 1\}$, and admissible for Lagrange interpolation from Π_{q-4}, and

- any d_{q-5} distinct points $z_{i,j}^{(1)}, \ldots, z_{i,j}^{(d_q-5)}$ lying in the interior of $T_{i,j}$ and admissible for Lagrange interpolation from Π_{q-5}, for each $j \in \{1, \ldots, \mu_i\} \setminus \{j'''\}$.

Theorem 2.7. [10] *The set of points $\mathcal{L}^{(q)} := \bigcup_{i=1}^{m} \mathcal{L}_i^{(q)}$ described above is a Lagrange interpolation set for $S_q^1(\Delta)$, $q \geq 5$.*

Remark 2.8. As in the case $q = 4$, our Lagrange interpolation scheme can be transformed into an appropriate Hermite interpolation scheme (cp. Remark 2.5). Moreover, Remark 2.6 about computation and approximation order of our interpolation method remains true in the case $q \geq 5$.

3   Interpolation by Cubic Splines

The Class of Triangulations. In this section we consider the following general type of triangulations Δ. The vertices of Δ are the vertices of closed simple polygons $P_0, P_1, \ldots, P_k$ which are nested and one vertex inside P_0. This means that $\Omega_{\mu-1} \subset \Omega_\mu$, where Ω_μ is the closed (not necessarily convex) polyhedron with boundary $P_\mu, \mu = 0, \ldots, k$, and Δ is a triangulation of $\Omega := \Omega_k$ (see Figure 3.1). To be more precise, we note that the vertices of P_μ are connected by line segments with the vertices of $P_{\mu+1}$, $\mu = 0, \ldots, k - 1$. On the other hand, for each closed simple polygon P_μ, there is no additional line segment connecting two vertices of P_μ, $\mu = 0, \ldots, k$. In order to construct interpolation points for $S_3^1(\Delta)$, we assume that the triangulation Δ has the following properties:

(C1) Each vertex of P_μ is connected with at least two vertices of $P_{\mu+1}, \mu = 0, \ldots, k - 1$.

(C2) There exist vertices w_μ of $P_\mu, \mu = 0, \ldots, k$, such that w_μ and $w_{\mu+1}$ are connected, and each vertex w_μ is connected with at least three vertices of $P_{\mu+1}, \mu = 0, \ldots, k - 1$.

Remark 3.1. (*i*) Since the polygons P_μ grow with increasing index μ, it is natural to assume that the number of vertices of $P_{\mu+1}$ is greater than the number of vertices of P_μ, $\mu = 0, \ldots, k - 1$. Then it is natural to connect the vertices of the polygons in such a way that the properties (C1) and (C2) are satisfied. (*ii*) Moreover, the properties (C1) and (C2) of Δ remain valid if Δ is **deformed**, i.e., the location of the vertices of Δ are changed but the connection of the vertices remain unchanged. (In other words, the graphs of the triangulation Δ and the deformed triangulation are the same.)

Decomposition of the Domain. In order to construct interpolation points, we decompose the domain Ω into finitely many sets $V_0 \subset V_1 \subset \ldots \subset V_m = \Omega$, where each set V_i, is the union of closed triangles of $\Delta, i = 0, \ldots, m$. Let V_0 be an arbitrary

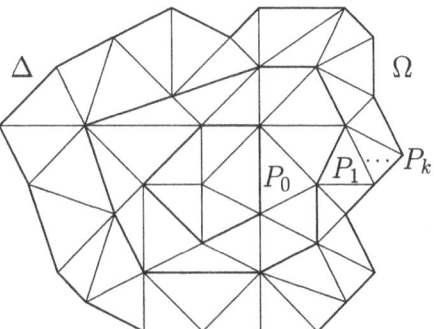

Fig. 3.1. Triangulation Δ (nested polygons).

closed triangle of Δ in Ω_0. We define the sets $V_1 \subset \ldots \subset V_m$ by induction according to the following `rule`: If V_{i-1} is defined, then we choose a vertex v_i of Δ with the following property: Let $T_{i,1}, \ldots, T_{i,n_i}$ $(n_i \geq 1)$ be all triangles of Δ with vertex v_i having a common edge with V_{i-1}. (Since Δ satisfies property (C1), we have $n_i \leq 2$.) We set $V_i = V_{i-1} \cup \overline{T}_{i,1} \cup \ldots \cup \overline{T}_{i,n_i}$. (Note that we choose the vertex v_i in such a way that at least one such triangle exists.) The vertices $v_i, i = 1, \ldots, m$, are chosen as follows. After choosing V_0 to be an arbitrary closed triangle of Δ in Ω_0, we pass through the vertices of P_0 in clockwise order by applying the above rule. (It is clear that the choice of these vertices is unique.) Now, we assume that we have passed through the vertices of $P_{\mu-1}$. Then w.r.t. clockwise order, we choose the first vertex of P_μ greater than w_μ which is connected with at least two vertices of $P_{\mu-1}$. Then we pass through the vertices of P_μ in clockwise order until w_μ^- and pass through the vertices of P_μ in anticlockwise order until w_μ^+ by applying the above rule. (Here w_μ^+ denotes the vertex next to w_μ in clockwise order and w_μ^- denotes the vertex next to w_μ in anticlockwise order.) Finally, we choose the vertex w_μ. (It is clear that the choice of the vertices is unique.) In this way, we obtain the sets $V_0 \subset V_1 \subset \ldots \subset V_m = \Omega$.

Construction of Interpolation Sets. Now, we construct interpolation sets for $S_3^1(\Delta)$ inductively as follows. First, we choose interpolation points on V_0 and then on $V_i \setminus V_{i-1}, i = 1, \ldots, m$. In the first step, we choose 10 different points (respectively 10 Hermite interpolation conditions) on V_0 which admit unique Lagrange interpolation by the space Π_3. (For example, we may choose four parallel line segments l_ν in V_0 and ν different points on each $l_\nu, \nu = 1, 2, 3, 4$.)

Now, we assume that we have already chosen interpolation points on V_{i-1}. Then we choose interpolation points on $V_i \setminus V_{i-1}$ as follows. By the above decomposition of Ω, $V_i \setminus V_{i-1}$ is the union of consecutive triangles $T_{i,1}, \ldots, T_{i,n_i}$ with vertex v_i having common edges with V_{i-1}. We denote the consecutive endpoints of these edges by $v_{i,0}, v_{i,1}, \ldots, v_{i,n_i}$. Moreover, the edges $[v_{i,j}, v_i]$ are denoted by $e_{i,j}, j = 0, \ldots, n_i$ (see Figure 3.2).

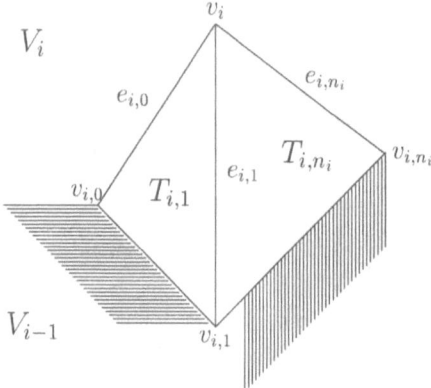

Fig. 3.2. The set $V_i \setminus V_{i-1}$.

The choice of interpolation points on $V_i \setminus V_{i-1}$ depends on the following properties of the subtriangulation $\Delta_i = \{T \in \Delta : T \subset V_i\}$ at the vertices $v_{i,0}, \ldots, v_{i,n_i}$:
(i) $e_{i,j}$ in non-degenerate at $v_{i,j}$. (ii) $e_{i,j}$ is non-degenerate at $v_{i,j}$ and in addition, $v_{i,j}$ is semi-singular w.r.t. Δ_i.

For $j \in \{0, n_i\}$, we set $c_{i,j} = 1$ if (ii) holds; and $c_{i,j} = 0$ otherwise. For j with $0 < j < n_i$, we set $c_{i,j} = 1$ if (i) holds; and $c_{i,j} = 0$ otherwise. Moreover, we set $c_i = \sum_{j=0}^{n_i} c_{i,j}$ and note that $0 \leq c_i \leq 3$. For **Lagrange interpolation**, we choose the following points on $V_i \setminus V_{i-1}$: If $c_i = 3$, then no point is chosen. If $c_i = 2$, then we choose v_i. If $c_i = 1$, then we choose v_i and one further point on some edge $e_{i,j}$ with $c_{i,j} = 0$. If $c_i = 0$, then we choose v_i and two further points on two different edges. For **Hermite interpolation**, we require the following interpolation conditions for $s \in S_3^1(\Delta)$ at the vertex v_i: If $c_i = 3$, then no interpolation condition is required at v_i. If $c_i = 2$, then we require $s(v_i) = f(v_i)$. If $c_i = 1$, then we require $s(v_i) = f(v_i)$ and $D_{e_{i,j}} s(v_i) = D_{e_{i,j}} f(v_i)$, where $e_{i,j}$ is some edge with $c_{i,j} = 0$. If $c_i = 0$, then we require $s(v_i) = f(v_i)$, $D_x s(v_i) = D_x f(v_i)$ and $D_y s(v_i) = D_y f(v_i)$. By the above construction, we obtain a set of points for Lagrange interpolation respectively a set of Hermite interpolation conditions.

Theorem 3.2. [12] *If the triangulation Δ satisfies the properties (C1) and (C2), then a unique spline in $S_3^1(\Delta)$ exists which satisfies the above Lagrange (respectively Hermite) interpolation conditions. In particular, the total number of interpolation conditions is equal to the dimension of $S_3^1(\Delta)$.*

Corollary 3.3. *Let Δ be a deformed Δ^1-partition. Then a unique spline in $S_3^1(\Delta)$ exists which satisfies the Lagrange (respectively Hermite) interpolation conditions obtained by our method.*

We note that the basic principle of passing through the vertices of the nested polygons of Δ can also be applied to the space $S_q^1(\Delta), q \geq 4$, in combination with the algorithm for constructing interpolation points in Section 2. Then, in contrast to Section 2, the choice of the vertices is unique as soon as the nested polygons, the starting triangle and the vertices w_μ have been identified.

4 Interpolation by Splines on Triangulations of Given Points

In this section, we construct a natural triangulation Δ for given points in the plane. The triangulation Δ is suitable for interpolation by $S_q^1(\Delta), q \geq 3$, respectively $S_q^2(\Delta), q \geq 5$.

Construction of the Triangulation. Let a set V of finitely many distinct points in $\mathbb{R}^2$ be given. We assume that V contains sufficiently many points. The triangulation Δ is constructed inductively as follows.

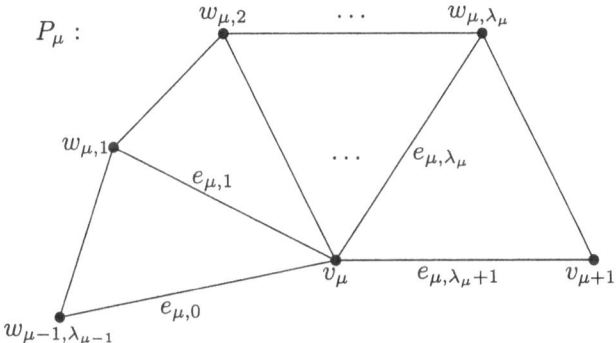

Fig. 4.1. Adding a polyhedron P_μ.

In the first step, we choose three points $v_1, v_2, v_3 \in V$ such that no point of V lies in the interior of the triangle formed by v_1, v_2, v_3. We assume that for a given subset $\tilde{V}$ of V, a simply connected triangulation $\tilde{\Delta}$ is already constructed with vertices in $\tilde{V}$. For simplicity, we denote the vertices on the boundary of $\tilde{\Delta}$ again by $v_1, \ldots, v_n$ (in clockwise order). For $\mu = 1, \ldots, n$, we choose points $w_{\mu,1}, \ldots, w_{\mu,\lambda_\mu} \in V \setminus \tilde{V}$, $\lambda_\mu \geq 1$ (in clockwise order) such that no point of $V \setminus \tilde{V}$ lies in the interior of the polyhedron P_μ formed by the points $v_\mu, w_{\mu-1,\lambda_{\mu-1}}, w_{\mu,1}, \ldots, w_{\mu,\lambda_\mu}, v_{\mu+1}$, where $w_{0,\lambda_0} := v_n$ and $v_{n+1} := w_{1,1}$. We connect the points $w_{\mu,1}, \ldots, w_{\mu,\lambda_\mu}$ with v_μ by line segments and denote the edges of P_μ with endpoint v_μ by $e_{\mu,0}, \ldots, e_{\mu,\lambda_\mu+1}$ (in clockwise order). We choose enough points $w_{\mu,1}, \ldots, w_{\mu,\lambda_\mu}$ such that $\lambda_\mu \geq 2$ if two edges in $\{e_{\mu,0}, \ldots, e_{\mu,\lambda_\mu+1}\}$ have the same slope. Analogously, we choose $\lambda_\mu \geq 3$ if an edge in $\{e_{\mu,1}, \ldots, e_{\mu,\lambda_\mu}\}$ has the same slope as $e_{\mu,0}$ and a further edge in $\{e_{\mu,1}, \ldots, e_{\mu,\lambda_\mu}\}$ has the same slope as $e_{\mu,\lambda_\mu+1}$.

P_μ:

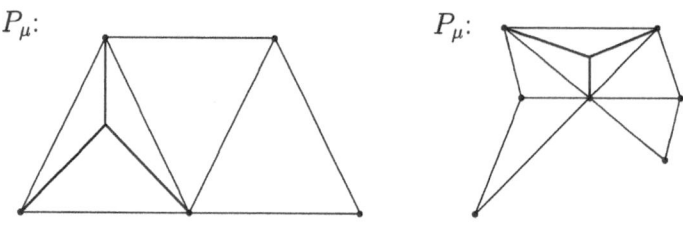

Fig. 4.2. Subdividing a triangle.

For the case when $r = 2$, exactly one triangle of P_μ has to be subdivided into three subtriangles if there do not exist four consecutive edges with different slopes in $\{e_{\mu,0}, \ldots, e_{\mu,\lambda_\mu+1}\}$. This means that we use a Clough-Tocher split only in this case. Here, we subdivide a triangle of P_μ which has an edge $e_{\mu,\nu}$ with slope different from all other edges in $\{e_{\mu,0}, \ldots, e_{\mu,\lambda_\mu+1}\}$, or an arbitrary triangle of P_μ, if there does not exist such an edge (see Figure 4.2). We subdivide this triangle such that we obtain four consecutive edges with endpoint v_μ which have different slopes.

If there exist sufficiently many points such that for each $\mu \in \{1, \ldots, n\}$ a polyhedron P_μ with the above properties can be added, we obtain a larger triangulation. If for some $\tilde{\mu} \in \{1, \ldots, n\}$, such a polyhedron cannot be added, we choose a point from $V \setminus \tilde{V}$ and add a triangle with vertex $v_{\tilde{\mu}}$ which has exactly one common edge with the given subtriangulation and so forth. By proceeding with this method, we finally obtain a triangulation Δ with the points of V as vertices. Note that the polyhedrons can be chosen such that a natural triangulation is obtained.

Construction of Interpolation Sets. In the following, we construct Hermite interpolation sets for $S_q^r(\Delta)$, where $q \geq 3$, if $r = 1$, and $q \geq 5$, if $r = 2$. The construction of Hermite interpolation sets is inductive and simultaneous with the construction of the triangulation.

We only have to describe some basic Hermite interpolation conditions. For doing this, as in Section 2, we denote by $D_e f$ the directional derivative along the edge e. Let $T \in \Delta$ be an arbitrary triangle with vertices z_1, z_2, z_3 and denote by e_k the edge $[z_k, z_{k+1}], k = 1, 2, 3$, where $z_4 = z_1$. For $r = 1$, we impose exactly one of the following conditions on the polynomial piece $p = s|_T \in \tilde{\Pi}_q$, where $s \in S_q^1(\Delta)$.

Condition Q: $D_x^\alpha D_y^\beta p(z_3) = D_x^\alpha D_y^\beta f(z_3)$, $0 \leq \alpha$, $0 \leq \beta$, $\alpha + \beta \leq q$.

Condition A_1: $D_x^\alpha D_y^\beta p(z_3) = D_x^\alpha D_y^\beta f(z_3)$, $0 \leq \alpha$, $0 \leq \beta$, $\alpha + \beta \leq q - 2$.

Condition B_1: $D_x^\alpha D_y^\beta p(z_3) = D_x^\alpha D_y^\beta f(z_3)$, $0 \leq \alpha$, $0 \leq \beta$, $\alpha + \beta \leq q - 3$, and
$\qquad\quad D_{e_2}^\alpha D_{e_3}^\beta p(z_3) = D_{e_2}^\alpha D_{e_3}^\beta f(z_3)$, $1 \leq \alpha$, $0 \leq \beta$, $\alpha + \beta = q - 2$.

Condition D_1: $D_{e_1}^\alpha D_{e_2}^\beta p(z_2) = D_{e_1}^\alpha D_{e_2}^\beta f(z_2)$, $\beta = 2, \ldots, q - 2 - \alpha$,
$\qquad\quad \alpha = 0, \ldots, q - 4$.

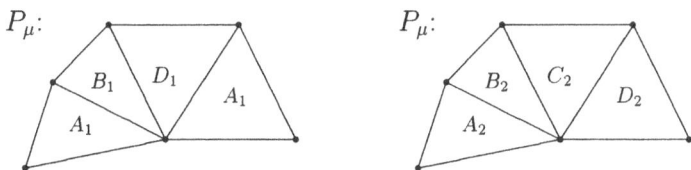

P_μ: P_μ:

Fig. 4.3. Interpolation sets for $r = 1$, respectively $r = 2$.

We determine the polynomial piece of the interpolating spline $s \in S_q^1(\Delta)$ on the first triangle by condition Q. Moreover, for each polyhedron P_μ of our inductive construction, we determine the polynomial pieces on the corresponding triangles as follows. By the construction of Δ, three edges $e_{\mu,\nu}, e_{\mu,\nu+1}, e_{\mu,\nu+2}$ with different slopes exist. The polynomial pieces on the triangles of P_μ which do not have $e_{\mu,\nu+1}$ as an edge are determined by condition A_1 and the C^1-property of s. The remaining polynomial pieces are determined by condition B_1, respectively C_1 (See Figure 4.3). Moreover, if for some μ such a polyhedron P_μ cannot be added, we determine the polynomial piece on the triangle which is added by condition A_1.

The resulting set of Hermite interpolation conditions is denoted by $\mathcal{H}_1$. Note that we impose Hermite interpolation conditions only at the points of V. Similarly to Lagrange interpolation sets we speak of a **Hermite interpolation set** if for each sufficiently differentiable function f there exists a unique spline satisfying corresponding Hermite interpolation conditions.

Theorem 4.1. [26] *The set $\mathcal{H}_1$ is a Hermite interpolation set for $S_q^1(\Delta), q \geq 3$.*

For $r = 2$, we impose one of the following conditions on the polynomial piece $p = s|_T \in \tilde{\Pi}_q$, where $s \in S_q^2(\Delta)$.

Condition Q: $D_x^\alpha D_y^\beta p(z_3) = D_x^\alpha D_y^\beta f(z_3)$, $0 \leq \alpha$, $0 \leq \beta$, $\alpha + \beta \leq q$.

Condition A_2: $D_x^\alpha D_y^\beta p(z_3) = D_x^\alpha D_y^\beta f(z_3)$, $0 \leq \alpha$, $0 \leq \beta$, $\alpha + \beta \leq q - 3$.

Condition B_2: $D_x^\alpha D_y^\beta p(z_3) = D_x^\alpha D_y^\beta f(z_3)$, $0 \leq \alpha$, $0 \leq \beta$, $\alpha + \beta \leq q - 4$, and
            $D_{e_2}^\alpha D_{e_3}^\beta p(z_3) = D_{e_2}^\alpha D_{e_3}^\beta f(z_3)$, $1 \leq \alpha$, $0 \leq \beta$, $\alpha + \beta = q - 3$.

Condition C_2: $D_x^\alpha D_y^\beta p(z_3) = D_x^\alpha D_y^\beta f(z_3)$, $0 \leq \alpha$, $0 \leq \beta$, $\alpha + \beta \leq q - 4$, and
            $D_{e_2}^\alpha D_{e_3}^\beta p(z_3) = D_{e_2}^\alpha D_{e_3}^\beta f(z_3)$, $2 \leq \alpha$, $0 \leq \beta$, $\alpha + \beta = q - 3$.

Condition D_2: $D_{e_1}^\alpha D_{e_2}^\beta p(z_2) = D_{e_1}^\alpha D_{e_2}^\beta f(z_2)$, $\beta = 3, \ldots, q - 3 - \alpha$,
            $\alpha = 0, \ldots, q - 6$.

If a triangle is subdivided, we need the following additional condition.

Condition $\tilde{C}_2$: $D_x^\alpha D_y^\beta p(z_3) = D_x^\alpha D_y^\beta f(z_3)$, $0 \leq \alpha$, $0 \leq \beta$, $\alpha + \beta \leq q - 4$, and
            $D_{e_2}^\alpha D_{e_3}^\beta p(z_3) = D_{e_2}^\alpha D_{e_3}^\beta f(z_3)$, $2 \leq \alpha, 2 \leq \beta, \alpha + \beta = q - 3$.

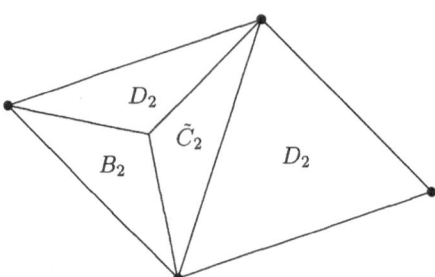

Fig. 4.4. Interpolation conditions for a subdivided triangle.

If no triangle of P_μ has to be subdivided, then by the construction of Δ, four edges $e_{\mu,\nu}, e_{\mu,\nu+1}, e_{\mu,\nu+2}, e_{\mu,\nu+3}$ with different slopes exist. In this case, the polynomial pieces of the interpolating spline $s \in S_q^2(\Delta)$ on the triangles of P_μ which do not have $e_{\mu,\nu+1}$, respectively $e_{\mu,\nu+2}$ as an edge are determined by condition A_2 and the C^2-property of s. The remaining polynomial pieces are determined by condition B_2, C_2 respectively D_2 (See Figure 4.3). If a triangle T with edges $e_{\mu,\nu}, e_{\mu,\nu+1}$ of P_μ is subdivided, then by construction of Δ the edges $e_{\mu,\nu}, e_{\mu,\nu+1}, e_{\mu,\nu+2}$ have different slopes. In this case, the polynomial pieces on the triangles of P_μ which do not have $e_{\mu,\nu+1}$ as an edge are determined by condition A_2. The four remaining polynomial pieces are determined by condition $B_2, \tilde{C}_2$ and D_2 (see Figure 4.4). Moreover, if for some μ such a polyhedron P_μ cannot be added, we determine the polynomial piece on the triangle which is added by condition A_2.

The resulting set of Hermite interpolation conditions is denoted by $\mathcal{H}_2$. Note that we only impose Hermite interpolation conditions at the points of V and the subdividing points.

Theorem 4.2. [26] *The set $\mathcal{H}_2$ is a Hermite interpolation set for $S_q^2(\Delta), q \geq 5$.*

Remark 4.3. Our method can also be used to construct Lagrange interpolation sets for $S_q^r(\Delta)$, where $q \geq 3$, if $r = 1$, and $q \geq 5$, if $r = 2$. For doing this, we choose distinct points lying on certain line segments in $T, T \in \Delta$. For details see [26].

Remark 4.4. By using Bézier-Bernstein techniques, we can show that the total number of interpolation conditions chosen by our method is equal to the dimension of $S_q^r(\Delta)$, where $q \geq 3$, if $r = 1$, and $q \geq 5$, if $r = 2$ (cf. [26]).

Remark 4.5. The interpolating spline is computed by passing from one triangle to the next and by solving several small systems instead of one large system. Therefore, the complexity of the algorithm is $\mathcal{O}(N)$, where N is the number of triangles.

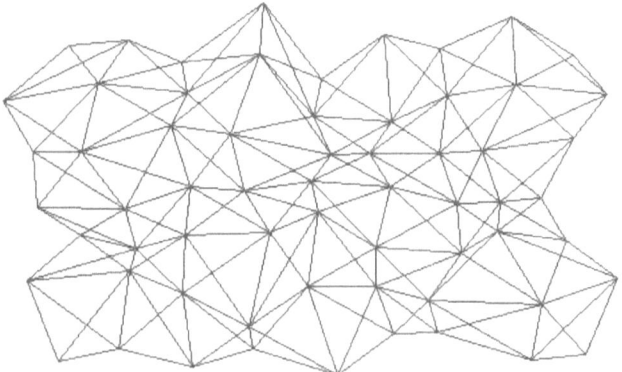

Fig. 4.5. A convex quadrangulation with diagonals.

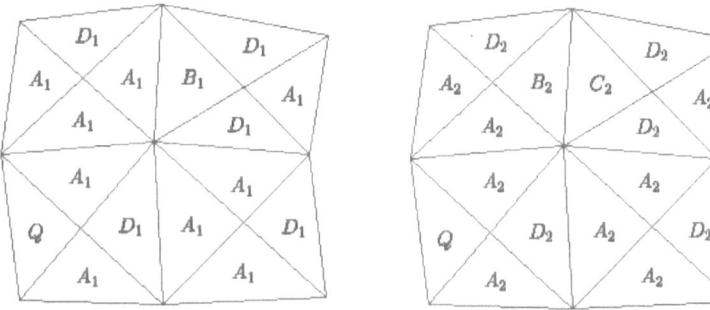

Fig. 4.6. Interpolation sets for splines on convex quadrangulations.

Remark 4.6. Our method can also be applied to certain classes of given triangulations, namely convex quadrangulations with diagonals. These are triangulations formed by closed convex quadrangles and their diagonals, where the intersection of any two quadrangles is empty, a common vertex or a common edge (see Figure 4.5). For such a triangulation, the distribution of interpolation conditions is indicated in Figure 4.6. In this case no triangle has to be subdivided (cf. [25]).

5 Hermite Interpolation with Optimal Approximation Order

We now describe a Hermite interpolation operator that assigns to every function $f \in C^{2r}(\Omega)$ a spline $s_f \in S_q^{r,\rho}(\Delta)$, where $S_q^{r,\rho}(\Delta)$ is the superspline subspace of

$S_q^r(\Delta)$, $q \geq 3r + 2$,

$$S_q^{r,\rho}(\Delta) := \{s \in S_q^r(\Delta) : s \in C^\rho(v) \text{ for all vertices } v \text{ of } \Delta\},$$

with $\rho = r + \left[\frac{r+1}{2}\right]$. (The dimension of $S_q^{r,\rho}(\Delta)$ is given in [14].) Since restrictions of a spline $s \in S_q^{r,\rho}(\Delta)$ to every triangle of Δ are polynomials, we are allowed to use derivatives of order greater than ρ, but in this case a particular triangle $T \in \Delta$ has to be chosen so that the derivative information comes from $s|_T$.

Let $f \in C^{2r}(\Omega)$. We impose on a spline $s_f \in S_q^{r,\rho}(\Delta)$ the following Hermite interpolation conditions, that fall into three groups corresponding to all vertices, edges and triangles of Δ, respectively.

1) Given any vertex v of Δ, let $T_v^1, \ldots, T_v^n$ be all triangles attached to v and numbered counterclockwise (starting from a boundary triangle if v is a boundary vertex). Denote by e_i the common edge of T_v^{i-1} and T_v^i, $i = 2, \ldots, n$. If v is an interior vertex, $e_1 = e_{n+1}$ denote the common edge of T_v^1 and T_v^n. Otherwise, e_1 and e_{n+1} are the boundary edges (attached to v) of T_v^1 and T_v^n respectively. As in Section 2, we denote by $D_{e_i} f(v)$ the directional derivative of f along edge e_i. If $\alpha + \beta > \rho$, then we set $D_{e_i}^\alpha D_{e_{i+1}}^\beta s_f(v) := D_{e_i}^\alpha D_{e_{i+1}}^\beta (s_f|_{T_v^i})(v)$. For every vertex v in Δ the following conditions are imposed on $s_f \in S_q^{r,\rho}(\Delta)$:

- $D_x^\alpha D_y^\beta s_f(v) = D_x^\alpha D_y^\beta f(v)$ for all $(\alpha, \beta) \in A_1$, where

$$A_1 := \{(\alpha, \beta) \in \mathbb{Z}^2 : \alpha \geq 0, \ \beta \geq 0, \ \alpha + \beta \leq \rho\},$$

- $D_{e_i}^\alpha D_{e_{i+1}}^\beta s_f(v) = D_{e_i}^\alpha D_{e_{i+1}}^\beta f(v)$ for all $(\alpha, \beta) \in A_2$, where

$$A_2 := \{(\alpha, \beta) \in \mathbb{Z}^2 : \alpha \leq r, \ \beta \leq r, \ \alpha + \beta \geq \rho + 1\},$$

  and for each $i \in \{1, \ldots, n\}$ such that e_i is nondegenerate at v,

- $D_{e_i}^\alpha D_{e_{i+1}}^\beta s_f(v) = D_{e_i}^\alpha D_{e_{i+1}}^\beta f(v)$ for all $(\alpha, \beta) \in A_3$, where

$$A_3 := \{(\alpha, \beta) \in \mathbb{Z}^2 : \alpha \geq r + 1, \ 2\alpha + \beta \leq 3r + 1, \ \alpha + \beta \geq \rho + 1\},$$

  and for each $i \in \{1, \ldots, n\}$ such that e_i is degenerate at v,

- $D_{e_1}^\alpha D_{e_2}^\beta s_f(v) = D_{e_1}^\alpha D_{e_2}^\beta f(v)$ and $D_{e_{n+1}}^\alpha D_{e_n}^\beta s_f(v) = D_{e_{n+1}}^\alpha D_{e_n}^\beta f(v)$ for all $(\alpha, \beta) \in A_3$ if v is a boundary vertex, and

- $D_{e_1}^\alpha D_{e_2}^\beta s_f(v) = D_{e_1}^\alpha D_{e_2}^\beta f(v)$ for all $(\alpha, \beta) \in A_2$ if v is a singular vertex.

2) On every edge e of Δ, with vertices v' and v'', we choose points

$$z_e^{\mu,i} := v' + \frac{i}{\kappa_\mu + 1}(v'' - v'), \quad i = 1, \ldots, \kappa_\mu, \quad \mu = 0, \ldots, r,$$

where $\kappa_\mu := q - 3r - 1 - (r - \mu) \bmod 2 = q - 2r - 1 - \mu - 2\left[\frac{r+1-\mu}{2}\right]$, and impose on $s_f \in S_q^{r,\rho}(\Delta)$ the following conditions:

- $D^\mu_{e^\perp} s_f(z_e^{\mu,1}) = D^\mu_{e^\perp} f(z_e^{\mu,1}), \ldots, D^\mu_{e^\perp} s_f(z_e^{\mu,\kappa_\mu}) = D^\mu_{e^\perp} f(z_e^{\mu,\kappa_\mu})$ for all $\mu = 0, \ldots, r$, where $D_{e^\perp}$ denotes differentiation in the direction orthogonal to e.

3) On every triangle $T \in \Delta$, with vertices v', v'' and v''', we choose uniformly spaced points

$$z_T^{i,j,k} := (iv' + jv'' + kv''')/q, \quad i + j + k = q,$$

and impose on $s_f \in S_q^{r,\rho}(\Delta)$ the following conditions:

- $s_f(z_T^{i,j,k}) = f(z_T^{i,j,k})$ for all i, j, k such that $i+j+k = q$ and $r < i, j, k < q-2r$.

Theorem 5.1. [13] *Let $r \geq 1$, $q \geq 3r + 2$ and $\rho = r + \left[\frac{r+1}{2}\right]$. Given $f \in C^{2r}(\Omega)$, there exists a unique spline $s_f \in S_q^{r,\rho}(\Delta)$ satisfying the above Hermite interpolation conditions. Moreover, if $f \in C^m(\Omega)$ ($m \in \{2r, \ldots, q+1\}$) and $T \in \Delta$, then*

$$\|D_x^\alpha D_y^\beta (f - s_f)\|_{L_\infty(T)} \leq K\, h_T^{m-\alpha-\beta} \max_{0 \leq m' \leq m} \|D_x^{m'} D_y^{m-m'} f\|_{C(T)},$$

for all $\alpha, \beta \geq 0$, $\alpha + \beta \leq m$, where h_T is the diameter of T and K is a constant which depends only on r, q and the smallest angle θ_Δ in Δ.

The following new characterization of C^r smoothness across a common edge of two polynomial patches plays an essential role in the proof of Theorem 5.1.

Theorem 5.2. [13] *Let T_1 and T_2 be two triangles sharing a common edge $e = [v_1, v_2]$, and let e_i be the edge of T_i attached to v_1 and different from e, $i = 1, 2$. Suppose a piecewise polynomial function s is defined on $T_1 \cup T_2$ as follows*

$$s|_{T_i} = p_i \in \Pi_q, \quad i = 1, 2.$$

Then $s \in C^r(T_1 \cup T_2)$, for some $r \leq q$, if and only if

$$\tau_1^\alpha D_{e_2}^\alpha D_e^{\gamma-\alpha} p_2(v_1) = \sum_{\beta=0}^\alpha (-1)^\beta \binom{\alpha}{\beta} \sin^{\alpha-\beta}(\theta_1 + \theta_2) \tau_2^\beta D_{e_1}^\beta D_e^{\gamma-\beta} p_1(v_1),$$

for all $\alpha = 0, \ldots, r$ and $\gamma = \alpha, \ldots, q$, where

$$\tau_i = \begin{cases} \sin \theta_i, & \text{if } e_1 \text{ and } e_2 \text{ are noncollinear}, \\ 1, & \text{otherwise}, \end{cases}$$

and θ_i is the angle between e and e_i, $i = 1, 2$.

It follows from Theorem 5.1 that the fundamental functions $s_1, \ldots, s_N$ of the above Hermite interpolation scheme form a basis for $S_q^{r,\rho}(\Delta)$. We note that a basis for this space has been constructed in [14] by using Bernstein-Bézier techniques. Although there exists some interrelation between two bases, particularly, the supports of basis functions are the same, the minimal determining set of [14] cannot be transformed by standard Bernstein-Bézier arguments into a Hermite interpolation scheme of our type.

The next theorem lists some of useful properties of our basis.

Theorem 5.3. [13] *The fundamental functions* $s_1, \ldots, s_N$ *form a basis for* $S_q^{r,\rho}(\Delta)$ *such that*

   1) $\{s_1, \ldots, s_N\}$ *is locally linearly independent, i.e., for every open* $B \subset \Omega$ *the subsystem* $\{s_i : B \cap \operatorname{supp} s_i \neq \emptyset\}$ *is linearly independent on* B,

   2) $\{s_1, \ldots, s_N\}$ *is least supported, i.e., for every basis* $\{b_1, \ldots, b_N\}$ *of* $S_q^{r,\rho}(\Delta)$ *there exists a permutation* π *of* $\{1, \ldots, N\}$ *such that*

$$\operatorname{supp} s_i \subset \operatorname{supp} b_{\pi(i)}, \quad \text{for all} \quad i = 1, \ldots, N,$$

   3) $\operatorname{supp} s_i$, $i = 1, \ldots, N$, *is either a triangle or the union of some triangles sharing one common vertex, and*

   4) *the corresponding normalized basis* $\{s_1^*, \ldots, s_N^*\}$, *with*

$$s_i^* := \|s_i\|_{L_\infty(\Omega)}^{-1} s_i, \quad i = 1, \ldots, N,$$

is stable in the sense that

$$K_1 \max_i |a_i| \leq \| \sum_{i=1}^N a_i s_i^* \|_{C(\Omega)} \leq K_2 \max_i |a_i|,$$

where K_1 *and* K_2 *depend only on* r, q, θ_Δ *and some measure of "near-degeneracy" of nondegenerate edges in* Δ.

Remark 5.4. Theorem 5.1 provides a new proof of the optimal approximation order of $S_q^r(\Delta)$, $q \geq 3r + 2$. Previous results on this subject were given in [3, 7, 8, 16]. As in [7, 16], the constant K that appears in Theorem 5.1 depends only on r, q and the smallest angle in Δ, and, therefore, does not grow for triangulations that contain near-singular vertices. Moreover, in contrast to quasi-interpolation methods of [7, 16], we show that optimal approximation order can be achieved by using Hermite interpolation.

Remark 5.5. According to Theorem 5.3, 2), our basis is best possible for the space $S_q^{r,\rho}(\Delta)$ in regard to the size of the supports of the basis functions. It shares this property with the basis constructed in [14]. The bases in [7, 16] fail to be least supported, but they have the advantage that stability constants K_1, K_2 depend only on the smallest angle in the triangulation while in our construction they also depend on the sums of pairs of adjacent angles.

6   Numerical Results

Finally, we give some numerical results for the interpolation methods of Section 3 and Section 4. We interpolate Franke's test function

$$f(x, y) = \tfrac{3}{4} e^{-\frac{(9x-2)^2+(9y-2)^2}{4}} + \tfrac{3}{4} e^{-\frac{(9x+1)^2}{49} - \frac{(9y+1)}{10}} + \tfrac{1}{2} e^{-\frac{(9x-7)^2+(9y-3)^2}{4}}$$
$$- \tfrac{1}{5} e^{-(9x-4)^2-(9y-7)^2}$$

Δ: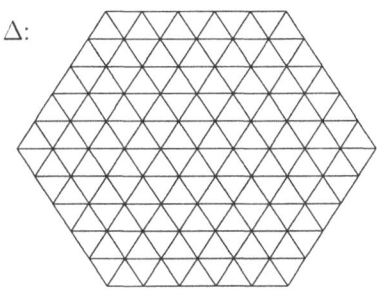

Fig. 6.1. Triangulation Δ.

by splines on a domain Ω with $[0,1] \times [0,1] \subseteq \Omega$. First, let Δ be a triangulation as in Figure 6.1. Obviously, Δ is of nested-polygon type.
Our results for the Hermite interpolating spline $s_f \in S_3^1(\Delta)$ are as follows :

$$[S_3^1 \mid 93 \mid 1.38 * 10^{-1}], \quad [S_3^1 \mid 291 \mid 1.64 * 10^{-2}],$$
$$[S_3^1 \mid 1011 \mid 1.20 * 10^{-3}], \quad [S_3^1 \mid 3747 \mid 1.61 * 10^{-4}],$$
$$[S_3^1 \mid 14403 \mid 2.03 * 10^{-5}],$$

where we set

$$[S_q^r \mid \text{number of interpolation conditions} \mid \text{error } \|f - s_f\|_\infty].$$

Now, let Δ be a triangulation that results from a given Δ^2-partition deformed by a randomizer (see Figure 4.5). Our results for the Hermite interpolating spline $s_f \in S_q^r(\Delta)$ are as follows :

$$[S_3^1 \mid 168 \mid 4.51 * 10^{-2}], \quad [S_3^1 \mid 583 \mid 1.05 * 10^{-2}],$$
$$[S_3^1 \mid 2163 \mid 8.00 * 10^{-4}], \quad [S_3^1 \mid 8323 \mid 9.06 * 10^{-5}],$$
$$[S_4^1 \mid 388 \mid 7.09 * 10^{-2}], \quad [S_4^1 \mid 1423 \mid 4.01 * 10^{-3}],$$
$$[S_4^1 \mid 5443 \mid 1.36 * 10^{-4}], \quad [S_4^1 \mid 21283 \mid 5.70 * 10^{-6}],$$
$$[S_5^1 \mid 708 \mid 4.23 * 10^{-2}], \quad [S_5^1 \mid 2663 \mid 2.12 * 10^{-3}],$$
$$[S_5^1 \mid 10323 \mid 2.61 * 10^{-5}], \quad [S_5^1 \mid 40643 \mid 7.17 * 10^{-7}],$$
$$[S_7^2 \mid 993 \mid 8.23 * 10^{-3}], \quad [S_7^2 \mid 3678 \mid 3.26 * 10^{-4}],$$
$$[S_7^2 \mid 14148 \mid 1.71 * 10^{-6}], \quad [S_7^2 \mid 55488 \mid 3.53 * 10^{-8}],$$
$$[S_8^2 \mid 1473 \mid 1.79 * 10^{-3}], \quad [S_8^2 \mid 5538 \mid 1.36 * 10^{-5}],$$
$$[S_8^2 \mid 21468 \mid 2.98 * 10^{-7}], \quad [S_8^2 \mid 84258 \mid 7.54 * 10^{-9}].$$

Acknowledgment. Oleg Davydov was supported in part by a research fellowship from the Alexander von Humboldt Foundation.

References

[1] M. H. Adam, *Bivariate Spline-Interpolation auf Crosscut-Partitionen*, Dissertation, Mannheim, 1995.

[2] P. Alfeld, B. Piper and L. L. Schumaker, *An explicit basis for C^1 quartic bivariate splines*, SIAM J. Numer. Anal. **24** (1987), 891–911.

[3] C. de Boor and K. Höllig, *Approximation power of smooth bivariate pp functions*, Math. Z. **197** (1988), 343–363.

[4] C. K. Chui and T. X. He, *On location of sample points in C^1 quadratic bivariate spline interpolation*, in "Numerical Methods of Approximation Theory," (L. Collatz, G. Meinardus and G. Nürnberger, Eds.), ISNM 81, Birkhäuser, Basel, 1987, 30–43.

[5] C. K. Chui and D. Hong, *Construction of local C^1 quartic spline elements for optimal-order approximation*, Math. Comp. **65** (1996), 85–98.

[6] C. K. Chui and D. Hong, *Swapping edges of arbitrary triangulations to achive the optimal order of approximation*, SIAM J. Numer. Anal. **34** (1997), 1472–1482.

[7] C. K. Chui, D. Hong and R.-Q. Jia, *Stability of optimal order approximation by bivariate splines over arbitrary triangulations*, Trans. Amer. Math. Soc. **347** (1995), 3301–3318.

[8] C. K. Chui and M. J. Lai, *On bivariate super vertex splines*, Constr. Approx. **6** (1990), 399–419.

[9] O. Davydov, *Locally linearly independent basis for C^1 bivariate splines of degree $q \geq 5$*, in "Mathematical Methods for Curves and Surfaces II," (M. Daehlen, T. Lyche and L. L. Schumaker, Eds.), Vanderbilt University Press, 1998, 71–78.

[10] O. Davydov and G. Nürnberger, *Interpolation by C^1 splines of degree $q \geq 4$ on triangulations*, preprint.

[11] O. Davydov, G. Nürnberger and F. Zeilfelder, *Approximation order of bivariate spline interpolation for arbitrary smoothness*, J. Comput. Appl. Math. **90** (1998), 117–134.

[12] O. Davydov, G. Nürnberger and F. Zeilfelder, *Interpolation by cubic splines on triangulations*, to appear in "Approximation Theory IX," (C. K. Chui and L. L. Schumaker, Eds.), Vanderbilt University Press.

[13] O. Davydov, G. Nürnberger and F. Zeilfelder, *Bivariate spline interpolation with optimal approximation order*, preprint.

[14] A. Ibrahim and L. L. Schumaker, *Super spline spaces of smoothness r and degree $d \geq 3r + 2$*, Constr. Approx. **7** (1991), 401–423.

[15] F. Jeeawock-Zedek, *Interpolation scheme by C^1 cubic splines on a non uniform type-2 triangulation of a rectangular domain*, C.R. Acad. Sci. Ser. I Math., **314** (1992), 413–418.

[16] M. J. Lai and L. L. Schumaker, *On the approximation power of bivariate splines*, Advances in Comp. Math. **9** (1998), 251–279.

[17] J. Morgan and R. Scott, *A nodal basis for C^1 piecewise polynomials of degree $n \geq 5$*, Math. Comp. **29** (1975), 736–740.

[18] E. Nadler, *Hermite interpolation by C^1 bivariate splines*, in "Contributions to the computation of curves and surfaces," (W. Dahmen, M. Gasca and C. Micchelli, Eds.), pp. 55–66, Academia de Ciencias, Zaragoza, 1990.

[19] G. Nürnberger, *Approximation by Spline Functions*, Springer-Verlag, Berlin, Heidelberg, New York, 1989.

[20] G. Nürnberger, *Approximation order of bivariate spline interpolation*, J. Approx. Theory **87** (1996), 117–136.

[21] G. Nürnberger, O. Davydov, G. Walz and F. Zeilfelder, *Interpolation by bivariate splines on crosscut partitions*, in "Multivariate Approximation and Splines," (G. Nürnberger, J. W. Schmidt and G. Walz, Eds.), ISNM, Birkhäuser, 1997, 189–203.

[22] G. Nürnberger and Th. Riessinger, *Lagrange and Hermite interpolation by bivariate splines*, Numer. Func. Anal. Optim. **13** (1992), 75–96.

[23] G. Nürnberger and Th. Riessinger, *Bivariate spline interpolation at grid points*, Numer. Math. **71** (1995), 91–119.

[24] G. Nürnberger and G. Walz, *Error analysis in interpolation by bivariate C^1-splines*, IMA J. Numer. Anal. **18** (1998), 485–507.

[25] G. Nürnberger and F. Zeilfelder, *Spline interpolation on convex quadrangulations*, to appear in "Approximation Theory IX," (C. K. Chui and L. L. Schumaker, Eds.), Vanderbilt University Press.

[26] G. Nürnberger and F. Zeilfelder, *Lagrange and Hermite interpolation by splines on triangulations*, in preparation.

[27] Z. Sha, *On interpolation by $S_2^1(\Delta_{m,n}^2)$*, Approx. Theory Appl. **1**, (1985), 71–82.

[28] Z. Sha, *On interpolation by $S_3^1(\Delta_{m,n}^2)$*, Approx. Theory Appl. **1**, (1985), 1–18.

Oleg Davydov
Universität Dortmund
Fachbereich Mathematik
44221 Dortmund, Germany
Email address: davydov@math.uni-dortmund.de

Günther Nürnberger and Frank Zeilfelder
Universität Mannheim
Fakultät für Mathematik und Informatik
68131 Mannheim, Germany
Email address: nuern@euklid.math.uni-mannheim.de,
zeilfeld@fourier.math.uni-mannheim.de

International Series of Numerical Mathematics
Vol. 132, © 1999 Birkhäuser Verlag Basel/Switzerland

On the Use of Quasi-Newton Methods
in DAE-Codes

Christoph Fredebeul, Christoph Weber

Abstract

The use of discretization methods for the numerical solution of differential algebraic equations (DAEs) leads to systems of nonlinear equations. These are usually solved with the help of Newton's method. To improve efficiency, the decomposed Jacobian remains unchanged for a fixed number of steps (at least as long as convergence is obtained). This may lead to poor convergence rates. Thus we consider the use of Quasi-Newton methods that allow a direct update of the factored Jacobian. A paper of J. M. Martinez [11] turns out to be helpful. However, implementing one of these methods in a DAE-code as DASSL written by L. R. Petzold [9] is not straightforward. Decisions and strategies are discussed and some numerical results are presented.

1 Derivation of the System of Equations to be solved

In this paper, we consider the differential-algebraic equation (DAE)

$$F(t, Y, Y') = 0, \quad Y(t_0) = Y_0, \tag{1}$$

where $F : I \times \mathbb{R}^d \times \mathbb{R}^d \to \mathbb{R}^d$, $I := [t_0, t_{end}]$, is a continuously differentiable function. Moreover, the solution $Y : I \to \mathbb{R}^d$ of (1) is supposed to exist uniquely.

Now, to solve (1) numerically, we take an implicit multistep formula

$$\sum_{i=0}^{k} a_{i,j} y_{j-i} - h_j \sum_{i=0}^{k} b_{i,j} y'_{j-i} = 0.$$

Reformulation yields

$$y'_j = \frac{1}{h_j b_{0,j}} \left(a_{0,j} y_j + \sum_{i=1}^{k} (a_{i,j} y_{j-i} - h_j b_{i,j} y'_{j-i}) \right)$$

$$=: \alpha_j y_j + \beta_j .$$

Hence, after discretization of the DAE, we arrive at

$$F(t_j, y_j, \alpha_j y_j + \beta_j) = 0. \tag{2}$$

Applying Newton's method gives

$$y_j^{m+1} = y_j^m - G^{-1}(y_j^m)\mathcal{F}(y_j^m), \quad m = 0, 1, 2, \ldots,$$

where

$$\begin{aligned}
\mathcal{F}(y_j^m) &= F(t_j, y_j^m, \alpha_j y_j^m + \beta_j), \\
G(y_j^m) &= (F_Y + \alpha_j F_{Y'})(t_j, y_j^m, \alpha_j y_j^m + \beta_j)
\end{aligned}$$

and the starting value y_j^0 is obtained from some explicit formula.

Now, to save work and to increase efficiency, some simplifications are applied. In DASSL we have

$$y_j^{m+1} = y_j^m - c_j\, G_l^{-1}\mathcal{F}(y_j^m).$$

Here, $G_l := G(y_l^0)$, $l \le j$, is saved (after Gaussian elimination) in factored form, i.e.

$$G_l = L_l U_l.$$

Note that G_l especially depends on α_l and, consequently, on the (time-)step size. Therefore, since G_l remains unchanged for some time steps, the scaling-factor

$$c_j := \frac{2\alpha_l}{\alpha_j + \alpha_l}$$

derived from an application to a linear model problem (see Petzold [1]) is introduced to improve the ratio of convergence.

A re-evaluation of G is done

- if $\mu := |(\alpha_j - \alpha_l)/(\alpha_j + \alpha_l)| >$ XRATE ($= 0.25$ in DASSL),

- if convergence appears to be too slow, i.e.

  - the estimated convergence ratio is too large,
  - too many iteration steps have been taken.

For more details (on testing convergence, on error control etc.) see [9] or [1].

Using all these strategies, DASSL works quite efficient when solving DAEs. Nevertheless, especially during the starting phase of DASSL, the (time-)step size may change rapidly and, as a consequence, a lot of re-evaluations and decompositions of the Jacobian are done. Hence, some advantages may be obtained when employing the strategies described in the following section.

2 A Family of Quasi-Newton Methods

In this section we briefly resume the approach to Quasi-Newton methods as presented in Martinez [11].

Let $F : \mathbb{R}^d \to \mathbb{R}^d$, $F \in C^1(\mathbb{R}^d)$ and $0 < \alpha < 1$. Solve

$$F(x) = 0$$

by a Quasi-Newton method

$$x^{k+1} = x^k - B_k^{-1} F(x^k),$$

$$B_k = A_k^{-1} R_k,$$

$A_k \in S_A$, $R_k \in S_R$. Suppose that (A_{k+1}, R_{k+1}) is obtained from (A_k, R_k) as the solution of the problem

$$\text{minimize} \quad \alpha \|A - A_k\|_F^2 + (1 - \alpha) \|R - R_k\|_F^2$$

$$\text{subject to} \quad Rs - At = 0,$$

where $s = x^{k+1} - x^k$, $t = F(x^{k+1}) - F(x^k)$ and $A \in S_A$, $R \in S_R$. $\| \cdot \|_F$ denotes the Frobenius norm.

- Note that $Rs - At = 0$ is equivalent to the well-known secant (or Quasi-Newton) equation.

- Originally, Martinez defined his methods using arbitrary norms, but in all relevant examples the Frobenius norm is applied.

- Methods of this type are known as *least-change secant updates*; those of Broyden, Dennis-Marvil, Johnson-Austria etc. are included as special cases.

- For more details and convergence results see [11].

Now, with respect to our intention of improving DASSL, let S_A (S_R) be the space of lower (upper) triangular matrices in $\mathbb{R}^{d \times d}$, and $a_{ii} = 1$, $i = 1, ..., d$, for any $A \in S_A$. Then, with

$$\lambda_i = \left(\sum_{j=i}^{d} r_{ij}^k s_j - \sum_{j=1}^{i-1} a_{ij}^k t_j - t_i \right) \bigg/ \left(\frac{1}{\alpha} \sum_{j=1}^{i-1} t_j^2 + \frac{1}{1-\alpha} \sum_{j=i}^{d} s_j^2 \right),$$

the update formulae look like

$$a_{ij} = a_{ij}^k + \frac{\lambda_i}{\alpha} t_j, \qquad j = 1, \ldots, i-1,$$

$$r_{ij} = r_{ij}^k - \frac{\lambda_i}{1-\alpha} s_j, \qquad j = i, \ldots, d.$$

Hence, if s and t are available, the update requires $\mathcal{O}(d^2)$ operations.

3   How to use Quasi-Newton Methods in DASSL

Although an update-formula is available, the implementation of Martinez' methods into DASSL [9] is not straightforward. There are (at least) four main questions to be posed (and, of course, answered):

1. How does $G_l = L_l U_l$ fit to $B_k = A_k^{-1} R_k$?

2. Which $\alpha \in (0, 1)$ should be taken?

3. What is a suitable direction for the secant update?

4. When should we do an update instead of (or additionally to) a re-evaluation?

3.1   How does $B_k = A_k^{-1} R_k$ fit to $G_l = L_l U_l$?

In DASSL the Jacobian is stored after decomposition in factored form. Obviously, we may take $L = A^{-1}$ and $U = R$. But now we are in trouble, since the update formula applies to $A = L^{-1}$, whereas Gaussian elimination yields $G = LU$ within $(d^3 - d)/3$ operations.

A way out is to rewrite the LINPACK subroutine DGEFA (see [3]) to give L^{-1} instead of L. As a consequence, the modified routine takes $(d^3 - d^2)/2$ operations whereas solving $Ly = z$ or $y = L^{-1}z$ requires the same effort.

Note that beside an increase in effort this approach has another drawback. Although L may have a special structure (e.g. banded), the inverse matrix may not possess the same one. Hence, storing L^{-1} instead of L may require additional memory.

Thus, we recommend our approach only if the Jacobian is to be considered as full or dense.

3.2   Which $\alpha \in (0, 1)$ should be taken?

The update-formula derived in Section 2 provides one degree of freedom, given by the parameter α. For some fixed values classical Quasi-Newton formulae may be obtained. However, depending on the DAE that yields equation (2), we have to scope with ill-conditioned iteration matrices. On the other hand, the condition number of L is constant whereas that of U reflects the one of the Jacobian. Hence, to let both parts of the decomposition enjoy the update, we set

$$\alpha = \frac{\|U_l\|}{\|L_l\| + \|U_l\|} \ .$$

3.3   What is a suitable direction for the secant update?

Note that, for t_j fixed, in DASSL only $1 - 2$ iteration steps are often sufficient to achieve convergence. Consequently, we are not allowed to benefit from the classical situation of many iterations where a direction for a secant update is somewhat naturally given by the difference of two iterated values. However, ignoring proceeding in time, we decide to take the direction indicated by the last (successful) iteration, i.e.

$$s = y_{j-1}^{\text{fin}} - y_{j-1}^0 \ .$$

With this we have

$$t = \mathcal{F}(y_j^0 + s) - \mathcal{F}(y_j^0).$$

As a consequence, we need one additional evaluation of F when an update should be done.

3.4 When should we do an update instead of (or additionally to) a re-evaluation?

In DASSL, the Jacobian is re-evaluated only at the beginning of an iteration. To keep the amount of changes in the code as low as possible, we have decided to apply a secant update only then, too. However, to gain some increase in efficiency, the value of XRATE (used to decide whether a re-evaluation should be done) is lowered and is now used to decide whether an update should be done. Hence, the conditions for an update (instead of a re-evaluation) are

$$\left| \frac{\alpha_j - \alpha_l}{\alpha_j + \alpha_l} \right| > 0.2$$

and, additionally, the Jacobian G is not too old, i.e. $j - l < 4$.

In case of divergence, we still prefer the re-evaluation.

4 Numerical evidence

We have applied the modified code, called DoDASSL (shortly "Do"), to 13 test examples ("E") of different differential indices, dimensions and tolerances and compared these results to DASSL (shortly "DA").

Example 0 is the Van-der-Pol equation (from [7]; $\varepsilon = 10^{-4}$), examples 1 and 2 are problems B5 and F4 (Stiff Detest [4]). Since they all are ODEs, their index is 0. Examples 3, 9 and 12 represent the planar pendulum by an index 1, 2 and 3 formulation, respectively. Example 4 is due to Koto [8], example 5 is due to Petzold [9]. Example 6 results from E5, multiplied by a linear time-dependend transformation. Examples 4 to 6 are of index 1. Example 7 is of Hessenberg-form and example 8 is a canonical subsystem, both of index 2. Examples 10 and 11 are due to [6] and [2], respectively. For more details, see [5].

In fact, the new approach is used mostly during the starting phase when the time step size is increased frequently. As a consequence, in (possible) contrast to F_Y and $F_{Y'}$, the modulus of α_l changes rapidly, too. Here, a secant update appears to be a proper way to avoid a re-evaluation (which is always connected with a decomposition).

To give an impression of the improvement, for a given tolerance ε we set $A_F(\varepsilon) := \#s + \#F + d * \#J (+ \#S$ in the case of DoDASSL), where $\#s$ denotes the number of steps, $\#F$ resp. $\#J$ the number of function resp. Jacobian evaluations, d the dimension and $\#S$ the number of secant updates applied to the Jacobian.

E	$A_F(10^{-3})$			$A_F(10^{-6})$			$A_F(10^{-9})$		
	DA	Do	Ratio	DA	Do	Ratio	DA	Do	Ratio
0	2557	2431	0.951	6457	6128	0.949	19095	18291	0.958
1	629	606	0.963	2163	2336	1.080	9393	5949	0.633
2	2592	2325	0.897	7157	6227	0.870	19546	17429	0.892
3	431	455	1.056	1269	1170	0.922	2596	2524	0.972
4	129	117	0.907	429	343	0.800	783	668	0.853
5	117	111	0.949	314	295	0.939	843	662	0.785
6	116	95	0.819	481	485	1.008	1533	1349	0.880
7	768	726	0.945	1641	1645	1.002	3863	3153	0.816
8	1075	637	0.593	2044	1607	0.786	5168	4775	0.924
9	938	935	0.997	2712	2561	0.944	8027	7296	0.909
10	439	409	0.932	1297	1049	0.809	3931	2779	0.707
11	136	135	0.993	254	227	0.894	620	408	0.658
12	1253	1238	0.988	3768	3681	0.977	10312	9989	0.969

In most cases, DoDASSL works more efficient than DASSL, in many cases it works significantly better.

However, in some cases other reasons influence on the performance, too. For the tolerance of 10^{-6}, the code DoDASSL becomes unstable when solving Example 1 (due to the highly oscillatory mode of the components of B5). For the tolerance of 10^{-9} the same happens to DASSL, too.

In other cases, we have not found such an explanation for outlayers.

References

[1] K. E. Brenan, S. L. Campbell, L. R. Petzold *"Numerical Solution of Initial-Value Problems in Differential-Algebraic Equations"* Elsevier, New York, 1989

[2] K. E. Brenan, L. R. Petzold *"The numerical solution of higher index differential/algebraic equations by implicit Runge-Kutta methods"* SIAM J. Numer. Anal., 26 (1989), 976-996

[3] J. J. Dongarra, J. R. Bunch, C. B. Moler, G. W. Stewart *"LINPACK User's Guide"* SIAM, Philadelphia, 1979

[4] W. H. Enright, T. E. Hull, B. Lindberg *"Comparing numerical methods for stiff systems of ordinary differential equations"* BIT, 15 (1975), 10-48

[5] Ch. Fredebeul, Ch. Weber *"Über den Einsatz von Quasi-Newton-Verfahren in DAGL-Lösern"* Technical Report No 141-T, Angewandte Mathematik, Uni-Dortmund, 1996

[6] C. W. Gear, L. R. Petzold *"ODE methods for the solution of differential algebraic systems"* SIAM J. Numer. Anal., 21 (1984), 716-728

[7] E. Hairer, G. Wanner *"Solving ordinary differential equations II: Stiff and differential-algebraic problems"* Springer 1991

[8] T. Koto *"Third-order semi-implicit Runge-Kutta methods for time-dependent index-one differential-algebraic equations"* Journal of Information Processing, 14 (1991)

[9] L. R. Petzold *"A description of DASSL: A differential/algebraic system solver"* Scientific Computing, eds. R. S. Stepleman et al., North-Holland, Amsterdam, 1983, 65-68

[10] J. E. Dennis, E. S. Marwil *"Direct secant updates of matrix factorizations"* Math. Comp., 38 (1982), 459-476

[11] J. M. Martinez *"A family of quasi-Newton methods for nonlinear equations with direct secant updates of matrix factorizations"* SIAM J. Numer. Anal., 27 (1990), 1034-1049

Christoph Fredebeul, Christoph Weber
Department of Mathematics
University of Dortmund
D-44221 Dortmund, Germany
Email address: Christoph.Fredebeul@math.uni-dortmund.de

International Series of Numerical Mathematics
Vol. 132, © 1999 Birkhäuser Verlag Basel/Switzerland

On the Regularity of Some Differential Operators

Karsten Kamber and Xinlong Zhou

Abstract

In the present paper we investigate the regularity of some differential operators which occur in the theory of orthogonal polynomials and in approximation theory.

1 Introduction

For some functions $\alpha_i(x) \in \text{Lip}\delta$, $0 < \delta \le 1$, $i = 0, 1, \ldots, r$ and $\alpha_r(x) \ne 0$, $x \in [0, 1]$, let $P(D)$ be the differential operator of order $2r$ given by

$$P(D) := \sum_{i=0}^{r} \alpha_i(x)(x(1-x))^i D^{r+i},$$

where $D := d/dx$. It is clear that for $r = 1$ and appropriate α_i the so-called Jacobi polynomials defined on $[0, 1]$ are the eigenfunctions of this differential operator (see [4] and [5]). In approximation theory this kind of operators may be regarded as a bridge between the approximation by trigonometric and algebraic polynomials (see [1] and the references therein). In fact, for a function f defined on $[0, 1]$ let $f(x) = f((1 - \cos \theta)/2) =: g(\theta)$; then g is periodic. Moreover, with appropriate α_i one may write

$$g^{(2r)}(\theta) = P(D)f(x) + Lf(x),$$

where for some β_i the operator L is given by $Lf(x) = \sum_{i=0}^{r-1} \beta_i(x)f^{(i)}(x)$. The regularity of $P(D)$ allows us to carry over the results from the periodic to the algebraic case and vice versa. For example, one of the important inequalities in approximation theory is the so-called Bernstein inequality concerning the derivatives of trigonometric polynomials. Thus, if T_n is a trigonometric polynomial of degree n, then

$$\int_{-\pi}^{\pi} |T_n(t)''(1 - \cos t)^\alpha (1 + \cos t)^\beta|^p dt \le Cn^{2p} \int_{-\pi}^{\pi} |T_n(t)(1 - \cos t)^\alpha (1 + \cos t)^\beta|^p dt.$$

On the other hand, for any $P_n \in \Pi_n$ (Π_n denotes the set of algebraic polynomials with degree $\le n$) the function defined by $T_n(\theta) := P_n((1 - \cos \theta)/2)$ is a trigonometric polynomial of degree n, and

$$T_n''(\theta) = x(1-x)P_n''(x) + \frac{1 - 2x}{2} P_n'(x) =: P(D)P_n(x).$$

The regularity of $P(D)$ (see Theorems 1.1 and 1.2 below) implies that the Bernstein inequality from above is equivalent to

$$\|w_{\alpha,\beta}\varphi^2 P_n''\|_p \leq Cn^2\|w_{\alpha,\beta}P_n\|_p.$$

For the investigation concerning the last inequality we refer the paper [3] and the references therein.

Throughout this paper we write $\varphi(x) := \sqrt{x(1-x)}$, and $w_{\alpha,\beta}(x) := x^\alpha(1-x)^\beta$ is the so-called Jacobi weight. Our interest here is to investigate the validity of the following inequality: for all $P_n \in \Pi_n$ and $n = 1, 2, \ldots$ there holds

$$\|w_{\alpha,\beta}\varphi^{2r} P_n^{(2r)}\|_{L_p[0,1]} \leq A(\|w_{\alpha,\beta}P(D)P_n\|_{L_p[0,1]} + \|w_{\alpha,\beta}P_n\|_{L_p[0,1]}), \quad (1)$$

where A does not depend on P_n and n. Furthermore, $\alpha, \beta > -1/p$ for $1 \leq p < \infty$ and $\alpha, \beta \geq 0$ for $p = \infty$.

To begin with, let us define the following functions:

$$\sigma_0(x) = \alpha_0(0) + \sum_{i=1}^{r} \alpha_i(0)x(x-1)\cdots(x-i+1),$$

and

$$\sigma_1(x) = \alpha_0(1) + \sum_{i=1}^{r}(-1)^i\alpha_i(1)x(x-1)\cdots(x-i+1).$$

Furthermore,

$$\bar{\sigma}_0 := \{\operatorname{Re} x : \sigma_0(x) = 0\}, \quad \text{and} \quad \bar{\sigma}_1 := \{\operatorname{Re} x : \sigma_1(x) = 0\}.$$

The following result is proved in [7]:

Theorem A   *Let $P(D)$, φ, $w_{\alpha,\beta}$ and $\bar{\sigma}_0$, $\bar{\sigma}_1$ be given as above where, for $1 \leq p \leq \infty$, we have $-1/p - \alpha \notin \bar{\sigma}_0$ and $-1/p - \beta \notin \bar{\sigma}_1$. Then (1) holds.*

We notice that for given α and β there are at most $2r$ different numbers p with $1 \leq p \leq \infty$ which satisfy $-1/p - \alpha$ in $\bar{\sigma}_0$ or $-1/p - \beta$ in $\bar{\sigma}_1$. Thus, up to at most $2r$ different numbers p, the differential operator $P(D)$ is regular in the sence of (1). On the other hand, it follows from [7] (see Lemma 2.3 and the proof of Theorem 1.2 therein) that, with $\alpha, \beta > -1/p$ for $1 \leq p < \infty$ and $\alpha, \beta \geq 0$ for $p = \infty$, we have for all $\alpha' > \alpha$ and $\beta' > \beta$ the estimate:

$$\|w_{\alpha',\beta'}\varphi^{2r} P_n^{(2r)}\|_{L_p[0,1]} \leq A(\|w_{\alpha,\beta}P(D)P_n\|_{L_p[0,1]} + \|w_{\alpha,\beta}P_n\|_{L_p[0,1]}), \quad (2)$$

where A does not depend on P_n and n. Therefore, it is natural to ask if the conditions of Theorem A are also necessary for (1). The aim of this paper is to give an answer of this problem. We have:

Theorem 1.1 *Let $1 \leq p < \infty$ and $\alpha, \beta > -1/p$. In order that (1) holds for all $P_n \in \Pi_n$ and $n = 1, 2, \ldots$ it is necessary and sufficient that $-1/p - \alpha \notin \overline{\sigma}_0$ and $-1/p - \beta \notin \overline{\sigma}_1$.*

The case $p = \infty$ is somewhat harder to treat. In order to state our next theorem we need the following two conditions:

(A) If $\alpha > 0$ then $-\alpha \notin \overline{\sigma}_0$. If $\alpha = 0$ then (i) $0 \notin \overline{\sigma}_0$ or (i') $y = 0$ is the single root of $\sigma_0(y)$, and the real part of any other roots is different from 0.

(B) If $\beta > 0$ then $-\beta \notin \overline{\sigma}_1$. If $\beta = 0$ then (ii) $0 \notin \overline{\sigma}_1$ or (ii') $y = 0$ is the single root of $\sigma_1(y)$, and the real part of any other roots is different from 0.

Our second result can now be presented as:

Theorem 1.2 *Let $p = \infty$ and $\alpha, \beta \geq 0$.*
For $r = 1$ the inequality (1) is valid if and only if

(i) $\mathrm{Re}\,(\alpha_0(0)/\alpha_1(0)) \neq \alpha$ *or* $\alpha_0(0) = 0$,

(ii) $\mathrm{Re}\,(\alpha_0(1)/\alpha_1(1)) \neq \beta$ *or* $\alpha_0(1) = 0$.

For $r \geq 2$ the inequality (1) is valid if and only if the conditions (A) and (B) are fulfilled.

Both theorems will be proved in the next section.

2 Proofs

Proof of Theorem 1.1.   In view of Theorem A, we only need to show that if $-1/p - \alpha \in \overline{\sigma}_0$ the inequality (1) cannot be true for all $P_n \in \Pi_n$, $n = 1, 2, \ldots$ To this end, assume the contrary. Thus, there exists y_0 such that $\sigma_0(y_0) = 0$ and $-1/p - \alpha = \mathrm{Re}(y_0)$. Let $f_0(x) = x^{y_0+r}$ and $f_\epsilon(x) = (x + \epsilon)^{y_0+r}$. It is clear that we have $f_\epsilon \in C^\infty[0, 1]$ for fixed $\epsilon > 0$, i.e., f_ϵ is infinitely often differentiable on $[0, 1]$. Hence (see, e.g., [2]), there exist polynomials $P_{n,\epsilon} \in \Pi_n$, such that for fixed $j = 0, 1, \ldots$ there holds

$$\|f_\epsilon^{(j)} - P_{n,\epsilon}^{(j)}\|_{L_\infty[0,1]} \longrightarrow 0 \quad (n \longrightarrow \infty).$$

We conclude from (1) that

$$\|w_{\alpha,\beta}\varphi^{2r} f_\epsilon^{(2r)}\|_{L_p[0,1]} \leq A(\|w_{\alpha,\beta}P(D)f_\epsilon\|_{L_p[0,1]} + \|w_{\alpha,\beta}f_\epsilon\|_{L_p[0,1]}), \qquad (3)$$

with a constant A independent of ϵ.

Let us consider the right hand side of (3). As

$$\|w_{\alpha,\beta}f_\epsilon\|_{L_p[0,1]}^p \leq C \int_0^1 w_{\alpha,\beta}^p(x)x^{-1-\alpha p+rp}\, dx \leq C,$$

the second term is bounded. We prove that the first term is bounded, too.

We can write

$$\|w_{\alpha,\beta}P(D)f_\epsilon\|_{L_p[0,1]} = \|w_{\alpha,\beta}P(D)f_\epsilon\|_{L_p[0,\frac{1}{2}]} + \|w_{\alpha,\beta}P(D)f_\epsilon\|_{L_p[\frac{1}{2},1]}.$$

Thus, as $\beta p > -1$ we have

$$\|w_{\alpha,\beta}P(D)f_\epsilon\|_{L_p[\frac{1}{2},1]}^p \leq C \sum_{i=0}^{r} \int_{\frac{1}{2}}^{1} (1-x)^{\beta p} x^{p(\alpha+i)} |f_\epsilon^{(r+i)}(x)|^p \, dx$$

$$\leq C \int_{\frac{1}{2}}^{1} (1-x)^{\beta p} \, dx \leq C.$$

On the other hand, we may write $P(D)$ as

$$P(D) = \sum_{i=0}^{r} \alpha_i(0) x^i D^{r+i} + \sum_{i=0}^{r} \{\alpha_i(x)(1-x)^i - \alpha_i(0)\} x^i D^{r+i} =: \overline{P}(D) + \overline{R}(D).$$

We notice that $\alpha_i \in \mathrm{Lip}\delta$. Hence, for $x \in [0,1/2]$ there holds for some $C > 0$

$$|\overline{R}(D)f_\epsilon(x)| \leq C \sum_{i=0}^{r} x^{\delta+i} |f_\epsilon^{(r+i)}(x)| \leq C x^{\delta-\alpha-\frac{1}{p}},$$

which implies $\|w_{\alpha,\beta}\overline{R}(D)f_\epsilon\|_{L_p[0,1/2]} \leq C$ for all $\epsilon > 0$. Furthermore, one has

$$\|w_{\alpha,\beta}\overline{P}(D)f_\epsilon\|_{L_p[0,\epsilon]} \leq C \left\{ \int_0^\epsilon (x+\epsilon)^{-1} \, dx \right\}^{\frac{1}{p}} \leq C.$$

Therefore, it follows from the above calculations that for some $C > 0$ there holds

$$\|w_{\alpha,\beta}P(D)f_\epsilon\|_{L_p[0,1]} \leq \|w_{\alpha,\beta}\overline{P}(D)f_\epsilon\|_{L_p[\epsilon,\frac{1}{2}]} + C.$$

We notice that by the choice of f_0 one has $\overline{P}(D)f_0 = 0$. Thus

$$\|w_{\alpha,\beta}\overline{P}(D)f_\epsilon\|_{L_p[\epsilon,\frac{1}{2}]} = \|w_{\alpha,\beta}\overline{P}(D)(f_\epsilon - f_0)\|_{L_p[\epsilon,\frac{1}{2}]}$$

$$\leq C \left(\int_\epsilon^{\frac{1}{2}} \sum_{i=0}^{r} x^{p(i+\alpha)} \left| \int_0^\epsilon (x+u)^{y_0-i-1} \, du \right|^p \, dx \right)^{\frac{1}{p}}$$

$$\leq C\epsilon \left(\int_\epsilon^{\frac{1}{2}} x^{-p-1} \, dx \right)^{\frac{1}{p}} \leq C.$$

Hence, the right hand side of (3) is bounded for all $\epsilon > 0$. To treat the left hand side of (3) we notice that for some $C > 0$ one has

$$\|w_{\alpha,\beta}\varphi^{2r}f_\epsilon^{(2r)}\|_{L_p[0,1]}^p \geq \|w_{\alpha,\beta}\varphi^{2r}f_\epsilon^{(2r)}\|_{L_p[\epsilon,\frac{1}{2}]}^p$$

$$\geq C \int_\epsilon^{\frac{1}{2}} \left(\frac{x}{x+\epsilon} \right)^{(\alpha+r)p} (x+\epsilon)^{-1} \, dx$$

$$\geq C(\ln(\frac{1}{2}+\epsilon) - \ln\epsilon).$$

Thus the left hand side of (3) is unbounded. This contradiction verifies our assertion.   ∎

Proof of Theorem 1.2. Like in the proof of Theorem 1.1 we use the notation

$$\overline{P}(D) := \sum_{i=0}^{r} \alpha_i(0) x^i D^{r+i} \quad \text{and} \quad \overline{\overline{P}}(D) := \sum_{i=0}^{r} \alpha_i(1)(1-x)^i D^{r+i}.$$

In [7] (see the proof of Theorem 1.1 therein) we proved for $\alpha, \beta \geq 0$ the following inequality:

$$\|w_{\alpha,\beta}\overline{P}(D)P_n\|_{[0,\frac{1}{2}]} + \|w_{\alpha,\beta}\overline{\overline{P}}(D)P_n\|_{[\frac{1}{2},1]} \leq M(\|w_{\alpha,\beta}P(D)P_n\|_{[0,1]} \tag{4}$$
$$+ \|w_{\alpha,\beta}P_n\|_{[0,1]}),$$

where $M > 0$ does not depend on $P_n \in \Pi_n$ and $\|f\|_{[a,b]} := \sup_{a \leq x \leq b} |f(x)|$. Moreover, we can rewrite σ_0 as

$$\sigma_0(x) = \alpha_r(0) \prod_{i=1}^{r}(x + y_i),$$

with some roots $-y_i$. Since $\alpha_r \neq 0$ we may therefore assume $\alpha_r(0) = 1$. Using the above notations and denoting $T_0 := P_n^{(r)}$, the following system of differential equations holds:

$$\begin{aligned}
xT_0' &+ y_1 T_0 &= T_1 \\
xT_1' &+ y_2 T_1 &= T_2 \\
&\cdots \\
xT_{r-1}' &+ y_r T_{r-1} &= \overline{P}(D)P_n.
\end{aligned}$$

We shall arrange y_i in the following way: $\mathrm{Re}(y_i) \neq \alpha$ for $i = k+1, \ldots, r$. It is clear that

$$\overline{P}(D) = \left(\prod_{i=1}^{r}(xD + y_i) \right) D^r. \tag{5}$$

Under the assumption that $\mathrm{Re}(y_i) \neq \alpha$ for $i = 1, 2, \ldots, r$ the following inequality was proved in [7] (see Lemma 2.2 and its proof): for $i = 0, 1, \ldots, r-j$, $j = 1, 2, \ldots, r$ we have

$$\|w_{\alpha,\beta}x^j T_i^{(j)}\|_{[0,\frac{1}{2}]} \leq C(\|w_{\alpha,\beta}\overline{P}(D)P_n\|_{[0,\frac{1}{2}]} + \|w_{\alpha,\beta}P_n\|_{[0,1]}), \tag{6}$$

where $C > 0$ does not depend on $P_n \in \Pi_n$.

For the case $k = 1$ we can prove in the same way for $i = 1, 2, \ldots, r-1$

$$\|w_{\alpha,\beta}T_i\|_{[0,\frac{1}{2}]} \leq C(\|w_{\alpha,\beta}\overline{P}(D)P_n\|_{[0,\frac{1}{2}]} + \|w_{\alpha,\beta}P_n\|_{[0,1]}), \tag{7}$$

and for $i = 1, 2, \ldots, r - j, \ j = 1, 2, \ldots, r - 1$

$$\|w_{\alpha,\beta} x^j T_i^{(j)}\|_{[0,\frac{1}{2}]} \le C(\|w_{\alpha,\beta} \overline{P}(D) P_n\|_{[0,\frac{1}{2}]} + \|w_{\alpha,\beta} P_n\|_{[0,1]}). \tag{8}$$

We are now in the position to prove the assertions of this theorem. In view of (4), it is clear that in order to verify the sufficiency of the conditions for (1), we only need to show

$$\|w_{\alpha,\beta} \varphi^{2r} P_n^{(2r)}\|_{[0,\frac{1}{2}]} \le C(\|w_{\alpha,\beta} \overline{P}(D) P_n\|_{[0,\frac{1}{2}]} + \|w_{\alpha,\beta} P_n\|_{[0,1]}) \tag{9}$$

and

$$\|w_{\alpha,\beta} \varphi^{2r} P_n^{(2r)}\|_{[\frac{1}{2},1]} \le C(\|w_{\alpha,\beta} \overline{\overline{P}}(D) P_n\|_{[\frac{1}{2},1]} + \|w_{\alpha,\beta} P_n\|_{[0,1]}). \tag{10}$$

We observe that it is enough to verify (9) under the assumption (i) for $r = 1$ and the condition (A) for $r \ge 2$. Then the inequality (10) follows from the transformation $t = 1 - x$ and the condition (ii) for $r = 1$ and (B) for $r \ge 2$, respectively.

Case 1 $(r = 1)$: Clearly, $y_1 = \alpha_0(0)$ as $\alpha_1(0) = 1$. Thus, if $\alpha_0(0) = 0$, then $y_1 = 0$ and $\overline{P}(D) = xD^2$. Hence, (9) holds trivially. While, for $\mathrm{Re}\, y_1 \ne \alpha$, the estimate (9) follows from (6).

Case 2 $(r \ge 2)$: For $\alpha > 0$ we get $-\alpha \notin \overline{\sigma}_0$, which implies $\mathrm{Re}\, y_i \ne \alpha$ for $i = 1, 2, \ldots, r$. Thus, (9) follows from (6). Otherwise $\alpha = 0$, and the condition means either $0 \notin \overline{\sigma}_0$ or $y = 0$ is the single root of σ_0 and for all other roots of σ_0 we have $\mathrm{Re}\, y \ne 0$. In the first case (9) follows from (6) again, since $\mathrm{Re}\, y_i \ne 0$ for $i = 1, 2, \ldots, r$. For the second one we deduce from inequalities (7) and (8) for $j = 0, 1, \ldots, r - 1$

$$\|w_{\alpha,\beta} x^j T_1^{(j)}\|_{[0,\frac{1}{2}]} \le C(\|w_{\alpha,\beta} \overline{P}(D) P_n\|_{[0,\frac{1}{2}]} + \|w_{\alpha,\beta} P_n\|_{[0,1]}).$$

As $T_1 = x T_0'$ the above implies for $j = 0, 1, \ldots, r - 1$

$$\|w_{\alpha\beta} (x^{j+1} T_0^{(j+1)} + j x^j T_0^{(j)})\|_{[0,\frac{1}{2}]} \le C(\|w_{\alpha,\beta} \overline{P}(D) P_n\|_{[0,\frac{1}{2}]} + \|w_{\alpha,\beta} P_n\|_{[0,1]}).$$

Using this inequality recursively we obtain

$$\|w_{\alpha,\beta} x^r T_0^{(r)}\|_{[0,\frac{1}{2}]} \le C_r(\|w_{\alpha,\beta} \overline{P}(D) P_n\|_{[0,\frac{1}{2}]} + \|w_{\alpha,\beta} P_n\|_{[0,1]}),$$

where C_r only depends on C and r. As $T_0 = P_n^{(r)}$, the last estimate implies (9) and the proof of the sufficient part is finished.

To prove that our assumptions are also necessary for (1), we assume the contrary. We know (see, e.g., [2]) that for any $f \in C^\infty[0,1]$ there exist $P_n \in \Pi_n$, such that for fixed $j = 0, 1, \ldots$ there holds

$$\|f^{(j)} - P_n^{(j)}\|_{[0,1]} \longrightarrow 0 \quad (n \longrightarrow \infty).$$

Thus (1) is equivalent to

$$\|w_{\alpha,\beta}\varphi^{2r}f^{(2r)}\|_{[0,1]} \le A(\|w_{\alpha,\beta}P(D)f\|_{[0,1]} + \|w_{\alpha,\beta}f\|_{[0,1]}) \quad \forall f \in C^\infty[0,1]. \quad (11)$$

To finish the proof we use the same technique as in the proof of Theorem 1.1, i.e., we show, that (11) cannot be true for some $f \in C^\infty[0,1]$ if at least one of our assumptions is not fulfilled.

Suppose now that (i) is not fulfilled in case $r = 1$ and $\alpha, \beta \ge 0$, that is, we have $\operatorname{Re}(\alpha_0(0)/\alpha_1(0)) = \alpha$ and $\alpha_0(0) \ne 0$ (note $\alpha_1(0) = 1$). In this case we replace f in (11) by

$$f_{\epsilon,\eta}(x) = \int_{x+\eta}^{\frac{1}{2}} t^{-y+\epsilon} \ln t \, dt$$

with $y = \alpha_0(0)$. It is clear that we have $f_{\epsilon,\eta} \in C^\infty[0,1]$ for any fixed $\epsilon, \eta > 0$. Computation shows that $\|w_{\alpha,\beta}f_{\epsilon,\eta}\|_{[0,1]}$ is bounded and

$$\lim_{\epsilon \to 0} \lim_{\eta \to 0} \|w_{\alpha,\beta}\varphi^2 f_{\epsilon,\eta}^{(2)}\|_{[0,1]} = \infty.$$

In what follows we shall prove

$$\overline{\lim_{\epsilon \to 0}} \; \overline{\lim_{\eta \to 0}} \; \|w_{\alpha,\beta}P(D)f_{\epsilon,\eta}\|_{[0,1]} < \infty.$$

Indeed,

$$
\begin{aligned}
P(D) &= \overline{P}(D) + (\alpha_1(x)(1-x) - \alpha_1(0))xD^2 + (\alpha_0(x) - \alpha_0(0))D \\
&=: \overline{P}(D) + R(D).
\end{aligned}
$$

Then, since $\alpha_0, \alpha_1 \in \operatorname{Lip}\delta$, one easily gets that $\|R(D)f_{\epsilon,\eta}\|_{[0,1]}$ is bounded for all $0 \le \epsilon, \eta \le 1$ and

$$\overline{\lim_{\epsilon \to 0}} \; \overline{\lim_{\eta \to 0}} \; \|w_{\alpha,\beta}\overline{P}(D)f_{\epsilon,\eta}\|_{[0,1]} < \infty.$$

Hence for $r = 1$ condition (i) is necessary for (1). In the same way one proves the necessity of (ii).

To show the necessity part for $r \ge 2$ we assume that (A) is not fulfilled. Hence, for $\alpha > 0$ we have $-\alpha \in \overline{\sigma}_0$ or equivalently $\sigma_0(y) = 0$ and $\operatorname{Re} y = -\alpha$. Thus we replace f in (11) by

$$f_{\epsilon,\eta}(x) = (x+\eta)^{y+r+\epsilon} \ln(x+\eta). \quad (12)$$

If $\alpha = 0$ then there exists one y with $\operatorname{Re} y = 0$ and $\sigma_0(y) = 0$. Moreover, if there is one and only one such y then $y = 0$ implies that y is at least a double root of σ_0. Thus, for $y \ne 0$ we use $f_{\epsilon,\eta}$ given by (12), otherwise define $f_{\epsilon,\eta}$ by $f_{\epsilon,\eta}^{(r)}(x) = (\ln(x+\epsilon))^2$. Clearly, both functions are in $C^\infty[0,1]$. Using the same method as above, we obtain a contradiction to (11) for this function $f_{\epsilon,\eta}$ when $\epsilon, \eta \to 0$.

Hence, (A) is necessary for (1). By using the transformation $t = 1 - x$ we deduce that the remaining part of our assumptions is also necessary for (1), and therefore the proof is finished. ∎

References

[1] P.L. Butzer, Legendre transform methods in the solution of basic problems in algebraic approximation, in: Functions, Series, Operators Vol. I, eds. B. Sz.-Nagy and J. Szabados, North-Holland, Amsterdam, 1983, 277–301.

[2] R.A. DeVore and G.G. Lorentz, Constructive Approximation, Springer, Berlin, 1993.

[3] S. Jansche, Norm inequalities involving ordinary and Jacobi derivatives, Appl. Math. Lett. **6** (1993), 13-19.

[4] P.G. Nevai, Orthogonal Polynomials, Memoirs Amer. Math. Soc. **213**, 1979.

[5] G. Szegö, Orthogonal Polynomials, Amer. Math. Soc. Coll. Publ. Vol. **23**, 1975.

[6] A.F. Timan, Theory of Approximation of Functions of a Real Variable, MacMillan, New York, 1963.

[7] H.J. Wenz and X.L. Zhou, Bernstein type inequalities associated with some differential operators, J. Mathematical Analysis and Applications **213** (1997), 250-261.

Department of Mathematics
University of Duisburg
D-47048 Duisburg, Germany
Email address: zhou@riemann.informatik.uni-duisburg.de

International Series of Numerical Mathematics
Vol. 132, © 1999 Birkhäuser Verlag Basel/Switzerland

Some Inequalities for Trigonometric Polynomials and their Derivatives

Hans-Bernd Knoop and Xinlong Zhou

Abstract

In trigonometric approximation one can show that smoothness of a function is equivalent to a quick decrease to zero of its error of approximation by trigonometric polynomials. The key steps to show this are the Jackson- and the Bernstein-inequality. In this paper we will investigate some Bernstein-type inequalities concerning conjugate functions and Laplacian. In a forth-coming paper we will show that these inequalities imply the equivalence of the order of approximation by the classical Jackson operator (in higher dimension) and a corresponding measure for the smoothness of the function.

1   Introduction

Let $\pi_{n,d}$ be the set of trigonometric polynomials of degree $\leq n$ with respect to each variable x_i, $i = 1, ..., d$. Denote $D_i := \partial/\partial x_i$ and $D^\alpha := D_i^{\alpha_1} \cdots D_d^{\alpha_d}$, where the multiindex $\alpha = (\alpha_1, \ldots, \alpha_d)$ has non-negative integers α_i. The length of α is $|\alpha| = \alpha_1 + \ldots + \alpha_d$. It is known that the so-called Bernstein-inequality asserts that for some C, which does not depend on n there holds (see [8])

$$\|D^\alpha T_n\|_p \leq C n^{|\alpha|} \|T_n\|_p, \quad T_n \in \pi_{n,d},$$

where the norm is given by

$$\|g\|_p := \left(\int_{[-\pi,\pi]^d} |g|^p \right)^{1/p}.$$

This type of inequalities plays an important role in Analysis and in particular in Approximation Theory. In fact, this kind of inequalities tells us in some sense the regularity of D^α (see [11]). It is thus useful in multivariate case to establish such inequality in use of Laplacian Δ on the right hand side, as Laplace operator is one of most frequently investigated differential operators. In [10] (for $p = \infty$) and [11] the second author shows

Theorem 1.1 *For fixed $1 \leq p \leq \infty$ and the dimension d there exists a positive constant C such that for all n and all $T_n \in \pi_{n,d}$*

$$\sup_{|\alpha|=3} ||D^\alpha T_n||_p \leq Cn||\Delta T_n||_p. \tag{1}$$

There is another kind of inequalities, which gives information on the conjugate function in case $d = 1$. Let for $f \in L^p_{2\pi}$ the conjugate function $\tilde{f}$ of f be defined by

$$\tilde{f}(x) := -\frac{1}{2\pi} \int_0^\pi \frac{f(x+t) - f(x-t)}{\tan \frac{t}{2}} dt$$

(cf. [12]). In case $1 < p < \infty$ one has (see [12])

$$||\tilde{T}_n||_p \leq C_p||T_n||_p, \quad \forall \quad T_n \in \pi_{n,1}. \tag{2}$$

The above inequality is not true for $p = 1$ and $p = \infty$. In fact, the following estimate is in general sharp :
For $p = 1$ or $p = \infty$ one has (see [12])

$$||\tilde{T}_n||_p \leq C \ln n ||T_n||_p, \quad \forall \quad T_n \in \pi_{n,1}.$$

On the other hand, it is known that for polynomials and their conjugate functions one has the so-called Szegö-inequality (see [12]):
For all $1 \leq p \leq \infty$ there holds

$$||\tilde{T}'_n||_p \leq Cn||T_n||_p, \quad \forall \quad T_n \in \pi_{n,1}.$$

A generalization of the Szegö-inequality is obtained in [3]:

Theorem 1.2 *For any $\epsilon > 0$ there exists a constant $C_k(\epsilon) > 0$ such that*

$$||\tilde{T}_n^{(k)}||_p \leq C_k(\epsilon)||T_n^{(k)}||_p + \epsilon n||T_n^{(k-1)}||_p, \quad \forall \quad T_n \in \pi_{n,1}.$$

With help of this inequality we proved in [3] the best lower estimate for the classical Jackson operator $J_n = J_{n,2}$ (see [4] and [7] for detail). Here $J_{n,r}$ is the convolution operator with the Jackson-type kernel $k_{n,r}$, defined by

$$k_{n,r}(t) = \alpha_{n,r} \cdot \frac{\sin^{2r}(nt/2)}{\sin^{2r}(t/2)}.$$

Thus the final version on the operator J_n after more than 80 years is:
There exists a positive constant $C > 0$ such that for all $1 \leq p \leq \infty$, for all $f \in L^p_{2\pi}$, and all $n \geq 1$

$$C^{-1}\omega_2(f, \frac{1}{n})_p \leq ||J_n f - f||_p \leq C\omega_2(f, \frac{1}{n})_p,$$

where $w_2(f, \cdot)_p$ is the modulus of smoothness of second order for f (see [7]). With the technique introduced in [5] it is shown that the above inequality holds for the operators $J_{n,r}$, and $r \geq 3$. Let us also remark that using this technique we establish in [6] the best lower and upper estimate for the Bernstein polynomial operator (see also [11] for the multivariate Bernstein operator). Now let $K_{n,r}$ be the tensor-product of the kernels $k_{n,r}$, i.e.

$$K_{n,r}(t) = \prod_{i=1}^{d} k_{n,r}(t_i), \quad t = (t_1, \ldots, t_d),$$

and let $J_{n,r}$ be the corresponding convolution operator with the kernel $K_{n,r}$. In [11] the following theorem is proved:
Let $r = 3, 4, \ldots$; there exists a positive constant $C_r > 0$ such that for all $1 \leq p \leq \infty$, for all $f \in L_{2\pi}^p$, and all $n \geq 1$

$$C_r^{-1} K_\Delta(f, \frac{1}{n})_p \leq ||J_{n,r} f - f||_p \leq C_r K_\Delta(f, \frac{1}{n})_p,$$

where $K_\Delta(f, \cdot)_p$ is the K-functional for f, defined by

$$K_\Delta(f, t)_p = \inf_g \{||f - g||_p + t^2 ||\Delta g||_p\}$$

with the Laplacian Δ.

The above equivalence for $r = 2$ is still missing. The reason is that we have not an analogue of the inequality in Theorem 1.2 for higher dimension with respect to the Laplacian. To deal with let us first give a generalization of the concept of conjugate function in multivariate case. Thus, let $L_i g$ be the conjugate function of g with respect to x_i and $\tilde{D}_i := D_i L_i$. In this way we define $\tilde{D}^\alpha = \tilde{D}_1^{\alpha_1} \ldots \tilde{D}_d^{\alpha_d}$. With these notations the Szegö-inequality in multivariable version is

$$||\tilde{D}^\alpha T_n||_p \leq C_\alpha n^{|\alpha|} ||T_n||_p, \quad \forall \quad T_n \in \pi_{n,d}. \tag{3}$$

The aim of this paper is to generalize the above inequalities in use of the Laplace operator. In a forthcoming paper we will investigate the best lower and upper estimates for the classical Jackson operator $J_{n,2}$ in higher dimension. We have

Theorem 1.3 *Let* $1 \leq p \leq \infty$, $k \in \mathbf{N}$, *and the dimension* $d \geq 1$ *be fixed. Then there exists a positive constant* $C_k > 0$ *such that for all* $\epsilon > 0$, *all* n *and all* $T_n \in \pi_{n,d}$

$$\sup_{|\alpha+\beta|=2k+1} ||D^\alpha \tilde{D}^\beta T_n||_p \leq C_k \left\{ \frac{1}{\epsilon^{2k+2}} \sup_{|\alpha|=2k+1} ||D^\alpha T_n||_p + \epsilon n ||\Delta^k T_n||_p \right\}. \tag{4}$$

It is clear that (4) implies (1). To see this one needs only to take infimum over all $\epsilon > 0$.

2   Proof of Theorem 1.3

Denote the kernel of the singular integral of (Gauß-)Weierstraß for $d = 1$ (see [1] and [2]) by

$$\eta_t(x) := \frac{1}{2\sqrt{\pi t}} \sum_{k=-\infty}^{\infty} \exp\left(-\frac{(2\pi k - x)^2}{4t}\right), \quad t > 0. \tag{5}$$

We have

Lemma 2.1 *For some absolute positive constant C there holds for any $t > 0$*

$$\int_0^{2\pi} |\tilde{\eta}'_t(x)| dx \le \frac{C}{\sqrt{t}}.$$

Proof. It is clear that, using integration by parts and the definition of conjugate function, one has

$$
\begin{aligned}
\tilde{\eta}'_t(x) &= -\frac{1}{2\pi} \int_0^\pi \frac{\eta'_t(x+u) - \eta'_t(x-u)}{\tan \frac{u}{2}} du \\
&= -\frac{1}{4\pi} \int_0^\pi \frac{\eta_t(x+u) + \eta_t(x-u) - 2\eta_t(x)}{\sin^2 \frac{u}{2}} du
\end{aligned}
$$

because of

$$\frac{d}{dx}\{\eta_t(x+u) - \eta_t(x-u)\} = \frac{d}{du}\{\eta_t(x+u) + \eta_t(x-u) - 2\eta_t(x)\}.$$

Replacing η_t in the above inequality by the expression of the right hand side of (5) and interchanging the integrals we conclude that

$$\int_0^{2\pi} |\tilde{\eta}'_t(x)| dx \le \frac{1}{8\pi^{\frac{3}{2}} t^{\frac{1}{2}}} \int_0^\pi \sum_{k=-\infty}^{\infty} \int_0^{2\pi} \frac{J(2k\pi - x, u)}{\sin^2 \frac{u}{2}} dx du,$$

where

$$J(y, u) = \left| \exp(-\frac{(y+u)^2}{4t}) + \exp(-\frac{(y-u)^2}{4t}) - 2\exp(-\frac{y^2}{4t}) \right|.$$

Changing the variable x by $2k\pi - x$, we obtain from above

$$
\begin{aligned}
\int_0^{2\pi} |\tilde{\eta}'_t(x)| dx &\le \frac{1}{8\pi^{\frac{3}{2}} t^{\frac{1}{2}}} \int_0^\pi \sum_{k=-\infty}^{\infty} \int_{2\pi(k-1)}^{2\pi k} \frac{J(x, u)}{\sin^2 \frac{u}{2}} dx du \\
&= \frac{1}{8\pi^{\frac{3}{2}} t^{\frac{1}{2}}} \int_0^\pi \int_{-\infty}^{\infty} \frac{J(x, u)}{\sin^2 \frac{u}{2}} dx du
\end{aligned}
$$

$$\leq \frac{\pi}{4\sqrt{\pi t}} \int_0^\infty \int_0^\infty \frac{J(x,u)}{u^2} \, dx \, du$$

$$\leq \frac{\pi}{4\sqrt{\pi t}} \int_0^\infty \int_0^\infty \left| \frac{e^{-(x+u)^2} + e^{-(x-u)^2} - 2e^{-x^2}}{u^2} \right| du \, dx.$$

Thus we need only to show that for some constant $C > 0$

$$\int_0^\infty \int_0^\infty \left| \frac{e^{-(x+u)^2} + e^{-(x-u)^2} - 2e^{-x^2}}{u^2} \right| du \, dx \leq C. \tag{6}$$

To this end we write the integral in (6) as

$$\int_0^1 \int_0^1 \cdots + \int_0^1 \int_1^\infty \cdots + \int_1^\infty \int_0^1 \cdots + \int_1^\infty \int_1^\infty =: I_1 + I_2 + I_3 + I_4.$$

It is clear that the mean value theorem yields

$$|I_1| \leq C_1 \quad \text{and} \quad |I_3| \leq C' \int_0^\infty e^{-x^2} dx \leq C_3,$$

while the boundedness of $\exp(-y^2)$ implies

$$|I_2| \leq 4 \int_1^\infty u^{-2} du \leq C_2.$$

To estimate I_4 we notice that

$$|I_4| \leq C'' + \int_1^\infty \int_1^\infty \frac{e^{-(x-u)^2}}{u^2} du \, dx \leq C_4.$$

The inequality (6) follows from the last four estimates. ∎

The kernel of the tensor product of the singular Weierstraß integral can be written as

$$K_t(x) := \frac{1}{(2\sqrt{\pi t})^d} \sum_{k \in \mathbf{Z}^d} \exp\left(-\frac{\sum_{i=1}^d (2\pi k_i - x_i)^2}{4t} \right).$$

Denote

$$W(t)f := K_t * f.$$

The following properties are well-known (see [1]):

$$\|f - W(t)f\|_p \leq Ct\|\Delta f\|_p, \quad \forall f, \Delta f \in L_p(T^d) \tag{7}$$

and

$$\|D^\alpha W(t)f\|_p \le \frac{C}{t^{\frac{|\alpha|}{2}}}\|f\|_p, \quad \forall f \in L_p(T^d). \tag{8}$$

Moreover, for f, $D^\alpha f \in L_p(T^d)$ there hold

$$D^\alpha W(t)f = W(t)D^\alpha f, \qquad \|D^\alpha W(t)f\|_p \le C\|D^\alpha f\|_p$$

and $W^j(t) = W(jt)$. Furthermore, Lemma 2.1 tells us

$$\|\tilde{D}_i W(t)f\|_p \le \frac{C}{\sqrt{t}}\|f\|_p, \quad \forall f \in L_p(T^d).$$

Consequently, we obtain

$$\|D^\alpha \tilde{D}^\beta W(t)f\|_p \le \frac{C}{t^{\frac{|\alpha+\beta|}{2}}}\|f\|_p, \quad \forall f \in L_p(T^d). \tag{9}$$

Now we are in the position to verify our main result Theorem 1.3.

Proof of Theorem 1.3: In the following, C_k will always denote some positive constant, independent of n and p. In each formula the constant may take a different value. Obviously, we have for any $t > 0$

$$\|D^\alpha \tilde{D}^\beta W(t)T_n\|_p \le \sum_{j=1}^{\infty} \|D^\alpha \tilde{D}^\beta W(jt)(I - W(t))T_n\|_p,$$

where I is the identity operator. Using this inequality repeatedly, we get from (9)

$$\|D^\alpha \tilde{D}^\beta W(t)T_n\|_p \le \sum_{j_1=1}^{\infty} \cdots \sum_{j_k=j_{k-1}}^{\infty} \|D^\alpha \tilde{D}^\beta W(j_k t)(I - W(t))^k T_n\|_p$$

$$\le \|(I - W(t))^k T_n\|_p \sum_{j_1=1}^{\infty} \cdots \sum_{j_k=j_{k-1}}^{\infty} \frac{C_k}{(j_k t)^{\frac{|\alpha+\beta|}{2}}}$$

$$\le C_k t^{-\frac{|\alpha+\beta|}{2}}\|(I - W(t))^k T_n\|_p.$$

Hence, recalling $|\alpha + \beta| = 2k + 1$, we deduce by (7)

$$\|D^\alpha \tilde{D}^\beta W(t)T_n\|_p \le C_k'' t^{-\frac{1}{2}}\|\Delta^k T_n\|_p. \tag{10}$$

On the other hand, we notice that, as $I - (I - W(t))^{k+1}$ is a linear combination of $W(jt)$ with $1 \le j \le k + 1$, one has by (7) and (10)

$$\|D^\alpha \tilde{D}^\beta T_n\|_p \le \|D^\alpha \tilde{D}^\beta (I - (I - W(t))^{k+1} T_n\|_p + \|D^\alpha \tilde{D}^\beta (I - W(t))^{k+1} T_n\|_p$$

$$\le C_k \left(t^{-\frac{1}{2}}\|\Delta^k T_n\|_p + t^{k+1}\|D^\alpha \tilde{D}^\beta \Delta^{k+1} T_n\|_p \right).$$

Finally, as $|\beta| \leq 2k+1$, the Bernstein-inequality and the Szegö-inequality imply

$$||D^\alpha \tilde{D}^\beta \Delta^{k+1} T_n||_p \leq C_k n^{2k+2} \sup_{|\gamma|=2k+1} ||D^\gamma T_n||_p.$$

Combining the last two estimates we obtain that

$$||D^\alpha \tilde{D}^\beta T_n||_p \leq C_k \left(t^{k+1} n^{2k+2} \sup_{|\gamma|=2k+1} ||D^\gamma T_n||_p + t^{-\frac{1}{2}} ||\Delta^k T_n||_p \right).$$

Choosing $t^{-1} = \epsilon^2 n^2$, the assertion (4) follows. ∎

References

[1] Butzer, P. L. and Berens, H., Semi-Groups of Operators and Approximation. Die Grundlehren der mathematischen Wissenschaften 145, Springer-Verlag, Berlin-Heidelberg-New York, 1967.

[2] Butzer, P. L. and Nessel, R. J., Fourier Analysis and Approximation. Vol. 1, Birkhäuser-Verlag, Basel 1970.

[3] Heckers, E., Knoop, H.-B. and Zhou, X. L., Approximation by convolution operators. Rendiconti del Circolo Matematico di Palermo **52** (1998), 523-536.

[4] Jackson, D., On the approximation by trigonometric sums and polynomials. *TAMS* **13** (1912), 491-515.

[5] Knoop, H.-B. and Zhou, X. L., The lower estimate for linear positive operators, I. *Constr. Approx.* **11** (1995), 53-66.

[6] Knoop, H.-B. and Zhou, X. L., The lower estimate for linear positive operators, II. *Results Math.* **25** (1994), 315-330.

[7] Lorentz, G. G., Approximation of Functions. Chelsea Publ. Comp., New York, 1986.

[8] Nikol'skii, S. M.: Approximation of Functions of Several Variables and Imbedding Theorems. Springer-Verlag, Berlin etc., 1975.

[9] Timan, A. F., Theory of Approximation of Functions of a Real Variable. Pergamon Press, Oxford etc., 1963.

[10] Zhou, X. L., Degree of approximation associated with some elliptic operators and its applications. *Approx. Theory Appli.* **2** (1995), 9-29.

[11] Zhou, X. L., Approximationsordnung und Regularität von Differentialoperatoren. Habilitationsschrift, Gerhard-Mercator-Universität Duisburg, 1996.

[12] Zygmund, A., Trigonometric Series. Vol. I and II, 2nd edn. Cambridge University Press, Cambridge etc., 1990.

Department of Mathematics
Gerhard-Mercator-University
D-47057 Duisburg, Germany
Email address: knoop@math.uni-duisburg.de
Email address: xzhou@informatik.uni-duisburg.de

International Series of Numerical Mathematics
Vol. 132, © 1999 Birkhäuser Verlag Basel/Switzerland

Inf-Convolution and Radial Basis Functions

Alain Le Méhauté

Abstract

In this paper, we show how it is possible to use the ideas of Inf-convolution, coming from convex optimisation, to the definition of new reproducing kernels and the explanation of some experimental constatations in the practical use of Radial Basis Functions.

1 Introduction

The purpose of this paper is to illustrate how to deal with the idea of inf-convolution of semihilbertian spaces in order to built (easily?) new reproducing kernels that can be used for the definition of radial basis functions.

Radial basis functions have been comprehensively reviewed in several recent papers (see [4] and references therein) and it is sufficient here to mention roughly how it works.

A radial function f on $\mathbb{R}^d$ is defined in such a way that there exists a univariate function $g : \mathbb{R}^+ \mapsto \mathbb{R}$ such that $f(x) = g(\|x\|) = g(r)$, for all $x \in \mathbb{R}^d$ ($r = \|x\|$ is the euclidian norm of x). In this paper we are dealing only with Lagrange interpolation, while Hermite or even Hermite-Birkhoff interpolation could be considered as well as approximation and smoothing [2,16]. In the former case, let us given $u \in \mathcal{C}(\mathbb{R}^d, \mathbb{R})$, $\mathcal{X} = \{x_1, x_2, \ldots, x_N\}$ a set of N distinct points in $\mathbb{R}^d$, $\mathbf{Z} = \{z_1, z_2, \ldots, z_N\}$ a set of data, $\mathbf{V}$ a linear subspace of $\mathcal{C}(\mathbb{R}^d, \mathbb{R})$ of finite dimension M and $\{v_1, v_2, \ldots, v_M\}$ as a basis of $\mathbf{V}$.

The interpolation problem we deal with is the following: find a radial function $\varphi \in \mathcal{C}(\mathbb{R}^d, \mathbb{R})$, with $\varphi(x) = \phi(\|x\|)$ and a function σ, of the form

$$\sigma(\bullet) = \sum_{i=1}^{N} \alpha_i \, \phi(\| \bullet - x_i\|) + \sum_{j=1}^{M} \beta_j \, v_j(\bullet) \tag{1}$$

such that $\sigma(x_k) = z_k, \; k = 1, \ldots, M$.

The coefficients $\{\alpha_i\}_{i=1}^{N}$ and $\{\beta_j\}_{j=1}^{M}$ are obtained from the *interpolation constraints*: $\sigma(x_k) = u(x_k) = z_k$, for $k = 1, \ldots, N$ and the *orthogonality constraints*: $\sum_{k=1}^{N} \alpha_k \, v_j(x_k) = 0$, for all $j = 1, \ldots, M$.

In matrix notation, this can be written as:

$$\begin{pmatrix} A & B \\ B^\top & 0 \end{pmatrix} \begin{pmatrix} \alpha \\ \beta \end{pmatrix} \begin{pmatrix} Z \\ 0 \end{pmatrix} \tag{2}$$

where $\boldsymbol{\alpha} = (\alpha_1, \alpha_2, \ldots, \alpha_N)^\top$, $\boldsymbol{\beta} = (\beta_1, \beta_2, \ldots, \beta_M)^\top$, $A = [\phi(\|x_i - x_j\|)]_{i,j=1}^N$, and $B = [v_j(x_i)], 1 \le i \le N, 1 \le j \le M$.

From the two sets of constraints, we obtain [6] that the linear system has a unique solution iff for any $\alpha \in (ImB)^\top \setminus \{0\}$ there exists $\gamma \in (ImB)^\top \setminus \{0\}$ such that $\gamma^\top A \gamma = 0$.

Let us remark that, practically, given any $\boldsymbol{Z}$, or equivalently $u \in C(\mathbb{R}^d, \mathbb{R})$ and thus $\boldsymbol{Z} = u^{\mathcal{X}}$, existence and unicity of a solution σ depends upon 1) the function ϕ, 2) the linear subspace $\boldsymbol{V}$, 3) the cardinality of $\mathcal{X}$, and 4) the location of the data points x_i in $\mathbb{R}^d$.

Various well-known Radial Basis Functions are widely used. Among them, gaussians RBF, where $\phi(r) = e^{-r^2}$; multiquadrics [10], with $\phi(r) = (h^2 + r^2)^{\pm\frac{1}{2}} (h > 0)$; thin plate splines [7,8] for which $\phi(r) = r^{2m+2s-d} log r$, whenever $2m + 2s - d \in 2\mathbb{N} \setminus \{0\}$ or $\phi(r) = r^{2m+2s-d}$ whenever $2m + 2s - d \in \mathbb{R}^+$ is not an even integer, and shifted versions of them [9]. Compactly supported Radial basis functions where introduced only very recently and we lack a satisfactory theory for them. It is the purpose of this paper to help providing a more general way to construct such bases. In the first constructions by Schaback & al [19,21,22], the main point was to have a piecewise polynomial basis, $i.e.$, more or less like in the univariate case. For example, Wendland introduces the compactly supported RBF (in $\mathbb{R}^3$): $\phi_{3,0}(r) = (1 - r)_+^2$, $\phi_{3,1}(r) = (1 - r)_+^4 (1 + 4r)$,

$$\phi_{3,2}(r) = (1 - r)_+^6 (3 + 18r + 35r^2), \phi_{3,3}(r) = (1 - r)_+^8 (1 + 8r + 25r^2 + 32r^3).$$

Another approach is nicely done by Buhmann [4], using a corollary of [18]: in order to use Bochner's theorem, and thus have a compactly supported ϕ such that $\hat{\phi} > 0$, he looks for a function such that $\phi(r) = \int_0^\infty \left(1 - \frac{r^2}{\beta}\right)_+^\lambda g(\beta) d\beta$, where g is continuous on $\mathbb{R}^+$ and compactly support ed. Then

$$\hat{\phi}(x) = C\|x\|^{-(d+2)} \int_0^\infty J_{\lambda + \frac{d}{2}} t^{\frac{d}{2} - \lambda + 1} g(t^2 \|x\|^{-2}) \, dt$$

and taking $g(\beta) = \left(1 - \beta^\mu\right)_+^\rho$ with $0 < \mu \le \frac{1}{2} \le \rho$ and, for $d = 1, \lambda \ge 1$, and for $d \ge 2, \lambda \ge \frac{1}{2}(d-1)$ he reach his goals, obtaining that, for instance with $\mu = \frac{1}{2}$ and ρ, λ being positive integers with $\rho \ge \lambda$, $\phi(x) = p(r) + q(r^2) log r$, for $r = \|x\| < 1$ and 0 for $r \ge 1$. In this expression, p and q are polynomials, $deg q = \lambda, p(1) = 0, q(0) = 0$. For example, for $r < 1$ and $\lambda = 1, \rho = 1$ we have $\phi(r) = \frac{1}{3} + r^2 - \frac{4}{3}r^3 + 2r^2 log r$, with $\lambda = 1, \rho = 4$ we have $\phi(r) = \frac{1}{15} + \frac{19}{6}r^2 - \frac{16}{3}r^3 + 3r^4 - \frac{16}{15}r^5 + \frac{1}{6}r^6 + 2r^2 log r$.

2 Semi-hilbertian Subspaces and Inf-convolution

2.1 Semi-hilbertian subspaces

Let us start this section with some definitions.

Let E, E' be two locally convex topological linear spaces in duality, (denoted as usual by $\langle x, x'\rangle, \forall x \in E, x' \in E'$).

A linear subspace $X \subset E$ is a *semi-hilbertian subspace* of E if there is a semi-scalar product $(\bullet, \bullet)$ on X and an associated semi-norm $|\bullet|$ such that:

1) the Null-space N of $|\bullet|$ is finite dimensional ;
2) the quotient-space X/N is complete;
3) $|x_n| \to 0$ implies that $\langle x_n, x'\rangle \to 0, \forall x' \in E'$ *i.e.*, X/N is topologically included in E/N for the weak topology.

Let $N^\circ = \{x' \in E' : \langle x, x'\rangle = 0, \forall x \in N\}$.

Let X be a semi-hilbertian subspace of E and h be such that

$$h(x) = \begin{cases} \frac{1}{2}|x|^2, & \text{if } x \in X \\ +\infty, & \text{elsewhere.} \end{cases}$$

h is said to be a *semi-hilbertian function* . (In fact h is a particular case of a *quadratic-convex function* [14])

A linear mapping $H \colon E' \mapsto E$ is a *semi-kernel* for X if for any $x' \in N^\circ \subset E'$ we have $H(x') \in X$ and $\langle x, x'\rangle = (x, H(x')), \forall x \in X$.

Properties [8,2,16]

1. If $N = \{0\}$, then H is a classical Schwartz kernel.
2. If $N \neq \{0\}$, a semi-kernel H is not unique: for any linear mapping $P \colon E' \mapsto N$, $H + P$ is also a semi-kernel.
3. Let $\mathcal{H} : N^\circ \mapsto X/N \subset E/N$ such that $x' \mapsto \mathcal{H}(x') = H(x') + N$. then $\mathcal{H}$ is the usual kernel in the space X/N.
4. Semi-symmetry: for all $x', y' \in N^\circ$, we have $\langle H(y'), x'\rangle = \langle H(x'), y'\rangle$.
5. Semi-positivity: for all $x' \in N^\circ$, we have $\langle H(x'), x'\rangle \geq 0$.

Let us define the following problem:

Problem 1. Given $L = \{\ell_1, \ell_2, \ldots, \ell_N\} \subset E'$ and $Z = \{z_1, z_2, \ldots, z_N\} \subset \mathbb{R}^N$, minimize $h(x)$ under the constraints : $\langle x, \ell_i\rangle = z_i, i = 1, 2, \ldots, N$.

From [12], we know that a function $\sigma \in X$, called a *spline function*, is a solution of this problem if and only if there exist $\lambda_1, \lambda_2, \ldots, \lambda_N$ such that

$$\langle x, \sigma\rangle = \langle x, \sum \lambda_i \ell_i\rangle, \quad \forall x \in N$$

Theorem 1. *[7] Let H be a semi-kernel for the semi-hilbertian subspace X of E. Then $\sigma \in X$ is a solution of (PB 1) if and only if $\sigma = \sum \lambda_i H(\ell_i) + p$, where $p \in N$ and $\sum \lambda_i \langle q, \ell_i\rangle = 0$, for all $q \in N$.*

A particular case consists in taking $E = \mathcal{C}(\mathbb{R}^d, \mathbb{R})$ and $\ell_i = \delta_{x_i}$ where the x_i are data points in $\mathbb{R}^d$. A semi reproducing kernel is then given by

$$K(t, t') = \langle H(\delta_{t'}), \delta_t \rangle$$

which gives $\sigma(x) = \sum \lambda_i \, K(x, x_i) + p, p \in \mathbf{N}$, subject to the compatibility conditions $\sum \lambda_i \, q(x_i) = 0$ for all $q \in \mathbf{N}$, specific spaces providing *thin plate splines* [7,8], *multiquadrics* [2], or more generally (m, ℓ, s)-splines [3,16].

2.2   Inf-convolution

Let us now introduce some results on *Inf- Convolution* [15]

Given two semi-hilbertian functions h_1 and h_2, let us denote by $\mathbf{X}_1 = Dom(h_1), \mathbf{X}_2 = Dom(h_2), \mathbf{N}_1 = kern(h_1), \mathbf{N}_2 = kern(h_2)$.

Definition 1. *The inf-convolution h of h_1 and h_2, denoted by $h = h_1 \nabla h_2$ is defined by*

$$h(x) = h_1 \nabla h_2(x) = Inf\{h_1(x_1) + h_2(x_2) : \ x = x_1 + x_2, x_1 \in \mathbf{X}_1, x_2 \in \mathbf{X}_2\}.$$

In [14] one can find the following properties:
• The domain and null-space of h are respectively $Dom(h) = \mathbf{X} = \mathbf{X}_1 + \mathbf{X}_2$ and $kern(h) = \mathbf{N} = \mathbf{N}_1 + \mathbf{N}_2$.

• **Theorem 2.** *The inf-convolution of two semi-hilbertian functions is itself a semi-hilbertian function.*

• From the definition of semi-hilbertian functions, $\mathbf{X}_i, i = 1, 2$, is a semi -hilbertian subspace of some E, with a semi-scalar product$(\bullet, \bullet)_i$, a semi-norm $| \bullet |_i$ and

$$h_i(x) = \begin{cases} \frac{1}{2}|x|_i^2, & \text{if } x \in X_i \\ +\infty, & \text{elsewhere.} \end{cases}$$

Thus, for all $x \in \mathbf{X} = \mathbf{X}_1 + \mathbf{X}_2$ we have

$$h(x) = h_1 \nabla h_2(x) = Inf\{\frac{1}{2}\left(|x_1|_1^2 + |x_2|_2^2\right) \text{ for } x = x_1 + x_2, x_i \in \mathbf{X}_i\}$$

and $h(x) = +\infty$ when ever $x \notin \mathbf{X}$.
• In the proof of Th.2, P.J.Laurent shows that there exist $x_1^* \in \mathbf{X}_1$ and $x_2^* \in \mathbf{X}_2$ such that $x = x_1^* + x_2^*$ and $h(x) = \frac{1}{2}\left(|x_1^*|_1^2 + |x_2^*|_1^2\right)$.

Thus we can define a semi-scalar product in $\mathbf{X}$: for any $x \in \mathbf{X}$, and the corresponding x_1^*, x_2^*, and any $y \in \mathbf{X}$, let $(x, y)_\mathbf{X} = (x_1^*, y_1)_1 + (x_2^*, y_2)_2$.

Theorem 3. *With the semi-scalar product $(\bullet, \bullet)_\mathbf{X}$ and the associated semi-norm $| \bullet |_\mathbf{X}$, the space $\mathbf{X}$ is a semi-hilbertian subspace of E.*

• From the construction, the null-space of the semi-norm is $\mathbf{N} = \mathbf{N}_1 + \mathbf{N}_2$.

2.3 Semi-kernel and inf-convolution

Theorem 4. *Let h_1 and h_2 be two semi-hilbertian functions on E, X_1 and X_2 the associated semi-hilbertian spaces, and H_1, H_2 two of the corresponding semi-kernels. Then $H = H_1 + H_2$ is a semi-kernel for the semi-hilbertian space $X = X_1 + H_1(x')$ which corresponds to the inf-convolution $h = h_1 \nabla h_2$.*

Proof: Let $x' \in (N_1 + N_2)^\circ$. For $x \in X$, let the optimal decomposition be $x = x_1^* + x_2^*$; then $\langle x, x' \rangle = \langle x_1^*, x' \rangle + \langle x_2^*, x' \rangle = (x_1^*, H_1(x'))_1 + (x_2^*, H_2(x'))_2$. But $H_1(x') \in X_1$ implies that $H_1(x') = H_1(x') + 0$ (where $0 \in H_1(x')$; thus $(x, H_1(x')) = (x_1^*, H_1(x'))_1 + (x_2^*, 0)_2 = (x_1^*, H_1(x'))_1$ and for the same reason $(x, H_2(x')) = (x_1^*, 0)_1 + (x_2^*, H_2(x'))_2 = (x_2, H_2(x'))_2$. Finally, for any $x \in X$, $\langle x, x' \rangle = (x, H_1(x')) + (x, H_2(x')) = (x, (H_1 + H_2)(x'))$ ∎

We are now in good position to define *inf-convolution splines* [15]. Let us consider the optimization problem:

Problem 2. *Minimize $(h_1 \nabla h_2)(x)$ under the contraints $\langle x, \ell_i \rangle = z_i$, for $i = 1, \ldots, N$.*

Theorem 5. *A function $\sigma \in X$ satisfying $\langle x, \ell_i \rangle = z_i$, for $i = 1, \ldots, N$ is a solution of (Pb 2) if and only if it can be written as*

$$\sigma = \sum_{i=1}^{N} \lambda_i \{ H_1(\ell_i) + H_2(\ell_i) \} + p_1 + p_2$$

where $p_1 \in N_1$, $p_2 \in N_2$ and the coefficients λ_i are such that $\sum \lambda_i \langle q, \ell_i \rangle = 0$ for all $q \in N_1 + N_2$.

This result ([15]) shows that an inf-convolution spline is obtained, as usual [12], from a linear combination of evaluations of the semi-kernel in the semi-hilbertian space, plus an element of the null space of the semi-norm; what is interesting for us here is the fact that the semi-kernel (resp. the null space) is a sum of semi-kernels (resp. null spaces), and also that this is a necessary and sufficient condition.

2.4 A particular case of Inf-convolution

This case was in fact the one studied by P.J.Laurent [13] (and also, but in a different way, by G.Wahba [20]).

Let h_1 be a semi-hilbertian function, $X_1 \in E, N_1$ and H_1 being the associated elements, and let V be a linear subspace of E, of finite dimension M, with $V \neq N_1$.

Let us define h_2 as

$$h_2(x) = \begin{cases} 0 & \text{for } x \in V \\ +\infty & \text{for } x \notin V \end{cases}.$$

It is easy to prove that h_2 is a semi-hilbertian function, with $Dom(h_2) = V$, $Ker(h_2) = V$ and a semi-kernel is H_2 such that $H_2(x') = 0$ $\forall x' \in E'$.

Now, (Pb 2), *i.e.*,"Minimize $(h_1 \nabla h_2)(x)$ under the contraints $\langle x, \ell_i \rangle = z_i, i = 1, \ldots, N$" becomes in this particular case:

Problem 3. *Minimize $h_1(x_1)$ under the contraints: $x_1 \in X_1$, and $\exists v \in V \langle x_1 + v, \ell_i \rangle = z_i, i = 1, \ldots, N$.*

and the characterization of the inf-convolution spline in this special case follows:

Theorem 6. *A function $\sigma \in X_1 + V$ satisfying $\langle \sigma, \ell_i \rangle = z_i, i = 1, \ldots, N$ is a solution of (Pb 3) if and only if*

$$\sigma = \sum_{i=1}^{N} \lambda_i \, H_1(\ell_i) + p_1 + v$$

where $p_1 \in N_1$, $v \in V$ and the coefficients λ_i are such that $\sum \lambda_i \langle q, \ell_i \rangle = 0$ for all $q \in N_1$ and $\sum \lambda_i \langle w, \ell_i \rangle = 0$ for all $w \in V$.

3   Inf-convolution of Radial Basis Functions

As already mentionned, in [4], M.D.Buhmann introduces some compactly supported radial basis functions which are of the form

$$\phi(r) = p(r) + q(r^2)logr, \quad \text{for } 0 \le r < 1,$$

with $p(1) = 0, q(0) = 0, degq = \lambda$.

Such a ϕ can easily be seen as a linear combination of reproducing semi-kernels, and thus related directly to those of J.Duchon by an inf-convolution process. Wendland's functions are of the form $p(r)$.

In order to be coherent with [21], we will assume that $d = 3$.

3.1   Considering the polynomial $q(r^2)logr$.

Let us start with the $q(r^2)logr$, which is obviously related to thin plate splines. From the definition, we have

$$q(r^2)logr = \sum_{k=0}^{\lambda} c_k \, r^{2k}logr = c_0logr + c_1r^2logr + \ldots + c_\lambda r^{2\lambda}logr.$$

Let us go back to the thin plate splines, or more generally the *Polyharmonic splines*. In the terminology of [7] and [2], they are characterized by:

- a semi-reproducing kernel $\mathcal{K}_{m,0,s}$ where

$$\mathcal{K}_{m,0,s}(r) = \begin{cases} c_{m,s}(-1)^{\frac{d}{2}+m+s+1}r^{2m+2s-d}logr & \text{when } 2m+2s-d \in 2\mathbb{N}, \\ d_{m,s}r^{2m+2s-d} & \text{for } 2m+2s-d \in \mathbb{R} \setminus 2\mathbb{N}, \end{cases}$$

  ($c_{m,s}$ and $d_{m,s}$ being positive constants). The parameters $m \in \mathbb{N}$ and $s \in \mathbb{R}$ are such that $-m - \frac{d}{2} < s < \frac{d}{2}$.
- a semi-hilbertian space: $X^{m,s} = D_0^{-m}(\widetilde{H}^s)$ where ($\mathcal{S}'$ is the set of tempered distributions on $\mathbb{R}^d$)

$$\widetilde{H}^s = \{u \in \mathcal{S}' : \int_{\mathbb{R}^d} |2\pi\xi|^{2s}|\widehat{u}(\xi)|^2 d\xi < \infty\}$$

$$X^{m,s} = \{v \in \mathcal{D}' : \forall \alpha \in \mathbb{N}^d \text{ with } |\alpha| = m, D^\alpha v \in \widetilde{H}^s\} \quad ;$$

- a semi-scalar product

$$(u,v)_{m,0,s} = \sum_{|\alpha|=m} \frac{m!}{\alpha!} \int_{\mathbb{R}^d} |2\pi\xi|^{2s} \widehat{D^\alpha u}(\xi)\overline{\widehat{D^\alpha v}(\xi)}d\xi;$$

- a semi-norm

$$|u|_{m,0,s} = \left[\sum_{|\alpha|=m} \frac{m!}{\alpha!} \int_{\mathbb{R}^d} |2\pi\xi|^{2s} |\widehat{D^\alpha u}(\xi)|^2 d\xi \right]^{\frac{1}{2}};$$

- a null-space $N_{m,0,s} = \mathbb{P}_{m-1}(\mathbb{R}^d)$;
- a semi-hilbertian function h_{m-1}, which is :

$$h_{m-1}(u) = \begin{cases} \left[\sum_{|\alpha|=m} \frac{m!}{\alpha!} \int_{\mathbb{R}^d} |2\pi\xi|^{2s} |\widehat{D^\alpha u}(\xi)|^2 d\xi \right], & \text{for } u \in X^{m,s} \\ +\infty & \text{if } u \notin X^{m,s} \end{cases}$$

- From the Sobolev inclusion theorem, in order to deal with functions instead of distributions, we require that $-m + \frac{d}{2} < s < \frac{d}{2}$ and thus $m + s > \frac{d}{2}$, which implies that $m > 1$. This explains that we cannot accept the index k to be 0 in the definition of the polynomial q (or equivalently, that we impose $q(0) = 0$, like in [4]).
- From [7], we know that $X^{m,s} \subset \mathcal{C}^n$ whenever $2m + 2s - d > 2n$, which is improved by [1]: a $X^{m,0}$-spline is in $\mathcal{C}^{2m-d-1}$ and whenever $2m + 2s - d$ is an odd integer , a $X^{m,s}$-spline is in $\mathcal{C}^n$ if $2m + 2s - d > n$.

Those results where established for a single kernel. Let us now do some inf-convolution.

From Th.4., the polynomial $q(r^2)logr = \sum_{k=1}^{\lambda} c_k r^{2k} logr$, which can be written as

$$H = H_1 + H_2 + \ldots + H_\lambda,$$

with $H_i = (-1)^{i+d+1} c_i r^{2i} log r$, is an inf-convolution of semi-kernels: it is related to the semi-hilbertian function

$$h = h_1 \nabla h_2 \nabla \dots \nabla h_\lambda.$$

The related semi-hilbertian subspace is

$$\boldsymbol{X} = X^{2,s} + X^{3,s} + \dots + X^{\lambda+1,s}.$$

The nullspace is

$$\boldsymbol{N} = \mathbb{P}_1 + \mathbb{P}_2 + \dots + \mathbb{P}_\lambda = \mathbb{P}_\lambda.$$

Let $\kappa = inf\{k : c_k \neq 0\}$, i.e., $q(r^2) = \sum_{k=\kappa}^\lambda c_k r^{2k} log r$; then $\boldsymbol{X}$ is a semi-hilbertian subspace of $C^{2\kappa-1}$.

The semi-scalar product in $\boldsymbol{X}$ is such that $(u,v)_{\boldsymbol{X}} = \sum_{k=\kappa}^\lambda (u_k^*, v_k^*)_{k+1,0,s}$, i.e.,

$$(u,v)_{\boldsymbol{X}} = \sum_{k=\kappa}^\lambda \left\{ \sum_{|\alpha|=k+1} \frac{(k+1)!}{\alpha!} \int_{\mathbb{R}^d} |2\pi\xi|^{2s} \widehat{D^\alpha u_k^*}(\xi) \overline{\widehat{D^\alpha v_k^*}(\xi)} d\xi \right\};$$

The associated semi-norm is

$$|u|_{\boldsymbol{X}} = \left[\sum_{k=\kappa}^\lambda \sum_{|\alpha|=k+1} \frac{(k+1)!}{\alpha!} \int_{\mathbb{R}^d} |2\pi\xi|^{2s} |\widehat{D^\alpha u_h^*}(\xi)|^2 d\xi \right]^{\frac{1}{2}};$$

and the semi-hilbertian function corresponds to

$$\mathcal{H}_1(x) = \begin{cases} \frac{1}{2}|x|_{\boldsymbol{X}}^2 & \text{if } x \in X^{\kappa+1,s} + X^{\kappa+2,s} + \dots + X^{\lambda+1,s} \\ +\infty & \text{elsewhere.} \end{cases}$$

(For $d = 3$, in order that $2m + 2s - d = 2i$ and from the Sobolev inclusion theorem, we have $m = i + 1$ and $s = \frac{1}{2}$).

3.2   We add a polynomial $p(r)$

This polynomial can be written as

$$p(r) = \sum_{k=0}^n d_k \, r^k$$

We will consider the odd and the even powers in two different ways; we separate them and write $p(r) = p_{odd} + p_{even}$, where

$$p_{odd}(r) = \sum_{i=0}^{\lfloor \frac{n}{2} \rfloor} d_{2i+1} r^{2i+1}, \text{ and } \quad p_{even}(r) = \sum_{i=0}^{\lfloor \frac{n}{2} \rfloor} d_{2i} r^{2i}$$

3.2.1 Considering p_{odd}

We have $p_{odd}(r) = \sum_{i=0}^{\lfloor \frac{n}{2} \rfloor} d_{2i+1} r^{2i+1} = d_1 r + d_3 r^3 + \cdots + d_{2\nu+1} r^{2\nu+1}$ with $\nu = \lfloor \frac{n}{2} \rfloor$; this sum corresponds to a sum of semi-reproducing kernels of polyharmonic splines, each of them being of the form $r^{2m+2s-d}$, with $2m + 2s - d = 2i + 1$.

- From the Sobolev condition, we get $s = 1$, and $m = i + 1$.
- The corresponding semi-hilbertian spaces: $X^{m,1} = X^{i+1,1} = D_0^{-m}(\tilde{H}^1)$
- with a semi-scalar product

$$(u, v)_{m,0,1} = \sum_{|\alpha|=m} \frac{m!}{\alpha!} \int_{\mathbb{R}^d} |2\pi\xi|^2 \widehat{D^\alpha u}(\xi) \overline{\widehat{D^\alpha v}(\xi)} d\xi;$$

- and a semi-norm

$$|u|_{m,0,1} = \left[\sum_{|\alpha|=m} \frac{m!}{\alpha!} \int_{\mathbb{R}^d} |2\pi\xi|^2 |\widehat{D^\alpha u}(\xi)|^2 d\xi \right]^{\frac{1}{2}};$$

- a null-space $N_{m,0,1} = \mathbb{P}_{m-1}(\mathbb{R}^d)$;
- a semi-hilbertian function $\tilde{h}_{m-1}$, which is (for $s = 1$):

$$\tilde{h}_{m-1}(u) = \begin{cases} \frac{1}{2} |u|_{m,0,1}^2, & \text{for } u \in X^{m,1} \\ +\infty & \text{if } u \notin X^{m,1} \end{cases}$$

- a semi-reproducing kernel

$$\tilde{H}_{m-1} = r^{2m-1}$$

These results are valid for a single kernel. As for $q(r^2) \log r$, the polynomial p_{odd} can be viewed from the Inf-convolution side.

From Th.4., p_{odd} corresponds to

$$\tilde{H}_{m-1} = \tilde{H}_1 + \tilde{H}_2 + \cdots + \tilde{H}_\nu$$

- this is a semi-kernel related to the inf-convolution

$$\tilde{h} = \tilde{h}_1 \nabla \tilde{h}_2 \nabla \ldots \nabla \tilde{h}_\nu.$$

- the semi-hilbertian subspace is

$$\tilde{X} = X^{1,1} + X^{2,1} + \cdots + X^{\nu+1,1}.$$

- the null-space is

$$\tilde{N} = \mathbb{P}_1 + \cdots + \mathbb{P}_\nu = \mathbb{P}_\nu.$$

- Let $\mu = \inf\{i : d_{2i+1} \neq 0\}$, i.e., p_{odd} is such that $p_{odd} = \sum_{i=\mu}^{\nu} d_{2i+1} r^{2i+1}$. Then $\tilde{X}$ is a semi-hilbertian subspace of $C^{2\nu}$.

- The semi-scalar product is

$$(u,v)_{\widetilde{X}} = \sum_{i=\mu}^{\nu} (u_i^*, v_i^*)_{i+1,0,1}.$$

- The associated semi-norm

$$|u|_{\widetilde{X}} = \left[\sum_{i=\mu}^{\nu} \sum_{|\alpha|=i+1} \frac{(i+1)!}{\alpha!} \int_{\mathbf{R}^d} |2\pi\xi| |\widehat{D^\alpha u_i^*}(\xi)|^2 d\xi \right]^{\frac{1}{2}}.$$

- and finally the semi-hilbertian function is

$$\mathcal{H}_2(x) = \begin{cases} \frac{1}{2}|x|_{\widetilde{X}}^2 & \text{if } x \in \widetilde{X} = X^{\mu,1} + X^{\mu+1,1} + \cdots + X^{\nu+1,1} \\ +\infty & \text{elsewhere.} \end{cases}$$

3.2.2   Considering p_{even}

There is still remaining $p_{even} = d_0 + d_2 r^2 + \cdots + d_{2\nu} r^{2\nu}$. As $r^2 = \|x\|^2$, p_{even} is a pure polynomial of degree 2ν in the coordinate of x.

We get here the opportunity to consider the special case of inf-convolution (Sect 2.4).

Let $V = \mathbb{P}_{2\nu}$, linear subspace of $\mathcal{C}^\mu$.

- The associated semi-hilbertian function is

$$\mathcal{H}_3(x) = \begin{cases} 0 & \text{for } x \in V = \mathbb{P}_{2\nu} \\ +\infty & \text{elsewhere.} \end{cases}$$

3.3   Considering finally $q + p_{odd} + p_{even}$

We collect all the results to obtain that a polynomial of the form introduced by Buhmann, $i.e., q(r^2) + p_{odd}(r) + p_{even}(r)$ has now an hilbertian space setting:

- it corresponds to the inf-convolution of the semi-hilbertian functions $\mathcal{H}_i$, $i = 1, 2, 3$, $i.e.$,

$$\mathcal{H} = \mathcal{H}_1 \nabla \mathcal{H}_2 \nabla \mathcal{H}_3$$

- the semi-kernel is

$$\sum_{k=\kappa}^{\lambda} c_k r^{2k} \log r + \sum_{i=\mu}^{\nu} d_{2i+1} r^{2i+1} + 0.$$

Problem 4. *The spline problem of the minimization of $\mathcal{H}$ under the constraints $\langle x, \ell_i \rangle = z_i$ is in fact the following : find*

$$u \in X + \widetilde{X} = X^{\kappa+1,0} + \cdots + X^{\lambda+1,0} + X^{\mu+1,\frac{1}{2}} + \cdots + X^{\nu+1,\frac{1}{2}}$$

minimizing

$$\frac{1}{2}\Big[\sum_{k=\kappa}^{\lambda}\sum_{|\alpha|=k+1}\frac{(k+1)!}{\alpha!}\int_{\mathbf{R}^d}|2\pi\xi|\,|\widehat{D^\alpha u_h^*}(\xi)|^2 d\xi$$

$$+\sum_{i=\mu}^{\nu}\sum_{|\alpha|=i+1}\frac{(i+1)!}{\alpha!}\int_{\mathbf{R}^d}|2\pi\xi|^2|\widehat{D^\alpha u_i^*}(\xi)|^2 d\xi\Big].$$

under the conditions

$$\bullet\, u = \overbrace{u_\kappa + u_{\kappa+1}+\cdots+u_\lambda}^{\in\,X}+\overbrace{u_\mu + u_{\mu+1}+\cdots+u_\nu}^{\in\,\tilde X}\in X+\tilde X$$

- there exists $v\in\mathbb{P}_{2\nu}$
- and $\langle u+v,\ell_i\rangle = z_i, i = 1,\ldots,N.$

Theorem 7. *A solution σ of this minimization has the following form:*

$$\sigma(x)=\sum_{i=1}^{N}\lambda_i\Big[\sum_{k=\kappa}^{\lambda}c_k\|x-x_i\|^{2k}log(\|x-x_i\|)$$

$$+\sum_{j=\mu}^{\nu}d_{2j+1}\|x-x_i\|^{2j+1}+\sum_{j=0}^{\nu}d_{2j}\|x-x_i\|^{2j}\Big]+\pi(x)$$

where $\sum_{j=0}^{\nu}d_{2j}\|x-x_i\|^{2j}=v\in V=\mathbb{P}_{2\nu}$ and $\pi\in\mathbb{P}_\lambda+\mathbb{P}_\nu=\mathbb{P}_{sup\{\lambda,\nu\}}$. This corresponds obviously to

$$\sigma(x)=\sum_{i=1}^{N}\lambda_i\big\{p(r_i)+q(r_i^2)logr_i\big\}+\pi(x)$$

where $r_i = \|x-x_i\|^{\frac{1}{2}}$.

It remain to choose the coefficients c_k and d_k.

4   The choice of the coefficients of the polynomials p and q

It is not actually well understood, but, up to the truncation at $r=1$, the Buhmann's functions, and also the Wendland's ones are easily obtained from an Hermite univariate interpolation process.

Solving the following problem: *find ϕ such that $\phi(r) = d_0 + d_1 r + d_2 r^2 + d_3 r^3 + c_1 r^2 log r$ subject to*

$$\phi(0)=\frac{1}{3},\phi'(0)=0,\phi(1)=0,\phi'(1)==0,\phi''(1)=0,$$

one finds the Buhmann's function $\phi(r) = \frac{1}{3} + r^2 - \frac{4}{3}r^3 + 2r^2 log r$; while solving the
problem: *find ϕ such that* $\phi(r) = d_0 + d_1 r + d_2 r^2 + \cdots + d_6 r^6 + c_1 r^2 log r$ *subject to*

$$\phi(0) = \frac{1}{3}, \phi'(0) = 0, \phi(1) = 0, \phi'(1) = 0, \phi''(1) = 0, \dots, \phi^{(5)}(1) = 0$$

one finds the function $\phi(r) = \frac{1}{15} + \frac{19}{6}r^2 - \frac{16}{3}r^3 + 3r^4 - \frac{16}{15}r^5 + \frac{1}{6}r^6 + 2r^2 log r$.
   The Wendland's functions are also subject to the same kind of interpolation:
$\phi_{3,0}(r) = (1-r)_+^2$ corresponds to $\phi \in \mathbb{P}_2$ and

$$\phi(0) = 1, \phi(1) = 0, \phi'(1) = 0,$$

$\phi_{3,1}(r) = (1-r)_+^4 \, (1+4r)$ corresponds to $\phi \in \mathbb{P}_5$ and

$$\phi(0) = \frac{1}{3}, \phi'(0) = 0, \phi(1) = 0, \phi'(1) = 0, \phi''(1) = 0, \phi^{(3)}(1) = 0$$

$\phi_{3,2}(r) = (1-r)_+^6 \, (3 + 18r + 35r^2)$ corresponds to $\phi \in \mathbb{P}_8$ and

$$\phi(0) = \frac{1}{3}, \phi'(0) = 0, \phi''(0) = 0, \phi(1) = 0, \phi'(1) = 0, \phi''(1) = 0, \dots, \phi^{(5)}(1) = 0$$

and $\phi_{3,3}(r) = (1-r)_+^8 \, (1 + 8r + 25r^2 + 32r^3)$ corresponds to $\phi \in \mathbb{P}_{11}$ and

$$\phi(0) = \frac{1}{3}, \phi'(0) = 0, \dots, \phi^{(3)}(0) = 0, \phi(1) = 0, \phi'(1) = 0, \dots, \phi^{(7)}(1) = 0.$$

Conclusion:
This work is still under investigation and forthcoming results would be published
elsewhere, particularly the fact that inf-convolution explains and provides a theoretical background to the domain decomposition method used by E. Kansa [11]
for multiquadrics, and also the fact that one can add polynomials to multiquadrics
in order to achieve polynomial precision, as was done by R. Carlson and T. Foley
in [5].

Acknowledgments. Supported in part by ECOS-CONICYT project C94E05.

References

[1] Arcangeli, R., *Etudes de problèmes de type elliptique ou parabolique avec conditions ponctuelles*, Thèse d'état, Université de Toulouse, 1974.

[2] Bouhamidi, A., *Interpolation et approximation par des fonctions splines radiales*, Ph.D. Dissertation, University of Nantes, 1992.

[3] Bouhamidi, A. and A. Le Méhauté, *Multivariate Interpolating (m, ℓ, s)-spline*, to appear, 1998.

[4] Buhmann, M.D., Radial functions on Compact Support, *Proc. Edinburgh Math. Soc.* **41** (1998),33–46.

[5] Carlson,R. and T.Foley, The parameter r^2 in multiquadrics interpolation, *Computers Math. Applic.* **21** (1991),29–42;

[6] Derrien F., *Distributions de type positif conditionnel et fonctions splines*, Ph.D. Dissertation, University of Nantes, 1997.

[7] Duchon, J., Interpolation des fonctions de deux variables suivant le principe de la flexion des plaques minces, *RAIRO Anal.Num.* **10** (1976), 5–12.

[8] Duchon, J., Splines minimizing rotation invariant semi-norms in Sobolev spaces, *Constructive Theory of functions if several variables*, Oberwolfach 1975, W.Schempp and K.Zeller, Lecture Notes in Math **571**, Springer Verlag, 1977, 85–100.

[9] Dyn, N., D.Levin and S.Rippa, Numerical procedures for global surfaces fitting of scattered data by radial functions, *SIAM J. Sci. Stat. Comp.* **7** (1986).

[10] Hardy, R.L., Multiquadric equations of topographic and other irregular surfaces, *J. Geophys. Res.* **76** (1971), 1905–1915.

[11] Kansa,E., A strictly conservative spatial approximation scheme for the governingengineering and physics equations over irregular regions and inhomogeneous scattered nodes, UCRI 101634, Lawrence Livermore National Laboratory (1989).

[12] Laurent,P.J.,*Approximation et Optimisation*, Hermann, Paris, 1972.

[13] Laurent,P.J., Inf-convolution splines pour l'approximation de données discontinues, Conf. Int.Workshop for the Approximation of functions, Haifa, 1980.

[14] Laurent,P.J., Quadratic Convex Analysis and Splines, *Proceed. Bombay Conference*, ISNM **76** Birkauser Verlag, (1986),17–43.

[15] Laurent,P.J., Inf-convolution splines, *Constr. Approx.* (1991), 469–484.

[16] Le Méhauté,A. and A.Bouhamidi, Splines in Approximation and Differential Operators: (m,ℓ,s) Interpolating Splines, *CRM Proceedings and Lecture Notes,* **17** (1998), 67–77.

[17] Madych,W.R. and S.A.Nelson, Multivariate interpolation and conditionnally positive definite functions, *J. of Approx. Theory and its Applications* **4** (1988), 77–89.

[18] Misiewicz,J.K. and D.S.P.Richards, Positivity of integrals of Bessel functions, *SIAM J. Math. Anal.***25** (1994), 596–601.

[19] Schaback,R. and H.Wendland, *Special cases of compactly supported radial basis functions,*University of Göttingen, 1994, preprint.

[20] Wahba,G., Partial splines and interactions splines for the semi-parametric estimation of functions of several variables, *Computer Sciences and Stat.* (1986)

[21] Wendland, H.,*Piecewise polynomial, positive definite and compactly supported radial basis functions of minimal degree*, University of Göttingen, 1995, preprint.

[22] Wu,Z., Multivariate compactly supported positive definite radial functions, *Adv. Comp. Math.***4** (1995), 283–292.

Alain Le Méhauté,
Département de Mathématiques,
Faculté des Sciences et Techniques,
Université de Nantes,
2 rue de la Houssinière,
BP 92208,
44322 Nantes Cedex 3
France.
Email-Address: alm@math.univ-nantes.fr

International Series of Numerical Mathematics
Vol. 132, © 1999 Birkhäuser Verlag Basel/Switzerland

On a Special Property of the Averaged Modulus for Functions of Bounded Variation

Burkhard Lenze

Abstract

In this little note, we take a look at the connection between the smoothness of a function of bounded variation and its variation function. Based on a number of well-known classical results we expect a very close relationship between the smoothness of these two functions. However, we immediately see that a near at hand generalization of the classical estimates doesn't yield correct results. Therefore, we will have to introduce a standard kind of averaging process in order to come to properly suited smoothness measures in this context. The precise averaging idea is exactly the same as the one used by Sendov and Popov to switch from the classical ω-modulus to the so-called τ-modulus.

1 Introduction

Without loss of generality we only consider functions of bounded variation on $[0, 1]$; any other compact interval may be treated in a similar way. Therefore, let $\mathrm{BV}[0, 1]$ be the space of all functions f of bounded variation on $[0, 1]$ canonically extended to the whole real line by defining $f(x) := f(0)$, $x < 0$, and $f(x) := f(1)$, $x > 1$. Moreover, let $V_{-\infty}^{\cdot}(f)$,

$$
\begin{aligned}
V_{-\infty}^{\cdot}(f): \quad & \mathbb{R} \to \mathbb{R} \,, \\
& x \mapsto V_{-\infty}^{x}(f) \,,
\end{aligned}
\tag{1}
$$

be the so-called nondecreasing variation function associated with f, where $V_{-\infty}^{x}(f)$ – as usual – denotes the total variation of f with respect to the interval $(-\infty, x]$ (cf. [15], pp. 15ff., for details). Finally, we define the integral p-norm resp. the supremum norm of f with respect to $[0, 1]$ as

$$
\|f\|_p := \left(\int_0^1 |f(x)|^p dx \right)^{\frac{1}{p}} \,, \quad 1 \le p < \infty \,,
\tag{2}
$$

$$
\|f\|_\infty := \sup\{|f(x)| : x \in [0, 1]\} \,.
\tag{3}
$$

If we now take a look at published contributions dealing with approximation of functions of bounded variation, we see that in many cases the approximation error is measured in terms of local differences of the corresponding variation function (for example, we mention the papers [1,2,3,4,6,7,9,10,11,12,13]). Therefore, it should be of some interest to study the connection between local differences of f and $V_{-\infty}^{\cdot}(f)$ or – more generally – between the smoothness of f and $V_{-\infty}^{\cdot}(f)$. Starting this investigation, a look into the classical literature about functions of bounded variation immediately provides us with quite a large number of results analyzing the intimate connection between the smoothness properties of f and $V_{-\infty}^{\cdot}(f)$. For example, it is well-known that for $f \in \mathrm{BV}[0,1]$ the derivatives of f and $V_{-\infty}^{\cdot}(f)$ exist almost everywhere and satisfy

$$(V_{-\infty}^{\cdot}(f))'(x) \;=\; |f'(x)| \;\; \text{for a.e. } x \in [0,1] \tag{4}$$

(cf. [15], Theorem (7.24), p. 113). Moreover, besides the well-known fact that f is continuous if and only if $V_{-\infty}^{\cdot}(f)$ is continuous, it can be shown that f is absolutely continuous if and only if $V_{-\infty}^{\cdot}(f)$ is absolutely continuous (cf. [15], Theorem (7.31), p. 117, and Exercise 8, p. 123). Finally, in view of quantitative results we mention the two equivalences

$$\left\{ \begin{array}{c} \|f(\cdot+\delta) - f(\cdot)\|_1 = O(\delta) \\ (\delta \to 0+) \end{array} \right\} \;\Longleftrightarrow\; \left\{ \begin{array}{c} \|V_{-\infty}^{\cdot+\delta}(f) - V_{-\infty}^{\cdot}(f)\|_1 = O(\delta) \\ (\delta \to 0+) \end{array} \right\}, \tag{5}$$

$$\left\{ \begin{array}{c} \|f(\cdot+\delta) - f(\cdot)\|_\infty = O(\delta) \\ (\delta \to 0+) \end{array} \right\} \;\Longleftrightarrow\; \left\{ \begin{array}{c} \|V_{-\infty}^{\cdot+\delta}(f) - V_{-\infty}^{\cdot}(f)\|_\infty = O(\delta) \\ (\delta \to 0+) \end{array} \right\}, \tag{6}$$

where the first one is an immediate consequence of the Hardy-Littlewood characterization of functions of bounded variation (cf. [5], p. 366) while the second one simply states that f is Lipschitz continuous of order 1 if and only if $V_{-\infty}^{\cdot}(f)$ is Lipschitz continuous of order 1 (easy exercise). Especially the last two equivalences give rise to the idea that the smoothness connection between f and $V_{-\infty}^{\cdot}(f)$ measured in integral p-norm is of similar type. However, this conjecture is wrong as the following simple example shows:
Assume that $1 < p < \infty$ and that f is defined as

$$f(x) := \left\{ \begin{array}{ll} 0, & x \neq \frac{1}{2}, \\ 1, & x = \frac{1}{2}. \end{array} \right. \tag{7}$$

An easy calculation yields that

$$\|f(\cdot+\delta) - f(\cdot)\|_p = 0 \;\; \text{for all } \delta > 0 , \tag{8}$$

while

$$\|V_{-\infty}^{\cdot+\delta}(f) - V_{-\infty}^{\cdot}(f)\|_p = 2\delta^{\frac{1}{p}} \;\; \text{for all } \delta \in (0, \tfrac{1}{2}) . \tag{9}$$

The last two identities show that we cannot yield analogous results to (5) and (6) in the integral p-norm setting. Roughly speaking, the reason for this difference lies in the fact that – for example – removable discontinuities of f do not affect an integral p-norm while they result in jump discontinuities for $V_{-\infty}^{\cdot}(f)$ which are of course p-norm relevant. Therefore, we have to introduce some modifications in measuring smoothness in order to get something like (5) and (6) for the whole scale of p-norms ($1 \leq p \leq \infty$). This will be done in the following section.

2 Main Results

First of all, for arbitrary $f \in \mathrm{BV}[0,1]$ (canonically extended on $\mathbb{R}$) we introduce the so-called local ω-modulus of f of first order and step size $\delta > 0$ at $x \in [0,1]$,

$$\omega(f, x, \delta) := \sup \left\{ |f(b) - f(a)| : a, b \in [x - \frac{\delta}{2}, x + \frac{\delta}{2}] \right\} . \tag{10}$$

This local modulus and its generalizations involving r-th order differences was intensively studied by Sendov and Popov. They used it in order to define a new kind of modulus of smoothness (they called it τ-modulus) which is more sensitive in case of local or even pointwise changes of f than the classical ω-modulus (cf. [8,14] for details). In the following lemma, we state a result showing the connection between the local ω-modulus of f and $V_{-\infty}^{\cdot}(f)$. From now on, n always denotes a natural number.

Lemma 2.1 *Let $f \in \mathrm{BV}[0,1]$ be given arbitrarily with canonical extension on $\mathbb{R}$. Then, for all $x \in [0,1]$ and $\delta \in (0,1)$ we have the inequality*

$$\omega(V_{-\infty}^{\cdot}(f), x, \delta) \leq \lim_{n \to \infty} 2 \sum_{i=0}^{2^n - 1} \omega(f, x - \frac{\delta}{2} + (i + \frac{1}{2})\frac{\delta}{2^n}, \frac{\delta}{2^n}) . \tag{11}$$

Proof. Let $x \in [0,1]$ and $\delta \in (0,1)$ be given arbitrarily. First of all, because of the monotonicity of $V_{-\infty}^{\cdot}(f)$ and well-known results for the variation, the left hand side of (11) reduces to

$$\omega(V_{-\infty}^{\cdot}(f), x, \delta) = V_{x - \frac{\delta}{2}}^{x + \frac{\delta}{2}}(f) . \tag{12}$$

Now, let

$$x - \frac{\delta}{2} =: x_0 < x_1 < \cdots < x_{k-1} < x_k := x + \frac{\delta}{2} \tag{13}$$

be an arbitrary partition of $[x - \frac{\delta}{2}, x + \frac{\delta}{2}]$. For all $n \in \mathbb{N}$ satisfying the inequality

$$\frac{\delta}{2^n} < \min\{(x_{i+1} - x_i) : 0 \leq i < k\} \tag{14}$$

we immediately obtain the estimate

$$\sum_{i=0}^{k-1} |f(x_{i+1}) - f(x_i)| \leq \sum_{i=0}^{2^n-1} 2\,\omega(f, x - \frac{\delta}{2} + (i + \frac{1}{2})\frac{\delta}{2^n}, \frac{\delta}{2^n}) \,. \tag{15}$$

The right hand side of the above inequality is nondecreasing for n increasing. Therefore, we can take the limit with respect to n,

$$\sum_{i=0}^{k-1} |f(x_{i+1}) - f(x_i)| \leq \lim_{n \to \infty} 2 \sum_{i=0}^{2^n-1} \omega(f, x - \frac{\delta}{2} + (i + \frac{1}{2})\frac{\delta}{2^n}, \frac{\delta}{2^n}) \,. \tag{16}$$

Now, the right hand side is independent of the initially chosen partition and we may take the supremum over all such partitions on the left hand side which yields

$$V_{x-\frac{\delta}{2}}^{x+\frac{\delta}{2}}(f) \leq \lim_{n \to \infty} 2 \sum_{i=0}^{2^n-1} \omega(f, x - \frac{\delta}{2} + (i + \frac{1}{2})\frac{\delta}{2^n}, \frac{\delta}{2^n}) \,. \tag{17}$$

∎

With the above result, we can now prove the following basic second lemma.

Lemma 2.2 *Let $f \in \mathrm{BV}[0,1]$ be given arbitrarily with canonical extension on $\mathbb{R}$ and let $1 \leq p \leq \infty$. Then, for all $\delta \in (0,1)$ the following two inequalities hold*

$$\|\omega(f, \cdot, \delta)\|_p \leq \|\omega(V_{-\infty}^{\cdot}(f), \cdot, \delta)\|_p \,, \tag{18}$$

$$\|\omega(V_{-\infty}^{\cdot}(f), \cdot, \delta)\|_p \leq 4 \liminf_{n \to \infty} 2^n \, \|\omega(f, \cdot, \frac{\delta}{2^n})\|_p \,. \tag{19}$$

Proof. The first inequality is rather trivial. We only have to use the fact that for all $a \leq b \leq c \leq d$ we have the inequality

$$|f(c) - f(b)| \leq V_a^d(f) \,. \tag{20}$$

In view of the second inequality, we only consider the integral case $1 \leq p < \infty$, because the case $p = \infty$ is even simpler. Now, for $\delta \in (0,1)$ given arbitrarily using our first lemma and Fatou's lemma we immediately obtain

$$\|\omega(V_{-\infty}^{\cdot}(f), \cdot, \delta)\|_p \leq \left\| \lim_{n \to \infty} 2 \sum_{i=0}^{2^n-1} \omega(f, \cdot - \frac{\delta}{2} + (i + \frac{1}{2})\frac{\delta}{2^n}, \frac{\delta}{2^n}) \right\|_p \tag{21}$$

$$\leq \liminf_{n \to \infty} 2 \sum_{i=0}^{2^n-1} \left(\int_0^1 (\omega(f, x - \frac{\delta}{2} + (i + \frac{1}{2})\frac{\delta}{2^n}, \frac{\delta}{2^n}))^p dx \right)^{\frac{1}{p}}$$

$$\leq \liminf_{n\to\infty} 2 \sum_{i=0}^{2^n-1} \left(\int_{-\frac{\delta}{2^{n+1}}}^{1+\frac{\delta}{2^{n+1}}} (\omega(f,t,\frac{\delta}{2^n}))^p dt \right)^{\frac{1}{p}}$$

$$\leq \liminf_{n\to\infty} 4 \cdot 2^n \left(\int_0^1 (\omega(f,t,\frac{\delta}{2^n}))^p dt \right)^{\frac{1}{p}}$$

$$= 4 \liminf_{n\to\infty} 2^n \left\| \omega(f,\cdot,\frac{\delta}{2^n}) \right\|_p .$$

For the last inequality we have used the fact that

$$\omega(f,t,\frac{\delta}{2^n}) \leq \omega(f,-t,\frac{\delta}{2^n}) , \quad t \in [-\frac{\delta}{2^{n+1}},0] \subset [-\frac{1}{4},0] , \tag{22}$$

and

$$\omega(f,1+t,\frac{\delta}{2^n}) \leq \omega(f,1-t,\frac{\delta}{2^n}) , \quad t \in [0,\frac{\delta}{2^{n+1}}] \subset [0,\frac{1}{4}] . \tag{23}$$

Summing up, we have proved the validity of (19). ■

Now, we are prepared to formulate our main result which gives a smoothness equivalence for functions of bounded variation and their corresponding variation function over the whole scale of usual p-norms.

Theorem 2.3 *Let* $f \in \mathrm{BV}[0,1]$ *be given arbitrarily with canonical extension on* $\mathbb{R}$ *and let* $1 \leq p \leq \infty$. *Then, the following equivalence holds*

$$\left\{ \begin{array}{c} \|\omega(f,\cdot,\delta)\|_p = O(\delta) \\ (\delta \to 0+) \end{array} \right\} \quad \Longleftrightarrow \quad \left\{ \begin{array}{c} \|\omega(V_{-\infty}^{\cdot}(f),\cdot,\delta)\|_p = O(\delta) \\ (\delta \to 0+) \end{array} \right\} . \tag{24}$$

Proof. The equivalence follows immediately using the inequalities (18) and (19). ■

Concluding Remarks.

(1) First of all, let us note that $\|\omega(f,\cdot,\delta)\|_p$, $\delta > 0$, $1 \leq p \leq \infty$, is nothing else but the p-norm τ-modulus of f of order one and step size δ as studied by Sendov and Popov (cf. [14]).

(2) Note that the equivalence in the above theorem is really a kind of saturation result since for $f \in \mathrm{BV}[0,1]$ and $1 \leq p \leq \infty$ we also have the small-o-equivalence, namely,

$$\left\{ \begin{array}{c} \|\omega(f,\cdot,\delta)\|_p = o(\delta) \\ (\delta \to 0+) \end{array} \right\} \quad \Longleftrightarrow \quad \left\{ \begin{array}{c} \|\omega(V_{-\infty}^{\cdot}(f),\cdot,\delta)\|_p = o(\delta) \\ (\delta \to 0+) \end{array} \right\} , \tag{25}$$

which simply implies that f is a constant function (cf. [5], Corollary 10.1.7, p. 361, and additionally use the stronger definition of the averaged local modulus ω to get rid of the almost everywhere restriction).

(3) Finally, let us note that the fractional cases $O(\delta^\alpha)$ for $0 < \alpha < 1$ and $\delta \to 0+$ are significantly different since there are Lipschitz continuous functions $f \in BV[0,1]$ of order α whose variation functions are known to be less smooth in view of their Lipschitz parameter. As an example, fix $\theta \in (1,2)$, set

$$x_k := \frac{1}{k} \quad \text{and} \quad y_{k,\theta} := \left(\frac{1}{k}\right)^\theta , \quad k \in \mathbb{N} , \tag{26}$$

and define the continuous piecewise linear function $f_\theta : [0,1] \to \mathbb{R}$ as follows:

$$f_\theta(x) := \begin{cases} 0, & \text{for } x = 0, \\ y_{k,\theta}, & \text{for } x = x_k \text{ and } k \in \mathbb{N} \text{ even}, \\ 0, & \text{for } x = x_k \text{ and } k \in \mathbb{N} \text{ odd}, \\ \text{linear}, & \text{for } x \in [x_{k+1}, x_k], k \in \mathbb{N}. \end{cases} \tag{27}$$

First of all, we show that f_θ is Lipschitz continuous on $[0,1]$ of order $\frac{\theta}{2}$. For $x, y \in [0,1]$, $x \neq y$, we consider a number of different cases.

Case 1: $x, y \in [(k_0 + 1)^{-1}, k_0^{-1}]$, $k_0 \in \mathbb{N}$ even:
Here, f_θ is explicitly given as

$$f_\theta(t) = \frac{\left(\frac{1}{k_0}\right)^\theta}{\frac{1}{k_0} - \frac{1}{k_0+1}} \left(t - \frac{1}{k_0 + 1}\right) , \quad t \in \left[\frac{1}{k_0 + 1}, \frac{1}{k_0}\right] , \tag{28}$$

which implies

$$\begin{aligned} |f_\theta(x) - f_\theta(y)| &= k_0^{-\theta} (k_0^{-1} - (k_0 + 1)^{-1})^{-1} |x - y| \\ &= k_0^{-\theta} (k_0(k_0 + 1)) |x - y|^{1 - \frac{\theta}{2}} |x - y|^{\frac{\theta}{2}} \\ &\leq k_0^{-\theta} (k_0(k_0 + 1))^{\frac{\theta}{2}} |x - y|^{\frac{\theta}{2}} \\ &\leq 2^{\frac{\theta}{2}} |x - y|^{\frac{\theta}{2}} . \end{aligned} \tag{29}$$

Case 2: $x, y \in [(k_0 + 1)^{-1}, k_0^{-1}]$, $k_0 \in \mathbb{N}$ odd:
An inequality similar to (29) may be obtained in a completely analogous way as in Case 1.

Case 3: $x \in [(k_0 + 1)^{-1}, k_0^{-1}]$, $y \in [k_0^{-1}, (k_0 - 1)^{-1}]$, $k_0 \in \mathbb{N}$, $k_0 \geq 2$ and even:
Using (28) and the representation

$$f_\theta(t) = \frac{\left(\frac{1}{k_0}\right)^\theta}{\frac{1}{k_0} - \frac{1}{k_0-1}} \left(t - \frac{1}{k_0 - 1}\right) , \quad t \in \left[\frac{1}{k_0}, \frac{1}{k_0 - 1}\right] , \tag{30}$$

we immediately obtain

$$
\begin{aligned}
& |f_\theta(x) - f_\theta(y)| \\
& = \left| k_0^{1-\theta}(k_0+1)\left(x - (k_0+1)^{-1}\right) + k_0^{1-\theta}(k_0-1)\left(y - (k_0-1)^{-1}\right)\right| \\
& = \left| k_0^{1-\theta}(k_0+1)\left((x - k_0^{-1}) + (k_0^{-1} - (k_0+1)^{-1})\right)\right. \\
& \qquad \left. + k_0^{1-\theta}(k_0-1)\left((y - k_0^{-1}) + (k_0^{-1} - (k_0-1)^{-1})\right)\right| \\
& = \left| k_0^{1-\theta}(k_0+1)\left(x - k_0^{-1}\right) + k_0^{1-\theta}(k_0-1)\left(y - k_0^{-1}\right)\right| \\
& \le 2^{\frac{\theta}{2}}\left|x - k_0^{-1}\right|^{\frac{\theta}{2}} + \left|y - k_0^{-1}\right|^{\frac{\theta}{2}} \\
& \le \left(2^{\frac{\theta}{2}} + 1\right)|x - y|^{\frac{\theta}{2}} .
\end{aligned}
\tag{31}
$$

<u>Case 4:</u> $x \in [(k_0+1)^{-1}, k_0^{-1}]$, $y \in [k_0^{-1}, (k_0-1)^{-1}]$, $k_0 \in \mathbb{N}$, $k_0 \ge 2$ and odd:
Simply change the rôles of x and y in Case 3 and obtain an inequality similar to (31).
The other cases ($x \in [(k_0+1)^{-1}, k_0^{-1}]$, $y = 0$, and $x \in [(k_0+1)^{-1}, k_0^{-1}]$, $y \in [r_0^{-1}, (r_0-1)^{-1}]$, with $k_0 > r_0$) are treated in a similar, even simpler way and are left to the reader. Summing up, we have shown that f_θ is Lipschitz continuous on $[0, 1]$ of order $\frac{\theta}{2}$.
In contrast to the smoothness of f_θ, the variation function of f_θ can be at most Lipschitz continuous of order $(\theta - 1)$, which is always smaller than $\frac{\theta}{2}$ for $\theta \in (1, 2)$. We draw this conclusion from the following estimate, where $k_0 \in \mathbb{N}$ is an arbitrarily given natural number:

$$
V_0^{k_0^{-1}}(f_\theta) \ge \sum_{k=k_0}^\infty \left(\frac{1}{2k}\right)^\theta \ge \int_{k_0}^\infty \left(\frac{1}{2x}\right)^\theta dx = \frac{2^{-\theta}}{\theta - 1}\left(\frac{1}{k_0}\right)^{\theta - 1}.
\tag{32}
$$

References

[1] R. Bojanic and F. H. Cheng, *Estimates for the rate of approximation of functions of bounded variation by Hermite-Fejér polynomials*, in: Second Edmonton Conference on Approximation Theory (Eds.: Z. Ditzian, A. Meir, S. D. Riemenschneider, A. Sharma), American Mathematical Society, Providence, 1983, 5–17.

[2] R. Bojanic and F. Cheng, *Rate of convergence of Bernstein polynomials for functions with derivatives of bounded variation*, J. Math. Anal. Appl. **141**, 1989, 136–151.

[3] R. Bojanic and F. Cheng, *Rate of convergence of Hermite-Fejér polynomials for functions with derivatives of bounded variation*, Acta Math. Hung. **59**, 1992, 91–102.

[4] R. Bojanic and M. Vuilleumier, *On the rate of convergence of Fourier-Legendre series of functions of bounded variation*, J. Approx. Theory **31**, 1981, 67–79.

[5] P. L. Butzer and R. J. Nessel, *Fourier Analysis and Approximation*, Vol. 1, Birkhäuser Verlag, Basel-Stuttgart, 1971.

[6] F. Cheng, *On the rate of convergence of Bernstein polynomials of functions of bounded variation*, J. Approx. Theory **39**, 1983, 259–274.

[7] F. Cheng, *On the rate of convergence of the Szász-Mirakyan operator for functions of bounded variation*, J. Approx. Theory **40**, 1984, 226–241.

[8] Z. Ditzian and V. Totik, *Moduli of Smoothness*, Springer-Verlag, Berlin-Heidelberg-New York, 1987.

[9] V. Gupta, *On the rate of pointwise convergence of modified Baskakov operators*, Soochow J. Math. **22**, 1996, 543–552.

[10] V. Gupta, *A note on the rate of convergence of Durrmeyer type operators for functions of bounded variation*, Soochow J. Math. **23**, 1997, 115–118.

[11] B. Lenze, *Operators for one-sided approximation by algebraical polynomials*, J. Approx. Theory **54**, 1988, 169–179.

[12] B. Lenze, *On one-sided spline approximation operators*, Numer. Funct. Anal. and Optimiz. **10**, 1989, 167–180.

[13] E. R. Love, G. Prasad, and A. Sahai, *An improved estimate of the rate of convergence of the integrated Meyer-König and Zeller operators for functions of bounded variation*, J. Math. Anal. Appl. **187**, 1994, 1–16.

[14] B. Sendov and V. A. Popov, *Averaged Moduli of Smoothness*, Bulgarian Mathematical Monographs Vol. 4, Bulgarian Academy of Sciences, Sofia, 1983 (in Bulgarian).

[15] R. L. Wheeden and A. Zygmund, *Measure and Integral*, Marcel Dekker, Inc., New York, 1977.

Burkhard Lenze
Fachbereich Informatik
Fachhochschule Dortmund
Postfach 105018
D-44047 Dortmund
Germany

Email address: lenze@fh-dortmund.de

International Series of Numerical Mathematics
Vol. 132, © 1999 Birkhäuser Verlag Basel/Switzerland

A Simple Approach to the Variational Theory
for Interpolation on Spheres

Jeremy Levesley, Will Light,[*] David Ragozin[†] and Xingping Sun[‡]

Abstract

In this paper we consider the problem of developing a variational theory for interpolation by radial basis functions on spheres. The interpolants have the property that they minimise the value of a certain semi-norm, which we construct explicitly. We then go on to investigate forms of the interpolant which are suitable for computation. Our main aim is to derive error bounds for interpolation from scattered data sets, which we do in the final section of the paper.

1   Introduction

Surface splines were introduced by Duchon [3], although one should not overlook earlier work of Atteia [1]. Given a function f in $C(\mathbb{R}^d)$ and a set of points $a_1, \ldots, a_m \in \mathbb{R}^d$, a surface spline is an interpolant to f on the set $\{a_1, \ldots, a_m\}$ which minimises a certain (rotational invariant) semi-norm on $\mathbb{R}^d$. This variational property, that is, the minimisation of a semi-norm, has many useful consequences. The first is that the interpolant is guaranteed to be unique. The second is that error estimates can be obtained quite easily. In this paper we want to develop interpolants where the points $a_1, \ldots, a_m$ are given in $\mathcal{S}^{d-1}$ and $f \in C(\mathcal{S}^{d-1})$. A number of authors have considered this setting already. The earliest work we are aware of is Wahba [24]. A series of important contributions has come from Freeden and his co-workers. The list of references from Freeden amounts to over 100 papers scattered throughout the literature, so we provide the reader with access to the bulk of this material via the survey paper [8]. Much of this work focusses on generating the semi-norm by reference to a differential or pseudo-differential operator. This is hardly surprising, since this was very much in the spirit of Duchon's investigations. The majority of the results cited concern $\mathcal{S}^2$. Again, this is a natural setting for such a study, because it has important practical applications in geodesy and

[*]Research partially supported by NATO grant CRG910885
[†]Research partially supported by EPSRC grant GR/K79710
[‡]Research partially supported by EPSRC grant GR/J19481

related fields. However, many of the results have straightforward generalisations to $\mathcal{S}^{d-1}$ for $d > 3$.

Our purposes in this paper are several. Firstly, we want to present as elementary an approach to the Freeden-Wahba circle of ideas as possible, while remaining reasonably complete and orderly in our treatment. It was something of a surprise to us as to how little prior knowledge is needed. In particular, we wish to develop suitable spaces in which the minimal norm problem can be posed without reference to differential operators. These spaces will be subspaces of $C(\mathcal{S}^{d-1})$, where $d \geq 2$. For these spaces we will derive a Sobolev type embedding theorem. This material forms Section 2. Section 3 deals with reproducing kernels and the closely related idea of representers for linear functionals for the Hilbert spaces constructed in Section 2. A variety of forms for the interpolant are given in Section 4. There is a strong parallel between the results obtained in this section and those developed by Schoenberg [21] and Micchelli [14] with reference to radial basis functions. The paper culminates in Section 5 with results on pointwise error estimates.

This paper is not a survey paper. In fact, it represents the fruits of a research project carried out over a summer at Leicester. Subsequent to that work, and over an extended period of time, we have come to understand something of how the work of others relates to the work described here, particularly in the Sections 2 to 4. We will end each of these Sections with an indication of some of these connections. In the final Section, which contains new material, we indicate related work as we go along.

2 The Hilbert Spaces

To define the spaces we wish to work with, it is necessary to understand a little about spherical harmonics. Let $\pi_k(\mathcal{S}^{d-1})$ denote the restriction of the space of polynomials of (total) degree at most k in $\mathbb{R}^d$ to the sphere $\mathcal{S}^{d-1}$. Let $\pi_{k-1}^{\perp}(\mathcal{S}^{d-1})$ denote the orthogonal complement of $\pi_{k-1}(\mathcal{S}^{d-1})$ in $\pi_k(\mathcal{S}^{d-1})$ where the inner product is the usual inner product on $\mathcal{S}^{d-1}$. The set of all $u \in \pi_k(\mathcal{S}^{d-1}) \cap \pi_{k-1}^{\perp}(\mathcal{S}^{d-1})$ forms a finite-dimensional linear space of polynomials of degree k, whose dimension we will denote by d_k. This number is computable, but its value need not enter into our discussions. We will use $\{Y_j^{(k)} : j = 1, 2, \ldots, d_k\}$ to denote any orthonormal basis for this linear space. These functions are the spherical harmonics of degree k. It is important from our point of view that the following summation formula (see Müller [15]) should hold:

$$\sum_{j=1}^{d_k} Y_j^{(k)}(x) Y_j^{(k)}(y) = \omega_{d-1}^{-1} d_k P_k(xy), \qquad \text{for all } x, y \in \mathcal{S}^{d-1}. \tag{1}$$

Here ω_{d-1} is the surface area of $\mathcal{S}^{d-1}$, xy is the inner product of x and y, and P_k is a certain univariate Gegenbauer polynomial which is normalised so that $P_k(1) = 1$.

To define our space, we begin with a non-negative integer κ and a sequence $\{\beta_k\}_{k=\kappa}^{\infty}$, where $\beta_k \geq 1$, $k \geq \kappa$. Then

$$X_\kappa = \left\{ \sum_{k=0}^{\infty} \sum_{j=1}^{d_k} a_{kj} Y_j^{(k)} : \sum_{k=\kappa}^{\infty} \beta_k^2 \sum_{j=1}^{d_k} a_{kj}^2 < \infty \right\}.$$

The condition imposed on the coefficients a_{kj} amounts to some sort of smoothness for the associated map from $\mathcal{S}^{d-1}$ to $\mathbb{R}$. A simple condition on the sequence $\{\beta_k\}_{k=\kappa}^{\infty}$ will enable us to show that the series $\sum_{k=0}^{\infty} \sum_{j=1}^{d_k} a_{kj} Y_j^{(k)}$ is uniformly convergent on $\mathcal{S}^{d-1}$, so $X_\kappa \subset C(\mathcal{S}^{d-1})$. We introduce the following semi-inner product on X_κ. If

$$f = \sum_{k=0}^{\infty} \sum_{j=1}^{d_k} a_{kj} Y_j^{(k)} \quad \text{and} \quad g = \sum_{k=0}^{\infty} \sum_{j=1}^{d_k} b_{kj} Y_j^{(k)},$$

then

$$\langle f, g \rangle = \sum_{k=\kappa}^{\infty} \beta_k^2 \sum_{j=1}^{d_k} a_{kj} b_{kj}.$$

This bilinear form enjoys all the usual properties of a real-valued inner product except that there is a kernel

$$N = \left\{ \sum_{k=0}^{\kappa-1} \sum_{j=1}^{d_k} a_{kj} Y_j^{(k)} : a_{kj} \in \mathbb{R} \right\}.$$

Let $\ell = d_0 + d_1 + \ldots + d_{\kappa-1}$. Choose $y_1, \ldots, y_\ell$ in $\mathcal{S}^{d-1}$ such that if $p \in N$ and $p(y_i) = 0$, $i = 1, 2, \ldots, \ell$ then $p = 0$. Now consider the new bilinear form $(\cdot, \cdot) : X_\kappa \times X_\kappa \to \mathbb{R}$ defined by

$$(f, g) = \sum_{i=1}^{\ell} f(y_i) g(y_i) + \langle f, g \rangle, \qquad f, g \in X_\kappa. \tag{2}$$

If f belongs to X_κ and

$$0 = (f, f) = \sum_{i=1}^{\ell} f^2(y_i) + \langle f, f \rangle,$$

then f is in N and $f(y_i) = 0$ for $i = 1, 2, \ldots, \ell$. Hence $f = 0$, and so $(\cdot, \cdot)$ is in fact an inner product on X_κ. The space X_κ turns out to be complete with respect to the norm induced on X_κ by this inner product. It is worthwhile adopting the notation $|\cdot|_\kappa = \sqrt{\langle \cdot, \cdot \rangle}$ for the seminorm induced by the semi-inner product. The symbol $\|\cdot\|$ will be reserved exclusively henceforward for the norm induced by the inner product $(\cdot, \cdot)$.

Lemma 2.1 *Suppose $\sum_{k=\kappa}^{\infty} \beta_k^{-2} d_k = A^2 < \infty$. For all $f \in X_\kappa$ of the form*

$$f = \sum_{k=\kappa}^{\infty} \sum_{j=1}^{d_k} a_{kj} Y_j^{(k)},$$

and for all x in $\mathcal{S}^{d-1}$, $|f(x)| \leq \omega_{d-1}^{-1/2} A |f|_\kappa$. Furthermore, $X_\kappa \subset C(\mathcal{S}^{d-1})$.

Proof. Fix $x \in \mathcal{S}^{d-1}$. Then using several applications of the Cauchy-Schwarz inequality, and the fact that $P_k(1) = 1$,

$$
\begin{aligned}
|f(x)| &\leq \sum_{k=\kappa}^{\infty} \left| \sum_{j=1}^{d_k} a_{kj} Y_j^{(k)}(x) \right| \\
&\leq \sum_{k=\kappa}^{\infty} \left(\sum_{j=1}^{d_k} a_{kj}^2 \right)^{1/2} \left(\sum_{j=1}^{d_k} \left(Y_j^{(k)}(x) \right)^2 \right)^{1/2} \\
&= \omega_{d-1}^{-1/2} \sum_{k=\kappa}^{\infty} \left(\sum_{j=1}^{d_k} a_{kj}^2 \right)^{1/2} d_k^{1/2} \left(P_k(|x|^2) \right)^{1/2} \\
&= \omega_{d-1}^{-1/2} \sum_{k=\kappa}^{\infty} \beta_k \left\{ \sum_{j=1}^{d_k} a_{kj}^2 \right\}^{1/2} \frac{\sqrt{d_k}}{\beta_k} \\
&\leq \omega_{d-1}^{-1/2} \left\{ \sum_{k=\kappa}^{\infty} \beta_k^2 \sum_{j=1}^{d_k} a_{kj}^2 \right\}^{1/2} \left\{ \sum_{k=\kappa}^{\infty} d_k \beta_k^{-2} \right\}^{1/2} \\
&= \omega_{d-1}^{-1/2} A \sqrt{\langle f, f \rangle}.
\end{aligned}
$$

This establishes the required inequality. The fact that $f \in C(\mathcal{S}^{d-1})$ follows immediately from the Weierstrass M-test. ∎

Lemma 2.2 *Suppose $\sum_{k=\kappa}^{\infty} \beta_k^{-2} d_k < \infty$. There exists a constant $C > 0$ such that $|f(x)| \leq C\|f\|$, for all $f \in X_\kappa$, and for all $x \in \mathcal{S}^{d-1}$.*

Proof. Suppose $f \in X_\kappa$ has the form

$$f = \sum_{k=0}^{\infty} \sum_{j=1}^{d_k} a_{kj} Y_j^{(k)}.$$

Set

$$g = f - \sum_{k=0}^{\kappa-1} \sum_{j=1}^{d_k} a_{kj} Y_j^{(k)}.$$

Then Lemma **2.1** shows that $|g(x)| \le \omega_{d-1}^{-1/2} A \sqrt{\langle g,g \rangle} = \omega_{d-1}^{-1/2} A \sqrt{\langle f,f \rangle}$. Hence,

$$
\begin{aligned}
|f(x)| &\le \left| \sum_{k=0}^{\kappa-1} \sum_{j=1}^{d_k} a_{kj} Y_j^{(k)}(x) \right| + \omega_{d-1}^{-1/2} A \sqrt{\langle f,f \rangle} \\
&\le \left\| \sum_{k=0}^{\kappa-1} \sum_{j=1}^{d_k} a_{kj} Y_j^{(k)} \right\|_\infty + \omega_{d-1}^{-1/2} A \sqrt{\langle f,f \rangle}.
\end{aligned}
$$

Now an alternative norm on $N = \operatorname{span}\{Y_j^{(k)} : 1 \le j \le d_k,\ 1 \le k \le \kappa - 1\}$ is defined for $p \in N$ by

$$
\left\{ \sum_{i=1}^{\ell} p^2(y_i) \right\}^{1/2}.
$$

On account of the equivalence of this and the supremum norm on the finite-dimensional space N, there exists a constant $B > 0$ such that

$$
\begin{aligned}
|f(x)| &\le B \left\{ \sum_{i=1}^{\ell} \left(\sum_{k=0}^{\kappa-1} \sum_{j=1}^{d_k} a_{kj} Y_j^{(k)}(y_i) \right)^2 \right\}^{1/2} + \omega_{d-1}^{-1/2} A \sqrt{\langle f,f \rangle} \\
&= B \left\{ \sum_{i=1}^{\ell} \left(f(y_i) - g(y_i) \right)^2 \right\}^{1/2} + \omega_{d-1}^{-1/2} A \sqrt{\langle f,f \rangle} \\
&\le B \left\{ \sum_{i=1}^{\ell} f^2(y_i) \right\}^{1/2} + B \left\{ \sum_{i=1}^{\ell} g^2(y_i) \right\}^{1/2} + \omega_{d-1}^{-1/2} A \sqrt{\langle f,f \rangle}.
\end{aligned}
$$

Using Cauchy-Schwartz and **2.1** again,

$$
\begin{aligned}
|f(x)| &\le B \left\{ \sum_{i=1}^{\ell} f^2(y_i) \right\}^{1/2} + \omega_{d-1}^{-1/2} \left(B \left(\ell A^2 \langle g,g \rangle \right)^{1/2} + A \sqrt{\langle f,f \rangle} \right) \\
&= B \left\{ \sum_{i=1}^{\ell} f^2(y_i) \right\}^{1/2} + \omega_{d-1}^{-1/2} A (1 + \sqrt{\ell} B) \sqrt{\langle f,f \rangle} \\
&\le \sqrt{2} \left(B + \omega_{d-1}^{-1/2} A (1 + \sqrt{\ell} B) \right) \left\{ \sum_{i=1}^{\ell} f^2(y_i) + \langle f,f \rangle \right\}^{1/2}.
\end{aligned}
$$

Setting $C = \sqrt{2} \left(B + \omega_{d-1}^{-1/2} A (1 + \sqrt{\ell} B) \right)$ gives $|f(x)| \le C \sqrt{\langle f,f \rangle}$ as required. ∎

We now wish to establish that X_κ is complete and is embeddable in $C(\mathcal{S}^{d-1})$. There are many ways of doing this. Our choice has been influenced by a desire to have an elementary discussion of the properties of X_κ. We will use ℓ_β to denote the Hilbert space of real sequences $a = \{a_{kj}\}_{j=1, k=\kappa}^{d_k, \infty}$ with finite norm

$$\|a\|_\beta = \Big(\sum_{k=\kappa}^{\infty} \beta_k^2 \sum_{j=1}^{d_k} a_{kj}^2 \Big)^{1/2}.$$

Theorem 2.3 *Let*

$$X_\kappa = \Big\{ \sum_{k=0}^{\infty} \sum_{j=1}^{d_k} a_{kj} Y_j^{(k)} : \sum_{k=\kappa}^{\infty} \beta_k^2 \sum_{j=1}^{d_k} a_{kj}^2 < \infty \Big\}$$

where $\sum_{k=\kappa}^{\infty} \beta_k^{-2} d_k < \infty$. With the norm $\|\cdot\|$ defined by the inner product (2), X_κ is complete, and X_κ is continuously embeddable in $C(\mathcal{S}^{d-1})$.

Proof. Define the mapping $J : X_\kappa \to \ell_\beta$ by $Jf = a$ where $f = \sum_{k=0}^{\infty} \sum_{j=1}^{d_k} a_{kj} Y_j^{(k)}$ and $a = \{a_{kj}\}_{j=1, k=\kappa}^{d_k, \infty}$. It is clear that J maps X_κ onto ℓ_β and that $|f|_\kappa = \|Jf\|_\beta$ for all $f \in X_\kappa$. Now let $\{f_q\}_{q=1}^{\infty}$ be a Cauchy sequence in X_κ. From the inequality

$$\|Jf_q - Jf_p\|_\beta = |f_q - f_p|_\kappa \le \|f_q - f_p\|$$

we see that $\{Jf_q\}$ is a Cauchy sequence in ℓ_β. Since ℓ_β is complete, $Jf_q \to b$ for some $b \in \ell_\beta$. Since J is onto, there exists $g \in X_\kappa$ such that $Jg = b$. Moreover, $|f_q - g|_\kappa = \|Jf_q - b\|_\beta \to 0$ as $q \to \infty$. Now let $L : X_\kappa \to N$ be the orthogonal projection. Then $\|Lf_q - Lf_p\| \le \|f_q - f_p\|$, showing that $\{Lf_q\}$ is a Cauchy sequence in N. Hence $Lf_q \to p$ for some $p \in \pi_{\kappa-1}$. Now,

$$
\begin{aligned}
\|f_q - g - p + Lg\| &\le \|f_q - g - Lf_q + Lg\| + \|Lf_q - p\| \\
&= \|f_q - g - L(f_q - g)\| + \|Lf_q - p\| \\
&= |f_q - g|_\kappa + \|Lf_q - p\| \\
&\to 0
\end{aligned}
$$

as $q \to \infty$. Thus $f_q \to g + p - Lg$ as $q \to \infty$, showing that X_κ is complete. Finally, from Lemma **2.2** it follows that $X_\kappa \subset C(\mathcal{S}^{d-1})$ and $\|f\|_\infty \le A\|f\|$, which demonstrates the existence of a continuous embedding. ∎

The definition of the space X in the generality described here can be found in the paper of Taijeron, Gibson and Chandler [23]. The above exposition first of all establishes that the point evaluation functionals are bounded linear functionals on X_κ. Then X_κ is shown to be complete. This shows that X_κ, in the language of Shapiro [20], is a *Hilbert function space*. As the introduction to the next section shows, every Hilbert function space has a reproducing kernel. In [23] none of the material contained in the lemmas and theorem of this section are present. Instead, they rely on the results of our Section 3 to infer the results presented here. Of course, that is a perfectly legitimate approach, but one which we don't find particularly appealing.

3 Reproducing Kernels

The space X_κ is a Hilbert space in which point evaluations are continuous linear functionals. Thus, to each $x \in \mathcal{S}^{d-1}$ there corresponds, via the Riesz representation theorem, an element $R_x \in X_\kappa$ such that, for all $f \in X_\kappa$, $f(x) = (f, R_x)$. The element R_x is often called the *representer* of x in X_κ. The mapping $H : \mathcal{S}^{d-1} \times \mathcal{S}^{d-1} \to \mathbb{R}$ given by $H(x, y) = R_x(y)$, is called the *reproducing kernel* for the Hilbert space X_κ, since each $f \in X_\kappa$ may be reproduced by H, using the rule $f(x) = (f, H(x, \cdot))$, $x \in \mathcal{S}^{d-1}$. One of the significant contributions of Duchon [3] was to identify the reproducing kernels for the "Beppo–Levi" type spaces. Our purpose in this section is to identify reproducing kernels for the spaces X_κ. This in turn will simplify the analysis of the interpolation process.

It will help to recall that

$$(f, g) \;=\; \sum_{i=1}^{l} f(y_i)g(y_i) + \langle f, g \rangle, \quad f, g \in X_\kappa.$$

The points $y_1, y_2, \ldots, y_l$ were chosen so that these point evaluations are linearly independent over the set $N = \operatorname{span}\{Y_j^{(k)} : 1 \le j \le d_k, \ 0 \le k \le \kappa\}$. Hence, we can find $q_1, q_2, \ldots, q_l \in N$ such that

$$q_i(y_j) = \delta_{ij}, \qquad 1 \le i, j \le l, \tag{3}$$

where δ is the usual Kronecker symbol.

Lemma 3.1 *The representer for y_j is q_j, $1 \le j \le l$. That is, $f(y_j) = (f, q_j)$ for all $f \in X_\kappa$ and $1 \le j \le l$.*

Proof. For any $f \in X_\kappa$,

$$\begin{aligned}
(f, q_j) \;&=\; \sum_{i=1}^{l} f(y_i)q_j(y_i) + \langle f, q_j \rangle \\
&=\; f(y_j) + \langle f, q_j \rangle,
\end{aligned}$$

using the biorthonormality property (3). The result follows on noting that N is the kernel of the semi-inner product $\langle \cdot, \cdot \rangle$. ∎

Note that Lemma **3.1** forces the sequence $\{q_j\}_{j=1}^{l}$ to be orthonormal since $(q_i, q_j) = q_i(y_j) = \delta_{ij}$, $1 \le i, j \le l$. The orthogonal projection $P : X_\kappa \to N$ turns out (somewhat unusually) to be an interpolatory projection:

$$Pf = \sum_{i=1}^{l} (f, q_i)q_i = \sum_{i=1}^{l} f(y_i)q_i,$$

where we recall that the q_i are the cardinal or Lagrange basis for interpolation at $y_1, y_2, \ldots, y_l$. To obtain the representer of the general point $x \in S^{d-1}$, we first of all obtain the representer of this point in $N^\perp$, the orthogonal complement of N in X_κ. What we mean by this is that we identify the function $R_x \in N^\perp$ such that $(f, R_x) = f(x)$, for all $f \in N^\perp$. It is then an easy step to move on to the representer for x on the whole of X_κ. Throughout this discussion we must assume that $\sum_{k=\kappa}^{\infty} \beta_k^{-2} d_k < \infty$.

Lemma 3.2 *Let R_x be the representer for x in $N^\perp$. Let $g \in X_\kappa$ be defined by*

$$g = \sum_{k=\kappa}^{\infty} \sum_{j=1}^{d_k} a_{kj} Y_j^{(k)},$$

where

$$a_{kj} = \beta_k^{-2} \left(Y_j^{(k)}(x) - (PY_j^{(k)})(x) \right), \quad 1 \le j \le d_k, \quad k \ge \kappa.$$

Then $R_x = g - Pg$, where $P : X_\kappa \to N$ is the orthogonal projection with respect to the inner product defined in Equation (2).

Proof. Take $f \in X_\kappa$. Then $f - Pf \in N^\perp$, and so

$$
\begin{aligned}
f(x) - (Pf)(x) &= (f - Pf, R_x) \\
&= \sum_{i=1}^{l} (f - Pf)(y_i) R_x(y_i) + \langle f - Pf, R_x \rangle \\
&= \langle f - Pf, R_x \rangle \\
&= \langle f, R_x \rangle,
\end{aligned}
$$

because P, as earlier remarked, is also the interpolatory projection onto N. Suppose

$$R_x = \sum_{k=0}^{\infty} \sum_{j=1}^{d_k} c_{kj} Y_j^{(k)},$$

and $f = Y_i^{(k)}$, where $1 \le i \le d_k$ and $k \ge \kappa$. Then

$$Y_i^{(k)}(x) - (PY_i^{(k)})(x) = \langle Y_i^{(k)}, R_x \rangle = c_{ki} \beta_k^2.$$

Solving for c_{ki} gives

$$c_{ki} = \beta_k^{-2} (Y_i^{(k)}(x) - (PY_i^{(k)})(x)), \quad 1 \le i \le d_k, \ k \ge \kappa,$$

and so we see that $c_{kj} = a_{kj}$, $1 \le j \le d_k$, $k \ge \kappa$. Now set

$$g = \sum_{k=\kappa}^{\infty} \sum_{j=1}^{d_k} a_{kj} Y_j^{(k)}.$$

Then, for all $f \in X_\kappa$, $f(x) - (Pf)(x) = \langle f, g \rangle$. However, $g \notin N^\perp$. This failing on the part of g is easily rectified by setting $R_x = g - Pg$. Then, for all $f \in X_\kappa$,

$$
\begin{aligned}
f(x) - (Pf)(x) &= \langle f, g \rangle \\
&= \langle f, g - Pg \rangle \\
&= \langle f - Pf, g - Pg \rangle \\
&= (f - Pf, g - Pg) \\
&= (f - Pf, R_x),
\end{aligned}
$$

as required. ∎

Theorem 3.3 *Let $x \in S^{d-1}$, and suppose $\sum_{k=\kappa}^{\infty} \beta_k^{-2} d_k < \infty$. Set*

$$
a_{kj} = \beta_k^{-2} \left(Y_j^{(k)}(x) - (PY_j^{(k)})(x) \right), \quad 1 \le j \le d_k, \quad k \ge \kappa.
$$

and let

$$
g = \sum_{k=\kappa}^{\infty} \sum_{j=1}^{d_k} a_{kj} Y_j^{(k)}.
$$

Then, R_x, the representer for x in X_κ, is given by $R_x = g - Pg + \sum_{i=1}^{l} q_i(x) q_i$.

Proof. Take $f \in X_\kappa$ and set $e = g - Pg$. Then $e(y_i) = 0$, $i = 1, 2, \ldots, l$. Hence, by Lemma **3.2**,

$$
\begin{aligned}
(f, e) &= (f - Pf, e) + (Pf, e) \\
&= f(x) - (Pf)(x) + \sum_{i=1}^{l} (Pf)(y_i) e(y_i) + \langle Pf, e \rangle \\
&= f(x) - (Pf)(x).
\end{aligned}
$$

Thus,

$$
\begin{aligned}
f(x) &= (f, e) + (Pf)(x) \\
&= (f, e) + \sum_{i=1}^{l} (f, q_i) q_i(x) \\
&= \left(f, e + \sum_{i=1}^{l} q_i(x) q_i \right).
\end{aligned}
$$

∎

Corollary 3.4 *Let H be the reproducing kernel for the Hilbert space X_κ. Then, in the notation of Theorem* **3.3**,

$$H(x,y) \;=\; g(y) - (Pg)(y) + \sum_{i=1}^{l} q_i(x)q_i(y), \quad x,y \in \mathcal{S}^{d-1}.$$

We want to conclude this section with a final result about *radiality*. The representer R_x, given in Theorem **3.3**, may be written as $R_x = g_x + p_x$, where

$$p_x \;=\; \sum_{k=0}^{\kappa-1}\sum_{j=1}^{d_k} b_{kj}(x)Y_j^{(k)}.$$

The function g_x can be written in the form

$$g_x(y) \;=\; \sum_{k=\kappa}^{\infty}\beta_k^{-2}\sum_{j=1}^{d_k}\left\{Y_j^{(k)}(x) - (PY_j^{(k)})(x)\right\}Y_j^{(k)}(y), \qquad y \in \mathcal{S}^{d-1}.$$

If we use the addition formula (1) then

$$g_x(y) \;=\; \sum_{k=\kappa}^{\infty}\gamma_k P_k(xy) - \sum_{k=\kappa}^{\infty}\beta_k^{-2}\sum_{j=1}^{d_k}(PY_j^{(k)})(x)Y_j^{(k)}(y),$$

where $\gamma_k = \beta_k^{-2}d_k\omega_{d-1}^{-1}$, and ω_{d-1} is the surface area of $\mathcal{S}^{d-1}$. Now,

$$\sum_{j=1}^{d_k}(PY_j^{(k)})(x)Y_j^{(k)}(y) \;=\; \sum_{i=1}^{l}q_i(x)\sum_{j=1}^{d_k}Y_j^{(k)}(y_i)Y_j^{(k)}(y)$$

$$=\; d_k\omega_{d-1}^{-1}\sum_{i=1}^{l}q_i(x)P_k(yy_i).$$

Hence, for $x,y \in \mathcal{S}^{d-1}$,

$$g_x(y) \;=\; \sum_{k=\kappa}^{\infty}\gamma_k P_k(xy) - \sum_{i=1}^{l}q_i(x)\sum_{k=\kappa}^{\infty}\beta_k^{-2}d_k\omega_{d-1}^{-1}P_k(yy_i)$$

$$=\; \phi(xy) - \sum_{i=1}^{l}q_i(x)\phi(yy_i),$$

where $\phi : [-1,1] \to \mathbb{R}$ is given by

$$\phi(t) \;=\; \sum_{k=\kappa}^{\infty}\gamma_k P_k(t), \qquad t \in [-1,1].$$

This is what we mean by radiality – the function g_x is a simple linear combination of 'shifts' of the univariate function ϕ. In $\mathbb{R}^d$ a radial basis function would have the form $y \mapsto \phi(\|y - a\|_2)$. On $\mathcal{S}^{d-1}$ one would expect the analogous form to be $y \mapsto \phi\big(d(y,a)\big)$ where $d(\cdot,\cdot)$ denotes the geodesic distance given by $d(u,v) = \arccos(uv)$. In our case the arccos function has been absorbed into the univariate function ϕ. We sum these arguments up in the next result.

Corollary 3.5 *Let* $x \in \mathcal{S}^{d-1}$ *and suppose* $\displaystyle\sum_{k=\kappa}^{\infty} \beta_k^{-2} d_k < \infty$. *Let* $\gamma_k = \beta_k^{-2} d_k \omega_{d-1}^{-1}$, $k = \kappa, \kappa + 1, \ldots,$ *where* ω_{d-1} *is the surface area of* $\mathcal{S}^{d-1}$. *Set*

$$\phi = \sum_{k=\kappa}^{\infty} \gamma_k P_k, \qquad and \qquad g_x(y) = \phi(xy) - \sum_{i=1}^{l} q_i(x)\phi(yy_i), \quad y \in \mathcal{S}^{d-1}.$$

Then the representer for x *in* X_κ *is given by*

$$R_x = g_x - Pg_x + \sum_{i=1}^{l} q_i(x)q_i.$$

The form of the reproducing kernel (again at the level of generality we are discussing) is found in [23]. The approach there is to present the reader with the reproducing kernel in closed form, when one need only check that it has the properties required of a reproducing kernel. Our approach we feel has the advantage that we already have prior knowledge from Section 2 that X_κ has a reproducing kernel, because X_κ is a Hilbert function space. This allows us to *calculate* the reproducing kernel, rather than to simply produce it 'out of a hat'. Of course, once one has the papers of Freeden [4] or Wahba [24], one can easily infer the form of the reproducing kernel at our level of generality. Very closely related material appears in the paper of Freeden and Hermann [7], and in fact, the existence and form of the reproducing kernel in certain special cases is given in that paper.

4   Interpolation

Our main purpose is to use the foregoing technical analysis to provide us with practical tools for interpolating scattered data on spheres. We now provide a sketch outline of how this is done. Details may be found in Golomb and Weinberger [10], or in an approach more carefully tailored to our present discussion in Light and Wayne [13].

We begin with m interpolation points $a_1, a_2, \ldots, a_m$. These points must be chosen so that $m \geq l$ and $a_1, a_2, \ldots, a_l$ (by reordering if necessary) are unisolvent with respect to N. That is, if $p \in N$ and $p(a_i) = 0$, $i = 1, 2, \ldots, l$, then $p = 0$. Suppose that data $c_1, c_2, \ldots, c_m$ are given at $a_1, a_2, \ldots, a_m$. Define

$$C = \{f \in X_\kappa \; : \; f(a_i) = c_i, \; i = 1, 2, \ldots, m\}.$$

Then C is a closed, convex subset of X_κ. As such it contains a unique point u of minimum norm. If we set

$$\Gamma \;=\; \{v \in X_\kappa \,:\, v(a_i) = 0,\ i = 1, 2, \ldots, m\},$$

then u is characterised by the fact that $u \in \Gamma^\perp$ and $u(a_i) = c_i$, $i = 1, 2, \ldots, m$. Let δ_y denote the point evaluation functional at y. Denote the representer of δ_{a_i} by R_{a_i}. Every element of $\Gamma^\perp$ is a linear combination of $R_{a_1}, R_{a_2}, \ldots, R_{a_m}$. Hence, there exists constants $\lambda_1, \lambda_2, \ldots, \lambda_m$ such that

$$u \;=\; \sum_{i=1}^{m} \lambda_i R_{a_i}.$$

The constants $\lambda_1, \lambda_2, \ldots, \lambda_m$, are determined by the equations

$$\sum_{i=1}^{m} \lambda_i R_{a_i}(a_j) \;=\; \sum_{i=1}^{m} \lambda_i (R_{a_i}, R_{a_j}) \;=\; c_j, \quad j = 1, 2, \ldots, m.$$

In general, Theorem **3.3** shows that the *minimal norm interpolant* u has quite a complex form, especially if the dimension of the kernel N is large. It is pertinent to observe that, since the values of any $v \in C$ are fixed at $a_1, a_2, \ldots, a_m$, minimising $\|v\|$ over $v \in C$ is equivalent to minimising the seminorm $|v|_\kappa$.

There is another way to specify this minimal norm interpolant. The inspiration for this alternative approach comes from the field of radial basis functions. The computational work involved is higher, but the algebraic form of the interpolant is easier to understand.

Definition 4.1 *Let $F \in C[-1, 1]$. Then F is said to be conditionally positive definite of order λ on S^{d-1} if for any $a_1, \ldots, a_m$ in S^{d-1}, and $c_1, \ldots, c_m$ in $\mathbb{R}$ satisfying $\sum_{i=1}^{m} c_i Y_j^{(k)}(a_i) = 0$ for $1 \le j \le d_k$, $0 \le k \le \lambda-1$, $\sum_{i,j=1}^{m} c_i c_j F(a_i a_j) \ge 0$. If, in addition, $\sum_{i,j=1}^{m} c_i c_j F(a_i a_j) > 0$ when at least one of the $c_1, \ldots, c_m$ is non-zero, then F is said to be strictly conditionally positive definite of order λ on S^{d-1}.*

Note that we have restricted the notion of conditional positive definiteness a little so as to suit our particular purposes. One can talk about such functions without confining oneself to the rather special class of functions which are compositions of a univariate function with the geodesic distance map.

Theorem 4.2 *Suppose that $\displaystyle\sum_{k=\kappa}^{\infty} \beta_k^{-2} d_k < \infty$ and set $\gamma_k = \beta_k^{-2} d_k \omega_{d-1}^{-1}$, $k = \kappa, \kappa +$
1, ... Let*

$$\phi = \sum_{k=\kappa}^{\infty} \gamma_k P_k,$$

where P_k is the kth Gegenbauer polynomial with respect to the weight function $w(t) = (1 - t^2)^{(d-3)/2}$, $t \in \mathbb{R}$, normalised so that $P_k(1) = 1$. Then ϕ is strictly conditionally positive definite of order κ on $\mathcal{S}^{d-1}$.

Proof. Let $a_1, a_2, \ldots, a_m \in \mathcal{S}^{d-1}$ and $c_1, c_2, \ldots, c_m \in \mathbb{R}$. Then, using (1) we have

$$
\begin{aligned}
\sum_{i,j=1}^{m} c_i c_j \phi(a_i a_j) &= \sum_{i,j=1}^{m} c_i c_j \sum_{k=\kappa}^{\infty} \gamma_k P_k(a_i a_j) \\
&= \sum_{i,j=1}^{m} c_i c_j \sum_{k=\kappa}^{\infty} \beta_k^{-2} \sum_{s=1}^{d_k} Y_s^{(k)}(a_i) Y_s^{(k)}(a_j) \\
&= \sum_{k=\kappa}^{\infty} \beta_k^{-2} \sum_{s=1}^{d_k} \left(\sum_{i=1}^{m} c_i Y_s^{(k)}(a_i) \right)^2 \\
&\geq 0.
\end{aligned}
$$

Suppose the above inequality is an equality. Recalling that $\beta_k^{-2} > 0$, we must have $\sum_{i=1}^{m} c_i Y_s^{(k)}(a_i) = 0$ for all $1 \leq s \leq d_k$ and $k = \kappa, \kappa + 1, \ldots$. If we also assume $\sum_{i=1}^{m} c_i Y_s^{(k)}(a_i) = 0$ for $1 \leq s \leq d_k$ and $k = 0, \ldots, \kappa - 1$, then we see that the functional which maps $f \in C(\mathcal{S}^{d-1})$ to $\sum_{i=1}^{m} c_i f(a_i)$ is zero on every spherical harmonic, and so we conclude that this functional is zero. That is, $c_1 = \ldots = c_m = 0$. Since $\phi \in C[-1, 1]$ it follows that ϕ is strictly conditionally positive definite of order κ.   ■

Theorem 4.3 *Let $a_1, a_2, \ldots, a_m \in \mathcal{S}^{d-1}$ be chosen so that $a_1, a_2, \ldots, a_l$ are unisolvent with respect to N. Let u be the minimal norm interpolant to data $d_1, d_2, \ldots, d_m$ specified at $a_1, a_2, \ldots, a_m$; that is, $u(a_i) = d_i$, $i = 1, 2, \ldots, m$, and $|u|_\kappa$ is minimal amongst all functions in X_κ with the interpolation property. Let ϕ be as given in Corollary 3.5. Then u has the form*

$$
u(y) = \sum_{i=1}^{m} \mu_i \phi(a_i y) + \sum_{k=0}^{\kappa-1} \sum_{j=1}^{d_k} \epsilon_{kj} Y_j^{(k)}(y), \qquad y \in \mathcal{S}^{d-1}.
$$

Furthermore, the parameters $\mu_1, \mu_2, \ldots, \mu_m$ and ϵ_{kj}, $1 \leq j \leq d_k$, $0 \leq k \leq \kappa - 1$, are uniquely determined by the equations

$$
u(a_s) = \sum_{i=1}^{m} \mu_i \phi(a_i a_s) + \sum_{k=0}^{\kappa-1} \sum_{j=1}^{d_k} \epsilon_{kj} Y_j^{(k)}(a_s) = d_s, \qquad s = 1, 2, \ldots, m,
$$

$$
\sum_{i=1}^{m} \mu_i Y_j^{(k)}(a_i) = 0, \qquad \begin{array}{l} j = 1, 2, \ldots, d_k, \\ k = 0, 1, \ldots, \kappa - 1. \end{array}
$$

Proof. We begin with the original form of u given by $u = \sum_{i=1}^{m} \lambda_i R_{a_i}$. From Corollary **3.5**, $R_{a_i} = g_{a_i} - P g_{a_i} + \sum_{j=1}^{l} q_j(a_i) q_j$, where

$$g_{a_i}(y) = \phi(a_i y) - \sum_{j=1}^{l} q_j(a_i) \phi(a_j y), \qquad y \in S^{d-1}.$$

This expression for R_{a_i} may be summarised by writing $R_{a_i} = g_{a_i} + n_{a_i}$, where $n_{a_i} \in N$, $i = 1, 2, \ldots, m$. Hence, for $y \in S^{d-1}$,

$$u(y) = \sum_{i=1}^{m} \lambda_i R_{a_i}(y) = \sum_{i=1}^{m} \lambda_i \left(\phi(a_i y) - \sum_{j=1}^{l} q_j(a_i) \phi(a_j y) \right) + \sum_{i=1}^{m} \lambda_i n_{a_i}(y)$$

$$= \sum_{i=1}^{m} \mu_i \phi(a_i y) + \sum_{k=0}^{\kappa-1} \sum_{j=1}^{d_k} \epsilon_{kj} Y_j^{(k)}(y),$$

where $\{\epsilon_{kj} : 1 \leq j \leq d_k, \ 0 \leq k \leq \kappa - 1\}$ are suitable real numbers and

$$\mu_i = \begin{cases} \lambda_i - \sum_{j=1}^{m} \lambda_j q_i(a_j), & i = 1, 2, \ldots, l, \\ \lambda_i, & i = l+1, l+2, \ldots, m. \end{cases}$$

Now, let $n \in N$. Then,

$$\sum_{i=1}^{m} \mu_i n(a_i) = \sum_{i=1}^{m} \lambda_i n(a_i) - \sum_{i=1}^{l} \sum_{j=1}^{m} \lambda_j q_i(a_j) n(a_i)$$

$$= \sum_{i=1}^{m} \lambda_i n(a_i) - \sum_{j=1}^{m} \lambda_j \sum_{i=1}^{l} n(a_i) q_i(a_j)$$

$$= 0,$$

recalling that $q_1, q_2, \ldots, q_l$ is the cardinal Lagrange basis for N.

It remains to be shown that u is uniquely determined by the equations given. Suppose that $\mu_1, \mu_2, \ldots, \mu_m$ and $\{\epsilon_{kj} : 1 \leq j \leq d_k, \ 0 \leq k \leq \kappa - 1\}$ are such that

$$\sum_{i=1}^{m} \mu_i \phi(a_i a_s) + \sum_{k=0}^{\kappa-1} \sum_{j=1}^{d_k} \epsilon_{kj} Y_j^{(k)}(a_s) = 0, \qquad s = 1, 2, \ldots, m,$$

$$\sum_{i=1}^{m} \mu_i Y_j^{(k)}(a_i) = 0, \qquad \begin{array}{l} j = 1, 2, \ldots, d_k, \\ k = 0, 1, \ldots, \kappa - 1. \end{array}$$

Then, by multiplying the sth equation by μ_s and summing over s, we see that

$$\sum_{i,s=1}^{m} \mu_i \mu_s \phi(a_i a_s) + \sum_{k=0}^{\kappa-1} \sum_{j=1}^{d_k} \epsilon_{kj} \sum_{s=1}^{m} \mu_s Y_j^{(k)}(a_s) = 0.$$

It follows therefore, that

$$\sum_{i,s=1}^{m} \mu_i \mu_s \phi(a_i a_s) = 0,$$

so that, by Theorem **4.2**, $\mu_1, \mu_2, \ldots, \mu_m = 0$. If

$$p = \sum_{k=0}^{\kappa-1} \sum_{j=1}^{d_k} \epsilon_{kj} Y_j^{(k)},$$

then $p \in N$ and $p(a_i) = 0$, $i = 1, 2, \ldots, l$. But since these points are unisolvent with respect to N, it follows that $p = 0$, i.e., $\epsilon_{kj} = 0$, $1 \le j \le d_k$, $0 \le k \le \kappa - 1$. Hence u is determined uniquely by the given equations. ∎

This Section has been included so that the reader will have a complete account of the theory of these minimal norm interpolants on spheres. There has been a considerable amount of work recently on the idea of conditionally positive definite functions on spheres, and we cannot even claim that this section provides any fresh or intrinsically superior approach! The interested reader can start with the survey of Cheney [2].

5   Error bounds

In this section we show that our interpolants have good behaviour with regard to error bounds. We will follow the approach of Schaback [18] and Powell [16] rather than the original material of Golomb and Weinberger [10], since this will produce a more immediate development of the results.

We maintain the same setup as in previous sections, so that $a_1, a_2, \ldots, a_m$ are the interpolation points and $a_1, a_2, \ldots, a_l$ are unisolvent with respect to N. The minimal norm interpolant can be written, using Theorem **4.3**, as

$$u(y) = \sum_{i=1}^{m} \mu_i \phi(a_i y) + n(y), \qquad y \in \mathcal{S}^{d-1},$$

where $n \in N$.

Theorem 5.1 *Let* $\mathcal{A} = \{a_1, \ldots, a_m\} \subseteq \mathcal{S}^{d-1}$. *For* $x \in \mathcal{S}^{d-1}$ *choose* $\alpha_1, \ldots, \alpha_m$ *such that*

$$n(x) = \sum_{i=1}^{m} \alpha_i n(a_i),$$

for all $n \in N$. *If* $f \in X_\kappa$ *and* u *is the minimal norm interpolant to* f *at* $a_1, \ldots, a_m$, *then*

$$|f(x) - u(x)| \le \left(\sum_{i,j=1}^{m} \alpha_i \alpha_j \phi(a_i a_j) - 2 \sum_{i=1}^{m} \alpha_i \phi(xa_i) + \phi(1) \right)^{\frac{1}{2}} |f|_\kappa.$$

Proof. Let Γ be the functional defined by

$$\Gamma f \;=\; f(x) - \sum_{i=1}^{m} \alpha_i f(a_i), \qquad f \in X_\kappa.$$

Then Γ is continuous by virtue of Lemma **2.2**. Because u is the minimal norm interpolant to f, $|f - u|_\kappa^2 = |f|_\kappa^2 + |u|_\kappa^2$. Hence,

$$|f(x) - u(x)| = |\Gamma(f - u)| \leq \|\Gamma\| \|f - u\| = \|\Gamma\| \|f - u\|_\kappa \leq \|\Gamma\| \|f\|_\kappa.$$

Now, let R be the representer for Γ in X_κ. Then, $\|\Gamma\|^2 = \|R\|^2 = (R, R) = \Gamma(R).$ From Corollary **3.5**,

$$R \;=\; R_x - \sum_{i=1}^{m} \alpha_i R_{a_i}.$$

In the notation of Corollary **3.5**, $R_x = g_x + n_x$ and $R_{a_i} = g_{a_i} + n_{a_i}$, where n_x and n_{a_i} are suitable elements of N. Since $\Gamma(n) = 0$ for all n in N,

$$\Gamma(R) \;=\; \Gamma(R_x) - \sum_{i=1}^{m} \alpha_i \Gamma(R_{a_i})$$

$$=\; \Gamma(g_x) - \sum_{i=1}^{m} \alpha_i \Gamma(g_{a_i}).$$

Now,

$$\Gamma(g_x) \;=\; \phi(xx) - \sum_{i=1}^{l} q_i(x)\phi(a_i x) - \sum_{i=1}^{m} \alpha_i \left(\phi(a_i x) - \sum_{i=1}^{l} q_j(x)\phi(a_i a_j) \right).$$

Similarly, for $i = 1, 2, \ldots, m$,

$$\Gamma(g_{a_i}) \;=\; g_{a_i}(x) - \sum_{s=1}^{m} \alpha_s g_{a_i}(a_s)$$

$$=\; \phi(a_i x) - \sum_{j=1}^{l} q_j(a_i)\phi(a_j x) - \sum_{s=1}^{m} \alpha_s \left(\phi(a_i a_s) - \sum_{i=1}^{l} q_j(a_i)\phi(a_s a_j) \right),$$

so that

$$\sum_{i=1}^{m} \alpha_i \Gamma(g_{a_i}) \;=\; \sum_{i=1}^{m} \alpha_i \phi(a_i x) - \sum_{j=1}^{l} \phi(a_j x) \sum_{i=1}^{m} \alpha_i q_j(a_i)$$

$$- \sum_{i,s=1}^{m} \alpha_i \alpha_s \phi(a_i a_s) + \sum_{s=1}^{m} \sum_{j=1}^{l} \alpha_s \phi(a_s a_j) \sum_{i=1}^{m} \alpha_i q_j(a_i)$$

$$= \sum_{i=1}^{m} \alpha_i \phi(a_i x) - \sum_{j=1}^{l} \phi(a_j x) q_j(x) - \sum_{i,s=1}^{m} \alpha_i \alpha_s \phi(a_i a_s) + \sum_{s=1}^{m} \sum_{j=1}^{l} \alpha_s \phi(a_s a_j) q_j(x).$$

Hence,

$$\Gamma(R) \;=\; \phi(1) - 2 \sum_{i=1}^{m} \alpha_i \phi(a_i x) + \sum_{i,s=1}^{m} \alpha_i \alpha_s \phi(a_i a_s).$$

■

It is natural to have expectations about the error bound given in Theorem **5.1**. If the points $x, a_1, a_2, \ldots, a_m$ are close together, then $|f(x) - u(x)|$ should be small. This is apparent from the form of the bound, since in this case $a_i a_j$ and $x a_i$ are close to 1, and the fact that $\sum_{i=1}^{m} \alpha_i = 1$ means that the term that is square rooted is small. The obvious way to investigate this is through Taylor series.

Theorem 5.2 *Let* $\{a_1, \ldots, a_m\} \subset S^{d-1}$ *be unisolvent with respect to* $\pi_{\kappa-1}(S^{d-1})$. *Let* $x \in S^{d-1}$ *and* $\alpha_1, \ldots, \alpha_m \in \mathbb{R}$ *be such that* $p(x) = \sum_{i=1}^{m} \alpha_i p(a_i)$ *for all* $p \in \pi_{n-1}(S^{d-1})$, *where* $n \geq \kappa$. *For* $f \in X_\kappa$, *let the minimal norm interpolant on* $a_1, \ldots, a_m$ *be given by*

$$(Uf)(y) = \sum_{i=1}^{m} \lambda_i \phi(a_i y) + p(y), \qquad y \in S^{d-1},$$

where $\lambda_1, \ldots, \lambda_m \in \mathbb{R}$, $p \in \pi_{\kappa-1}(S^{d-1})$ *and* $\phi \in C^{(n)}[1 - \epsilon, 1]$, *where* $0 < \epsilon < 1$. *Suppose* $1 - a_i a_j < h < \epsilon$ *whenever* $\alpha_i \alpha_j \neq 0$, $i, j = 1, \ldots, m$, *and* $1 - x a_i < h < \epsilon$ *whenever* $\alpha_i \neq 0$, $i = 1, \ldots, m$. *Then*

$$|f(x) - (Uf)(x)|^2 \leq \frac{h^n}{n!} \max_{z \in [1-\epsilon,1]} |\phi^{(n)}(z)| \left(\sum_{i=1}^{m} |\alpha_i| \left(\sum_{j=1}^{m} |\alpha_j| + 2 \right) \right) |f|_\kappa^2,$$

for all $f \in X_\kappa$.

Proof. By expanding ϕ as a Taylor series about 1, there exists ξ_{ij} and ξ_i in $(1-\epsilon, 1)$ such that

$$\sum_{i,j=1}^{m} \alpha_i \alpha_j \phi(a_i a_j) - 2 \sum_{i=1}^{m} \alpha_i \phi(x a_i)$$

$$= \sum_{i,j=1}^{m} \alpha_i \alpha_j \left\{ \sum_{s=0}^{n-1} \frac{(1 - a_i a_j)^s}{s!} \phi^{(s)}(1) + \frac{(1 - a_i a_j)^n}{n!} \phi^{(n)}(\xi_{ij}) \right\}$$

$$- 2 \sum_{i=1}^{m} \alpha_i \left\{ \sum_{s=0}^{n-1} \frac{(1 - x a_i)^s}{s!} \phi^{(s)}(1) + \frac{(1 - x a_i)^n}{n!} \phi^{(n)}(\xi_i) \right\}.$$

Set $p_s(y) = \sum_{j=1}^{m} \alpha_j (1 - a_j y)^s$ and $q_s(y) = (1 - xy)^s$ for $y \in S^{d-1}$ and $s = 0, \ldots, n-1$. Then,

$$
\sum_{i,j=1}^{m} \alpha_i \alpha_j \phi(a_i a_j) - 2 \sum_{i=1}^{m} \alpha_i \phi(x a_i) = \sum_{s=0}^{n-1} \frac{\phi^{(s)}(1)}{s!} \left(\sum_{i=1}^{m} \alpha_i \left[p_s(a_i) - 2q_s(a_i) \right] \right)
$$

$$
+ \sum_{i=1}^{m} \alpha_i \left(\sum_{j=1}^{m} \alpha_j \frac{(1 - a_i a_j)^n}{n!} \phi^{(n)}(\xi_{ij}) - 2 \frac{(1 - a_i x)^n}{n!} \phi^{(n)}(\xi_i) \right)
$$

$$
= \sum_{s=0}^{n-1} \frac{\phi^{(s)}(1)}{s!} \left(p_s(x) - 2q_s(x) \right)
$$

$$
+ \sum_{i=1}^{m} \alpha_i \left(\sum_{j=1}^{m} \alpha_j \frac{(1 - a_i a_j)^n}{n!} \phi^{(n)}(\xi_{ij}) - 2 \frac{(1 - a_i x)^n}{n!} \phi^{(n)}(\xi_i) \right).
$$

Note that if $s \neq 0$, then $q_s(x) = 0$ for all $x \in S^{d-1}$, while $q_0(x) = 1$. Similarly,

$$
p_s(x) = \sum_{i=1}^{m} \alpha_j (1 - a_j x)^s = \sum_{j=1}^{m} \alpha_j q_s(a_j) = q_s(x).
$$

Hence $p_s(x) = 1$ if $s = 0$ and is zero otherwise. Therefore,

$$
\left| \phi(1) + \sum_{i,j=1}^{m} \alpha_i \alpha_j \phi(a_i a_j) - 2 \sum_{i=1}^{m} \alpha_i \phi(x a_i) \right| =
$$

$$
= \left| \phi(1) - \phi(1) + \sum_{i=1}^{m} \alpha_i \left(\sum_{j=1}^{m} \alpha_j \frac{(1 - a_i a_j)^n}{n!} \phi^{(n)}(\xi_{ij}) - 2 \frac{(1 - a_i x)^n}{n!} \phi^{(n)}(\xi_i) \right) \right|
$$

$$
\leq \max_{z \in [1-\epsilon, 1]} |\phi^{(n)}(z)| \frac{h^n}{n!} \left| \sum_{i=1}^{m} |\alpha_i| \left(\sum_{j=1}^{m} |\alpha_j| + 2 \right) \right|.
$$

This establishes the result.　■

Versions of Theorems **5.1** and **5.2**, with $n = 1$, are the basis for the error estimates found in a variety of works by Freeden and his co-authors. However, the error bounds are usually given by reference to the reproducing kernel and not to the function ϕ. It seems to us to be rather easier to understand what order of convergence is likely, simply by knowing the continuity of ϕ around the value 1. In particular Freeden and Hermann [7] take advantage of the fact that in Theorem **5.2** when $n = 1$, $\alpha_1 = 1$ is the only available choice, and $p(x) = \sum_{i=1}^{m} \alpha_i p(x_i)$ holds good only for polynomials of degree zero. We will recover their precise error estimate in a moment, but first some discussion of the idea of density is necessary. The geodesic distance between two points $x, y \in S^{d-1}$ is given by $d_g(x, y) = \arccos(xy)$. This is a natural measure of distance between two points and corresponds to the 'great circle' distance on the earth. If $\mathcal{A}$ is a subset of S^{d-1}, then the quantity $\max_{x \in S^{d-1}} \min_{a \in \mathcal{A}} d_g(x, a)$ will be called the geodesic density of $\mathcal{A}$ in S^{d-1}. We are going to show how an asymptotic error estimate can be obtained using this density. The estimate will be valid only for the cases $d = 3$ and $\kappa \leq 2$. The reason for this is that the computations are already difficult to perform even for this special case. Arbitrary d and κ are very complicated to handle, but a complete treatment can be found in [9].

Theorem 5.3 *For each $h > 0$, let $\mathcal{A}_h$ be a finite subset of S^2 with geodesic density h. Let $\kappa \leq 2$. For each $f \in X_\kappa$, let $U_h f$ be the minimal norm interpolant to f on $\mathcal{A}_h$, where*

$$(U_h f)(x) = \sum_{a \in \mathcal{A}_h} \mu_a \phi(xa) + n(x), \qquad x \in S^2.$$

Here $\mu_a \in \mathbb{R}$, $n \in \pi_{\kappa-1}(S^2)$ and $\phi \in C^{(2)}[1 - \epsilon, 1]$ for some $0 < \epsilon < 1$. Then there exists $h_0 > 0$ and $C > 0$, both independent of h, such that for all x in S^2, all $f \in X_\kappa$ and for all $h < h_0$,

$$|f(x) - (U_h f)(x)| \leq Ch^2 |f|_\kappa.$$

Proof. Fix $f \in X_\kappa$ and $x \in S^2$. Without loss of generality we may assume $x = y_1 = (1, 0, 0)$, so that x is the 'north pole'. Define y_2, y_3, y_4 by the two conditions $y_2 y_3 = y_3 y_4 = y_4 y_2$ and $\arccos(y_i y_1) = 8h$, $i = 2, 3, 4$. Define $B_i = \{u \in S^2 : \arccos(uy_i) < h\}$, $i = 1, 2, 3, 4$. Since $\mathcal{A}_h$ is of geodesic density h in S^2, it follows that each ball B_i contains a point $a_i \in \mathcal{A}_h$, $i = 1, 2, 3, 4$. We want to apply Theorem **5.1**, using the points $y_1, a_1, \ldots, a_4$. Thus we have to find $\alpha_1, \ldots, \alpha_4$ such that $n(y_1) = \sum_{i=1}^4 \alpha_i n(a_i)$ for all $n \in \pi_1(S^2)$. This constrains the $a_1, \ldots, a_4$ to be unisolvent with respect to $\pi_1(S^2)$. This, in turn, implies that $\alpha_1, \ldots, \alpha_4$ are uniquely determined numbers. Let us suppose for a moment that this is the case. From Theorem **5.1**,

$$|f(x) - (U_h f)(x)|^2 \leq \left\{ \sum_{i,j=1}^4 \alpha_i \alpha_j \phi(a_i a_j) - 2 \sum_{i=1}^4 \alpha_i \phi(xa_i) + \phi(1) \right\}^2 |f|_\kappa. \quad (4)$$

Now the points y_1, a_1, a_2, a_3 and a_4 are all in the 'spherical cap'

$$\{u \in S^2 : \arccos(uy_1) \leq 9h\}.$$

Any two points in this spherical cap have inner product at least $\cos(9h)$, and hence at least $1 - 41h^2$. We now assume that h is sufficiently small so that $41h^2 < \epsilon$. It then follows that all the arguments of ϕ appearing in Equation (4) are at least $1 - \epsilon$. Moreover, $1 - a_i a_j \leq 41h^2$ and similarly $1 - y_1 a_i \leq 41h^2$. Hence, by Theorem **5.2**,

$$|f(x) - (U_h f)(x)|^2 \leq \frac{1}{2} \max_{z \in [1-\epsilon, 1]} |\phi^{(2)}(z)| (41h^2)^2 \left(\sum_{i=1}^4 |\alpha_i| \left(\sum_{j=1}^4 |\alpha_j| + 2 \right) \right) |f|_\kappa^2. \quad (5)$$

The proof will be completed by showing that there are numbers h_0, C_1 independent of h such that $\sum_{i=1}^{4} |\alpha_i| \leq C_1$ for all $h < h_0$. We then further restrict h_0 if necessary so that $41h_0^2 < \epsilon$. Setting

$$C^2 = (41/2)(C_1^2 + 2C_1)\|\phi^{(2)}\|_{C[1-\epsilon,1]},$$

and substituting in Equation (5) gives

$$|f(y_1) - (U_h f)(y_1)| \leq Ch^2 |f|_2,$$

for all $h < h_0$. The existence of the constant C_1 is the subject of the next few Lemmas.   ∎

We have to discuss interpolation by linear polynomials on S^2, denoted by $\pi_1(S^2)$, and also by linear polynomials on $\mathbb{R}^3$, denoted by $\pi_1(\mathbb{R}^3)$. Of course, the distinction between these two interpolants is a fine one, since the linear interpolant in $\pi_1(S^2)$ is simply the restriction of the linear interpolant in $\pi_1(\mathbb{R}^3)$ to the surface S^2. It will help to make some preliminary observations about the balls $B_1, \ldots, B_4$ introduced in the proof of Theorem **5.3**. Firstly, points in B_1 have first coordinate at least $\cos(h)$. Secondly, the first coordinate of any point in B_2, B_3 and B_4 lies between $\cos(7h)$ and $\cos(9h)$.

Lemma 5.4 *For $h > 0$ define $y_1 = (1,0,0)$, and y_2, y_3, y_4 such that $y_2 y_3 = y_3 y_4 = y_4 y_2$ and $\arccos(y_i y_1) = 8h$, $i = 2,3,4$. Let $B_i = \{u \in S^2 : \arccos(u y_i) < h\}$, $i = 1,2,3,4$. Then the balls $B_1, \ldots, B_4$ are pairwise disjoint.*

Proof. By the third remark preceding the Lemma, B_1 is disjoint from the other three balls. The centres of the balls B_2, B_3, B_4 lie on the "lattitude" of points with first coordinate $\cos(8h)$. This latitude has radius $2\pi \sin(8h) > 2\pi(2/\pi)8h = 32h$. Since the centres of B_2, B_3, B_4 are equally spaced around this latitude, and these balls have diameter $2h$, they do not intersect.   ∎

The linear interpolant can be realised explicitly as follows. For $i = 1,2,3,4$ define the hyperplane

$$L_i = \left\{ \sum_{\substack{j=1 \\ j \neq i}}^{4} \mu_j b_j : \sum_{\substack{j=1 \\ j \neq i}}^{4} \mu_j = 1 \right\}.$$

Let $p_i^h \in \pi_1(\mathbb{R}^3)$, $i = 1,2,3,4$, be the cardinal functions for Lagrange interpolation at b_1, b_2, b_3, b_4 respectively. Thus $p_i^h(b_j) = 1$ if $i = j$, and is zero otherwise, $i, j = 1, \ldots, 4$. Then

$$|p_i^h(x)| = \frac{\text{dist}(x, L_i)}{\text{dist}(b_i, L_i)}, \qquad i = 1, \ldots, 4, \; x \in \mathbb{R}^3.$$

Here dist denotes the usual Euclidean distance in $\mathbb{R}^3$.

To complete the argument needed for Theorem **5.3**, one looks at the definition of $\alpha_1, \ldots, \alpha_4$. Since $n(y_1) = \sum_{i=1}^{4} \alpha_i n(a_i)$ it follows that $\alpha_i = p_i^h(y_1)$, and so,

$$\sum_{i=1}^{4} |\alpha_i| = \sum_{i=1}^{4} |p_i^h(y_1)|.$$

Note that the construction of $p_1^h, \ldots, p_4^h$ is only possible if the points $b_1, \ldots, b_4$ are unisolvent with respect to $\pi_1(\mathcal{S}^2)$. The important question now is how $|p_i^h(y_1)|$ depends on h as $h \to 0$. Note that the dependence on h comes from the fact that although not made explicit in the notation, the points $b_1, \ldots, b_4$ depend on h. The next three results show how to bound $\sum_{i=1}^{4} |p_i^h(y_1)|$ for all sufficiently small h. This establishes simultaneously the existence of the constant C_1 needed in Theorem **5.3**, and the unisolvency of $b_1, \ldots, b_4$.

Lemma 5.5 *We have* $|p_1^h(y_1)| \leq 81$.

Proof. Observe that $L_1 = \{\sum_{j=2}^{4} \mu_j b_j : \sum_{j=2}^{4} \mu_j = 1\}$. By the second remark preceding Lemma **5.4**, any point in $\mathcal{C} = \text{co}\{b_2, b_3, b_4\}$ has first coordinate lying between $\cos(9h)$ and $\cos(7h)$. Now,

$$\begin{aligned}
\text{dist}(y_1, L_1) &= \min\left\{\sqrt{(1-x)^2 + y^2 + z^2} : (x, y, z) \in \mathcal{C}\right\} \\
&\leq \min\left\{\sqrt{(1-\cos(9h))^2 + y^2 + z^2} : (x, y, z) \in \mathcal{C}\right\}.
\end{aligned}$$

Because there exists a point in $\mathcal{C}$ of the form $(a, 0, 0)$, $\text{dist}(y_1, L_1) \leq 1 - \cos(9h)$. Now suppose $b_1 = (u, v, w)$. Then by the first remark preceding Lemma **5.4**, $u \geq \cos(h)$. Thus,

$$\begin{aligned}
\text{dist}(b_1, L_1) &= \min\{\sqrt{(u-x)^2 + (v-y)^2 + (w-z)^2} : (x, y, z) \in L_1\} \\
&\geq \cos(h) - \cos(7h).
\end{aligned}$$

Now,

$$\frac{\text{dist}(y_1, L_1)}{\text{dist}(b_1, L_1)} \leq \frac{1 - \cos(9h)}{\cos(h) - \cos(7h)} \leq \frac{1 - (1 - 81h^2/2)}{(1 - (h^2/2)) - 1} \leq 81.$$

∎

Because of the symmetry of the problem, it suffices now to bound $|p_4^h(y_1)|$ as $h \to 0$. The next two Lemmas achieve this. It helps to adopt a simple orientation for the points y_2, y_3, y_4. This is $y_4 = (\cos(8h), -\sin(8h), 0)$, and $y_3 = (\cos(8h), \sin(8h)\cos(2\pi/3), \sin(8h)\sin(2\pi/3))$. Then y_2 is identical with y_3 except that the last coordinate is $-\sin(8h)\sin(2\pi/3)$.

Lemma 5.6 *Suppose* y_1 *lies above* L_4. *Then for all sufficiently small* $h > 0$, $\text{dist}(y_1, L_4) \leq 25h^2$.

Proof. Let ℓ_1 be the line in R^3 which passes through b_2 and b_3. Let $p_1 \in S^2 \cap L_4$ be such that $\arccos(y_1 p_1) \leq h$. Let ℓ_2 be the line which passes through p_1, lies in L_4 and is perpendicular to ℓ_1. The slope of ℓ_2 is at least

$$\frac{\cos(h) - \cos(7h)}{\sin(h) + \sin(9h)} = \frac{24h}{10} + \mathcal{O}(h^2).$$

Now let ℓ_3 be the line lying on L_4, perpendicular to ℓ_1 and passing through $y = z = 0$. Then this line crosses the axis $y = z = 0$ at x, where

$$\begin{aligned} x \geq \cos(9h) + \sin(7h)\left(\frac{24h}{10} + \mathcal{O}(h^2)\right) &= 1 - \frac{81h^2}{2} + 7h\frac{24h}{10} + \mathcal{O}(h^3) \\ &\geq 1 - 24h^2 + \mathcal{O}(h^3). \end{aligned}$$

Hence, for all sufficiently small $h > 0$, $\mathrm{dist}(y_1, L_4) \leq 1 - x \leq 25h^2$. ∎

Lemma 5.7 *For all sufficiently small $h > 0$, $|p_4^h(y_1)| \leq 13$.*

Proof. If y_1 lies below L_4 then $\mathrm{dist}(y_1, L_4) \leq \mathrm{dist}(b_4, L_4)$ and so $|p_4^h(y_1)| \leq 1$. If y_1 lies above L_4, then for all sufficiently small h, $\mathrm{dist}(y_1, L_4) \leq 25h^2$ by the previous Lemma. Now we consider two sets. Firstly, if (x, y, z) is a point on L_4 with $y \geq \sin(h)$, then such a point is at distance at least $\sin(5h) - \sin(h)$ from b_4, and so is $4h + \mathcal{O}(h^2)$ away from b_4. If (x, y, z) in L_4 is such that $y \leq \sin(h)$, then we need only consider the points of this form which lie on lines perpendicular to the line joining b_2 to b_3 (so they pass through the convex hull of these two points). The x-coordinate of such points is at least

$$\begin{aligned} \cos(9h) + \left(\frac{24h}{10} + \mathcal{O}(h^2)\right)\left(\sin(h) + \sin(7h)\right) &= 1 - \frac{81h^2}{2} + \frac{24.8.h^2}{10} + \mathcal{O}(h^3) \\ &\geq 1 - 22h^2 + \mathcal{O}(h^3). \end{aligned}$$

The x-coordinate of b_4 is at most

$$\cos(7h) = 1 - \frac{49h^2}{2} + \mathcal{O}(h^3) \leq 1 - 24h^2$$

for suitably small h. Hence for these values of h, $\mathrm{dist}(b_4, L_4) \geq 2h^2$. Finally, we obtain $|p_4^h(y_1)| \leq 25/2 \leq 13$. ∎

The error estimate of Freeden and Hermann is weaker than Theorem **5.3**, but is consequently much easier to derive.

Theorem 5.8 *(Freeden and Hermann [7]) For each $h > 0$, let $\mathcal{A}_h$ be a finite subset of S^{d-1} with geodesic density h. Let $\kappa \leq 1$. For each $f \in X_\kappa$, let $U_h f$ be the minimal norm interpolant to f on $\mathcal{A}_h$, where*

$$(U_h f)(x) = \sum_{a \in \mathcal{A}_h} \mu_a \phi(xa) + n(x), \qquad x \in S^{d-1}.$$

Here $\mu_a \in \mathbb{R}$, $n \in \pi_{\kappa-1}(\mathcal{S}^{d-1})$ and $\phi \in C^{(1)}[1-\epsilon, 1]$ for some $0 < \epsilon < 1$. Then there exists $h_0 > 0$ and $C > 0$, both independent of h, such that for all x in $\mathcal{S}^{d-1}$, all $f \in X_\kappa$ and for all $h < h_0$,

$$|f(x) - (U_h f)(x)| \leq Ch|f|_\kappa.$$

Proof. As in Theorem **5.3**, fix $f \in X_\kappa$ and $x \in \mathcal{S}^{d-1}$. Because $\mathcal{A}_h$ has geodesic density h in $\mathcal{S}^{d-1}$, it follows that there exists a point $a \in \mathcal{A}_h$ such that $d_g(x, a) < h$. Now $n(x) = n(a)$ for all $n \in \pi_0(\mathcal{S}^{d-1})$. Moreover, $1 - xa < 1 - \cos(h) < h^2/2$. Take $h < \epsilon$. Then by Theorem **5.2**,

$$|f(x) - (U_h f)(x)|^2 \leq \max_{z \in [1-\epsilon, 1]} |\phi^{(1)}(z)|(3h^2/2)|f|_\kappa^2 \leq C^2 h^2 |f|_\kappa,$$

where $C^2 = (3/2) \max_{z \in [1-\epsilon, 1]} |\phi^{(1)}(z)|$. ∎

Some remarks are now in order. Firstly, the 'space index' κ, which measures the dimension of the kernel of the seminorm on X_κ is restricted in Theorem **5.2** so that $\kappa \leq n$. This restriction is implicit already in Theorem **5.1**. Equation (5) will not hold in general without this restriction. However, the proof of Theorem **5.1** can still be executed in exactly the same manner without this restriction. This leads to an error bound containing not only the function ϕ but also some polynomial terms. If such a bound is desired, it is better expressed in terms of the reproducing kernel, with no attempt being made to discard polynomial terms. Secondly, readers who are familiar with radial basis functions in $\mathbb{R}^d$ will find that Theorems **5.3** and **5.8** appear to give *double* the expected order of convergence. For example, with a Euclidean density of h and a radial basis function $\phi \in C^{(1)}(\mathbb{R})$, one would expect to obtain $|f(x) - (U_h f)(x)| = \mathcal{O}(h^{1/2})$ and not $\mathcal{O}(h)$ as given. The simple explanation is that our 'radial' function (in common with those of most other authors) is of the form $x \mapsto \phi(xa)$, $x \in \mathcal{S}^{d-1}$, and so is not a function of the geodesic distance. If we wrote $\phi(xa) = \psi(\arccos(xa))$ for $x \in \mathcal{S}^{d-1}$, and assumed $\psi \in C^{(1)}[-1, 1]$, then our error estimate would return to being $\mathcal{O}(h)$. Thirdly, we attribute Theorem **5.8** to Freeden and Hermann. However, the proof is couched in a rather different language from [7]. The idea of using a single interpolation point has a long history. See for example [8] for many references, or the very recent paper by Schreiner [22]. This is hardly surprising, in view of the technical difficulties of obtaining a practical error bound when more than one interpolation point is used. (Witness the effort expended in the subsidiary Lemmas to Theorem **5.3**.) Our final objective is to show how the order of convergence may be doubled using a technique exploited in the $\mathbb{R}^d$-theory by Schaback [19].

Theorem 5.9 *Let $\mathcal{A} = \{a_1, a_2, \ldots, a_m\} \subseteq \mathcal{S}^{d-1}$. For $x \in \mathcal{S}^{d-1}$ choose $\alpha_1, \ldots, \alpha_m$ such that $n(x) = \sum_{i=1}^m \alpha_i n(a_i)$ for all $n \in N$. For $f \in X_\kappa$, let Uf be the minimal norm interpolant to f on $\mathcal{A}$. Suppose $f = \sum_{k=0}^\infty \sum_{j=1}^{d_k} a_{kj} Y_j^{(k)}$ and that the* *quantity* $|||f|||^2 = \sum_{k=\kappa}^{\infty} \beta_k^4 \sum_{j=1}^{d_k} a_{kj}^2 < \infty$. *Then,*

$$\|f - Uf\|_\infty \le \omega_{d-1}^{1/2}\Big\{\phi(1) - 2\sum_{i=1}^{m}\alpha_i\phi(xa_i) + \sum_{i,j=1}^{m}\alpha_i\alpha_j\phi(a_ia_j)\Big\}|||f|||.$$

Proof. From Equation (5) in the proof of Theorem **5.1**, we obtain,

$$\|f - Uf\|_\infty = \sup_{x \in \mathcal{S}^{d-1}} |f(x) - (Uf)(x)| \le \|\Gamma\|\,\|f - Uf\|_\kappa.$$

Because Uf is the minimal norm interpolant to f,

$$|f - Uf|_\kappa^2 = \langle f - Uf, f - Uf \rangle = \langle f - Uf, f \rangle.$$

Now suppose $f = \sum_{k=0}^{\infty}\sum_{j=1}^{d_k} a_{kj}Y_j^{(k)}$ and $f - Uf = \sum_{k=\kappa}^{\infty}\sum_{j=1}^{d_k} b_{kj}Y_j^{(k)}$. Then, using the Cauchy-Schwartz inequality,

$$
\begin{aligned}
\langle f - Uf, f \rangle &= \sum_{k=\kappa}^{\infty}\beta_k^2\sum_{j=1}^{d_k} a_{kj}b_{kj} \le \Big(\sum_{k=\kappa}^{\infty}\sum_{j=1}^{d_k} b_{kj}^2\Big)^{1/2}\Big(\sum_{k=\kappa}^{\infty}\beta_k^4\sum_{j=1}^{d_k} a_{kj}^2\Big)^{1/2}\\
&= \Big(\sum_{k=\kappa}^{\infty}\sum_{j=1}^{d_k} b_{kj}^2\Big)^{1/2}|||f|||.
\end{aligned}
$$

Note that the use of the Cauchy-Schwartz inequality is legitimate, since, because $X_\kappa \subset C(\mathcal{S}^{d-1})$, and $\{Y_j^{(k)}\}_{j=1,k=0}^{d_k,\infty}$ is an orthonormal basis for $L^2(\mathcal{S}^{d-1})$,

$$\sum_{k=\kappa}^{\infty}\sum_{j=1}^{d_k} b_{kj}^2 = \|f - Uf\|_2^2 \le \|f - Uf\|_\infty^2 < \infty.$$

Furthermore,

$$\langle f - Uf, f \rangle \le \Big(\sum_{k=\kappa}^{\infty}\sum_{j=1}^{d_k} b_{kj}^2\Big)^{1/2}|||f||| = \|f - Uf\|_2|||f||| \le \omega_{d-1}^{1/2}\|f - Uf\|_\infty|||f|||,$$

where ω_{d-1} is the surface area of $\mathcal{S}^{d-1}$. Putting all this together, we obtain

$$\|f - Uf\|_\infty^2 \le \|\Gamma\|^2|f - Uf|_\kappa^2 = \|\Gamma\|^2\langle f - Uf, f \rangle \le \omega_{d-1}^{1/2}\|\Gamma\|^2\|f - Uf\|_\infty|||f|||.$$

Now the proof of Theorem **5.1** contains a calculation of $\|\Gamma\|^2$, and this gives

$$\|f - Uf\|_\infty \le \omega_{d-1}^{1/2}\Big\{\phi(1) - 2\sum_{i=1}^{m}\alpha_i\phi(xa_i) + \sum_{i,j=1}^{m}\alpha_i\alpha_j\phi(a_ia_j)\Big\}|||f|||.$$

This Theorem can be used to improve the error in either Theorem **5.3** or Theorem **5.8**. We give the former as an example.

Corollary 5.10 *Assume, in addition to the hypotheses of Theorem* **5.3**, *that* $f = \sum_{k=0}^{\infty} \sum_{j=1}^{d_k} a_{kj} Y_j^{(k)}$, *where* $|||f|||^2 = \sum_{k=\kappa}^{\infty} \beta_k^4 \sum_{j=1}^{d_k} a_{kj}^2 < \infty$. *Then there exists* $h_0 > 0$ *and* $C > 0$, *both independent of* h, *such that for all* $f \in X_\kappa$,

$$\|f - U_h f\|_\infty \leq C h^4 |||f|||.$$

It is worth noting two points in connection with Corollary **5.10**. Firstly, the sequence $\{\beta_k\}_{k=\kappa}^{\infty}$ has the property that every member has value greater than or equal to unity. The larger the values of this sequence, the smoother the functions in X_κ are. Because $||| \cdot |||$ involves β_k^4, the requirement that $|||f||| < \infty$ is therefore seen to be demanding considerably more smoothness on the function f. In fact, the smoothness needed is, in quite a precise sense, double that which is required for $|f|_\kappa < \infty$. Secondly, the use of this 'error doubling trick' in $\mathbb{R}^d$ usually involves some boundary conditions, which are rather difficult to understand. Of course, the sphere has no boundary, and therefore we encounter no such problems.

There seems little hope that the techniques involved in the proof of Lemmas **5.4** to **5.7** can be generalised. In von Golitschek and Light [9] Theorem **5.3** is shown to hold independent of the value of κ. In conclusion, we should also mention the related paper of Jetter, Stöckler and Ward [11]. This last paper contains a similar result to [9], but the proof appears to be quite different.

References

[1] M. Atteia, *Existence et determination des fonctions "spline" a plusiers varaiables*, C.R. Acad. Sci. Paris **262** (1966), 575-578.

[2] E.W. Cheney, *Approximation using positive definite functions* in *Approximation Theory VII*, C.K. Chui and L.L. Schumaker (eds.) (1995), 145-168.

[3] J. Duchon, *Splines minimizing rotation-invariant seminorms in Sobolev spaces*, in *Constructive Theory of Functions of Several Variables*, Lecture Notes in Mathematics **571**, eds. W. Schempp and K. Zeller, Springer-Verlag (Berlin), 1977, pp. 85-100.

[4] W. Freeden, *On spherical spline interpolation and approximation*, Math. Meth. in Appl. Sci., **3** (1981), 551-575.

[5] W. Freeden, *Spherical spline approximation: basic theory and computational aspects*, J. Comput. Appl. Math., **11** (1984), 367-375.

[6] W. Freeden, *On approximation by harmonic splines*, Manuscr. Geod. **6** (1981), 193-244.

[7] W. Freeden and P. Hermann, *Uniform approximation by spherical spline interpolation*, Math. Z. **193** (1986), 265-275.

[8] W. Freeden, M. Schreiner and R. Franke, *A survey on spherical spline approximation*, Preprint.

[9] M. von Golitschek and W. Light, *Interpolation by polynomials and radial basis functions on spheres*, Constructive Approximation (accepted).

[10] M. Golomb and H.F. Weinberger, *Optimal Approximation and Error Bounds*, in *On Numerical Approximation*, ed. R.E. Langer, University of Wisconsin Press (Madison), 1959, pp. 117-190.

[11] K. Jetter, J. Stöckler and J.D. Ward, *Error estimates for scattered data interpolation on spheres*, Math. Comp. (to appear).

[12] J. Levesley, W.A. Light, D. Ragosin and X. Sun *Variational theory for interpolation on spheres*, Department of Mathematics and Computer Science Technical Report, 1996/21, University of Leicester, England.

[13] W.A. Light and H.S.J. Wayne, *Error estimates for approximation by radial basis functions*, in *Approximation theory, wavelets and applications*, ed. S.P. Singh, Kluwer Academic, Dordrecht, 1995, 215-246.

[14] C.A. Micchelli, *Interpolation of scattered data: distance matrices and conditionally positive definite functions*, Constr. Approx. **2** (1986), 1-22.

[15] C. Müller, *Spherical Harmonics*, Lecture Notes in Mathematics **17**, Springer-Verlag, Berlin, 1966.

[16] M.J.D. Powell, *The uniform convergence of thin plate spline interpolation in two dimensions*, Numerische Math. **68** (1) (1994), 107-128.

[17] W. Rudin, *Functional Analysis*, 2nd ed., McGraw-Hill, New York, 1973.

[18] R. Schaback, *Multivariate interpolation and approximation by translates of a basis function*, in *Approximation theory VIII*, Vol I, eds C.K. Chui and L.L. Schumaker, World Scientific, Singapore, 1995, 491-514.

[19] R. Schaback, *Private Communication*.

[20] H. Shapiro, *Topics in approximation theory*, Springer Lecture Notes in Mathematics, Vol **187**, Springer-Verlag, Berlin, 1981.

[21] I.J. Schoenberg, *Positive definite functions on spheres*, Duke Math. J. **9** (1942), 96-108.

[22] Schreiner, M. *Locally supported kernels for spherical spline interpolation*, J. Approx. Th. **89**(2) (1997), 172-194.

[23] Taijeron, H.J., A.G. Gibson and C. Chandler, *Spline interpolation and smoothing on hyperspheres*, SIAM J. Sci. Comput. **15**(5) (1994), 1111-1125.

[24] G. Wahba, *Spline interpolation and smoothing on the sphere*, SIAM J. Sci. Stat. Comput. **2** (1981), 5-16.

Jeremy Levesley and Will Light
Department of Mathematics and Computer Science
University of Leicester
Leicester LE1 7RH, England

David Ragozin
Department of Mathematics
University of Washington, Box 354350
Seattle, Washington 98195-4350, USA

Xingping Sun
Department of Mathematics
Southwest Missourri State University
Springfield, Missourri, USA

International Series of Numerical Mathematics
Vol. 132, © 1999 Birkhäuser Verlag Basel/Switzerland

Constants in Comonotone Polynomial
Approximation – A Survey

L. Leviatan and I.A. Shevchuk

Abstract

We survey the Jackson and Jackson type estimates for comonotone polynomial approximation of continuous and r-times continuously differentiable
functions which change their monotonicity finitely many times in a finite
interval, say $[-1, 1]$, with special attention to the constants involved in the
estimates. We describe four possibilities ranging from the existence of estimates involving absolute constants and which are valid for polynomials of
all degrees except the very few first ones, through estimates where either the
constants or the degrees of the polynomials depend on the location of the
points of change of monotonicity, then through estimates where either the
constants or the degrees of the polynomials depend on the specific function,
and finally cases where there are no Jackson type estimates possible.

1 Introduction

We are going to survey what is known about the degree of approximation of a continuous function f on a finite interval, say $[-1, 1]$, which changes its monotonicity
finitely many time, say at s fixed points, in that interval, by polynomials that
keep the shape of the function in the sense that they change monotonicity exactly
at these points. Such approximants are said to be comonotone with the function.
We will concentrate on the constants in the Jackson-type estimates which present
new and interesting phenomena. We are seeking estimates with constants C and
N which are independent of the degree n of the polynomials. The constants may
depend on parameters like the number r of derivatives that f possess, the number
s of changes of monotonicity and their locations, and the order k of the various
moduli of smoothness of the function f, or its derivative as the case may be. We
will indicate which parameters the constants depend on and by default this will
mean independence of all others.

We will illustrate the results by truth tables. For completeness we present
also the truth tables for the corresponding estimates in the case of monotone
approximation, most of which are well known and follow from results by Lorentz,
Zeller, DeVore, Shvedov, Yu, Wu, Zhou, Kopotun, Listopad, Dzyubenko, and the
authors, (see [9–12] and the references therein). We will mention the monotone

case only briefly, in Remarks 2.1, 2.11, 2.12 and 2.16, especially in relation to some recent new results.

Since shape constraints limit the degree of approximation, we will relax the constraints of shape in small parts of the interval and see that we can improve the guaranteed degree of approximation. However, this can be done only up to a limit as we demonstrate in Section 3.

2   Truly Comonotone Approximation

Suppose that $f \in \mathbb{C}[-1, 1]$ changes monotonicity at $Y_s : -1 < y_s < \cdots < y_1 < 1$ and assume that in $[y_1, 1]$, the function is nondecreasing. We will denote by $\Delta^{(1)}(Y_s)$, the collection of all such functions. The degree of comonotone approximation by polynomials p_{n-1} of degree not exceeding $n - 1$ will be denoted by

$$E_n^{(1)}(f, Y_s) := \inf_{p_{n-1} \in \Delta^{(1)}(Y_s)} \|f - p_{n-1}\|.$$

Here and below $\| \cdot \|$ denotes the uniform norm over $[-1, 1]$, or if needed, the L_∞-norm.

The first Jackson type estimates for comonotone polynomial approximation were obtained by Newman [15] (see also Iliev [6]) who proved that

$$E_n^{(1)}(f, Y_s) \le C(s)\omega\left(f, \frac{1}{n}\right), \qquad n \ge 1,$$

where $\omega(f, t)$ is the modulus of continuity of f, that is, $\omega(f, t) := \omega_1(f, t)$, where we write

$$\omega_k(f, t) := \sup_{h \in [0, t]} \max_{x \in [-1, 1-kh]} \left| \sum_{j=0}^{k} (-1)^{k-j} \binom{k}{j} f(x + jh) \right|,$$

for the moduli of smoothness of f.

If $f \in \mathbb{C}^1[-1, 1]$, then Beatson and Leviatan [1] showed that

$$E_n^{(1)}(f, Y_s) \le C(s)\frac{1}{n}\omega\left(f', \frac{1}{n}\right), \qquad n \ge 2.$$

At the same time Shvedov [16] showed that it is not possible to obtain the Jackson estimate with higher moduli of smoothness when we just assume that $f \in \mathbb{C}[-1, 1]$ unless the constant depends on Y_s. Namely, he proved that given $s \ge 1$, for any $A > 0$ and each $n \ge 2$, a function $g = g_{A,n}$ which changes monotonicity s times exists, such that if p_n is a polynomial comonotone with g, then

$$\|g - p_n\| > A\omega_2(g, 1).$$

On the other hand he showed that

$$E_n^{(1)}(f, Y_s) \leq C(Y_s)\omega_2\left(f, \frac{1}{n}\right), \qquad n \geq 1.$$

Thus, Gilewicz and Shevchuk [5] have investigated the question when does one have for each $g \in \Delta^{(1)}(Y_s) \cap C^r[-1, 1]$, the estimates,

$$E_n^{(1)}(g, Y_s) \leq \frac{C}{n^r}\omega_k\left(g^{(r)}, \frac{1}{n}\right) \qquad n \geq N. \tag{1}$$

They proved that (1) holds with $C = C(k, r, s)$ and $N = N(k, r, s)$, only when either $k = 1$, or $r > s$, or in the particular case $k = 2$ and $r = s$, moreover in these cases one can always take $N = k + r$. If $k = 2$ and $0 \leq r < s$, or $k = 3$ and $1 \leq r \leq s$, or if $k > 3$ and $2 \leq r \leq s$, then the estimates hold either with $C = C(k, r, Y_s)$ and $N = k + r$, or with $C = C(k, r, s)$ and $N = N(k, r, Y_s)$, and they fail to hold with $C = C(k, r, s)$ and $N = N(k, r, s)$.

In all other cases, namely, when either $r = 0$ and $k \geq 3$, or $r = 1$ and $k > 3$, it is impossible to achieve any such estimate even with constants C and N which depend on g. In fact, Wu and Zhou [17] proved that there is a function $g \in C[-1, 1]$, which changes monotonicity exactly at the y_i's, satisfying

$$\limsup_{n \to \infty} \frac{E_n^{(1)}(g, Y_s)}{\omega_3(g, 1/n)} = \infty,$$

which shows that for $r = 0$ and $k \geq 3$, we cannot have an estimate of type (1), even with C and N depending on g. Dzyubenko, Gilewicz and Shevchuk [4] proved the same for $r = 1$ and $k \geq 4$.

We illustrate the results in the following tables. The sign "+" in entry (k, r) means that (1) is valid for this pair of numbers (and the proper s for which the table is given) with $C = C(k, r, s)$ and $N = k + r$. The sign "⊕" means that (1) is valid with some of the constants depending on the actual location of the points of change of monotonicity, and the estimates are invalid with $C = C(k, r, s)$ and $N = N(k, r, s)$. The sign "−" means that no such estimate exists, even with constants C and N which depend on g. Finally, in rare cases (1) holds with constants C and N, one of which at least depends on g (see Remark 2.1 below). We designate them with the sign "⊖".

r	⋮	⋮	⋮	⋮	⋮	⋰
2	+	+	+	+	+	⋯
1	+	+	+	+	+	⋯
0	+	+	⊖	−	−	⋯
	1	2	3	4	5	k

Fig. 1, $s = 0$ (the monotone case).

r					
3	+	+	+	+	+	$\cdots$
2	+	+	+	+	+	$\cdots$
1	+	+	$\oplus$	$-$	$-$	$\cdots$
0	+	$\oplus$	$-$	$-$	$-$	$\cdots$
	1	2	3	4	5	k

Fig. 2, $s = 1$.

r					
4	+	+	+	+	+	$\cdots$
3	+	+	+	+	+	$\cdots$
2	+	+	$\oplus$	$\oplus$	$\oplus$	$\cdots$
1	+	$\oplus$	$\oplus$	$-$	$-$	$\cdots$
0	+	$\oplus$	$-$	$-$	$-$	$\cdots$
	1	2	3	4	5	k

Fig. 3, $s = 2$.

r					
$s+2$	+	+	+	+	+	$\cdots$
$s+1$	+	+	+	+	+	$\cdots$
s	+	+	$\oplus$	$\oplus$	$\oplus$	$\cdots$
$s-1$	+	$\oplus$	$\oplus$	$\oplus$	$\oplus$	$\cdots$
$\vdots$	$\vdots$	$\vdots$	$\vdots$	$\vdots$	$\vdots$	$\vdots$
2	+	$\oplus$	$\oplus$	$\oplus$	$\oplus$	$\cdots$
1	+	$\oplus$	$\oplus$	$-$	$-$	$\cdots$
0	+	$\oplus$	$-$	$-$	$-$	$\cdots$
	1	2	3	4	5	k

Fig. 4, $s > 2$.

Remark 2.1 *Obviously, in the case of monotone approximation there can be no difference between the signs "+" and "$\oplus$" ($Y_s = \emptyset$). It turns out that there is one more advantage to the monotone case, namely, in Fig. 1, we have a case where "$\ominus$" occurs. That (1) is invalid in that case, if both C and N are absolute constants is due to Shvedov [16]. However, we have recently shown [12] that in that special case, (1) is valid with $C = C(g)$ and $N = 3$, or with an absolute constant C and $N = N(g)$. (See similar phenomena in Fig. 8 below).*

For later reference we note that it follows from the above that if $f \in W_\infty^r$, then

$$E_n^{(1)}(f, Y_s) \leq C(r, s)\frac{\|f^{(r)}\|}{n^r}, \quad n \geq r. \tag{2}$$

It is possible to extend some of these estimates to pointwise estimates. With $\rho_n(x) := \sqrt{1 - x^2}/n + 1/n^2$, Dzyubenko, Gilewicz and Shevchuk [4] investigated the possibility of having a polynomial p_{n-1}, which is comonotone with $g \in \mathbb{C}^r[-1, 1]$, such that

$$|g(x) - p_{n-1}(x)| \leq C[\rho_n(x)]^r \omega_k(g^{(r)}, \rho_n(x)), \quad n \geq N. \tag{3}$$

Again we summarize these results in tables and we compare them to the table of the monotone case. Here too we have a new result, the "−" sign in the case $r = 0$ and $k = 3$ in Fig. 5 (see [12]).

r	⋮	⋮	⋮	⋮	⋮	⋰
2	+	+	+	+	+	⋯
1	+	+	+	+	+	⋯
0	+	+	−	−	−	⋯
	1	2	3	4	5	k

Fig. 5, $s = 0$ (the monotone case).

r	⋮	⋮	⋮	⋮	⋮	⋰
3	+	+	+	+	+	⋯
2	+	+	+	+	+	⋯
1	+	+	⊕	−	−	⋯
0	+	⊕	−	−	−	⋯
	1	2	3	4	5	k

Fig. 6, $s = 1$.

r	⋮	⋮	⋮	⋮	⋮	⋰
3	⊕	⊕	⊕	⊕	⊕	⋯
2	⊕	⊕	⊕	⊕	⊕	⋯
1	⊕	⊕	⊕	−	−	⋯
0	⊕	⊕	−	−	−	⋯
	1	2	3	4	5	k

Fig. 7, $s > 1$.

(Note that Fig. 6 is the same table as in Fig. 2, and note the major difference between the case $s = 1$ and $s > 1$.)

Another way to take into account the different behavior of polynomials near the endpoints of the interval is the Ditzian–Totik moduli of smoothness which are defined in the following way. With $\varphi(x) := \sqrt{1 - x^2}$ and $k \geq 1$, we let

$$\Delta_{h\varphi}^k f(x)$$
$$:= \begin{cases} \sum_{i=0}^k (-1)^{k-i} \binom{k}{i} f(x + (i - \frac{k}{2})h\varphi(x)), & x \pm \frac{k}{2} h\varphi(x) \in [-1, 1] \\ 0, & \text{otherwise.} \end{cases}$$

We define

$$\omega_k^\varphi(f, t) := \sup_{0 < h \leq t} \|\Delta_{h\varphi}^k f\|, \qquad \omega^\varphi(f, t) := \omega_1^\varphi(f, t).$$

Recall that the advantage of these moduli over the ordinary ones is two-fold. First, they are smaller (sometimes by order of magnitude) and second, they lend themselves to inverse results, see (10) below.

Recently we proved,

Theorem 2.2 ([9]) *We have the inequality*

$$E_n^{(1)}(f, Y_s) \leq C(s)\omega^\varphi\left(f, \frac{1}{n}\right), \qquad n \geq 1.$$

Let $\mathbb{B}^r$, $r \geq 1$, be the space of functions $g \in \mathbb{C}[-1, 1]$, which possess a locally absolutely continuous $(r - 1)$st derivative in $(-1, 1)$, such that

$$\|\varphi^r g^{(r)}\| < \infty.$$

Then Theorem 2.2 implies

$$E_n^{(1)}(f, Y_s) \leq C(s)\frac{\|\varphi f'\|}{n}, \qquad n \geq 1,$$

provided $f \in \mathbb{B}^1$.

In view of (2), we begin with the seemingly easier case and prove

Theorem 2.3 ([11]) *If $f \in \mathbb{B}^r$, with either $s = 1$ and $r = 3$, or $r > 2s + 2$, then*

$$E_n^{(1)}(f, Y_s) \leq C(r)\frac{\|\varphi^r f^{(r)}\|}{n^r}, \qquad n \geq r. \tag{4}$$

Unlike (2), in all other cases estimates of type (4) are invalid. We have

Theorem 2.4 ([9]) *Let the constant $A > 0$ be arbitrary and let $2 \leq r \leq 2s + 2$, excluding the case $r = 3$ and $s = 1$. Then for any n, there exists a function $g = g_{r,s,n} \in \mathbb{B}^r$, which changes monotonicity s times in $[-1, 1]$, for which*

$$e_n^{(1,s)}(g) \geq A\|\varphi^r g^{(r)}\|.$$

Here we use the notation

$$e_n^{(1,s)}(g) := \inf E_n^{(1)}(g, Y_s^*),$$

where the infimum is taken over all collections Y_s^*, such that g changes monotonicity at Y_s^*. Since g may be constant in some subinterval (and this is indeed the case for some of the functions $g_{r,s,n}$ of Theorem 2.4), then there may be many admissible Y_s^*'s. **Thus we encounter a new phenomenon**.

Still we can salvage something, namely

Theorem 2.5 ([11]) *If $f \in \mathbb{B}^r$, $r \geq 1$, then*

$$E_n^{(1)}(f, Y_s) \leq C(r, Y_s) \frac{\|\varphi^r f^{(r)}\|}{n^r}, \qquad n \geq r,$$

and

$$E_n^{(1)}(f, Y_s) \leq C(r, s) \frac{\|\varphi^r f^{(r)}\|}{n^r}, \qquad n \geq N(r, Y_s).$$

In an effort to explain the new phenomenon, we go back to estimates involving the Ditzian–Totik moduli. Thus, we are interested in the truth tables for the estimates

$$E_n^{(1)}(g, Y_s) \leq \frac{C}{n^r} \omega_k^\varphi \left(g^{(r)}, \frac{1}{n} \right), \qquad n \geq N, \tag{5}$$

for each $g \in \Delta^{(1)} \cap C^r[-1,1]$. First Kopotun and Leviatan proved,

Theorem 2.6 ([7]) *The inequalities*

$$E_n^{(1)}(f, Y_s) \leq C(Y_s) \omega_2^\varphi \left(f, \frac{1}{n} \right), \qquad n \geq 2,$$

and

$$E_n^{(1)}(f, Y_s) \leq C(s) \omega_2^\varphi \left(f, \frac{1}{n} \right), \qquad n \geq N(Y_s),$$

hold.

At the same time since for all pairs (k, r), $\omega_k^\varphi(g^{(r)}, t) \leq \omega_k(g^{(r)}, t)$, so that any "$-$" sign in tables 1–4, implies a similar one for the D–T tables. Furthermore repeating the arguments in [11] one can prove that the truth tables of (5), coincides with the truth tables of the estimate (1), (see Figs. 1– 4).

On the other hand such truth tables imply Theorem 2.3 for $r = 1$ and Theorem 2.5 for $r = 1$ and $r = 2$, but they do not imply any other case.

Therefore we need a modification of the D–T moduli. We set

$$\varphi_\delta(x) := \sqrt{\left(1 - x - \frac{\delta}{2}\varphi(x)\right)\left(1 + x - \frac{\delta}{2}\varphi(x)\right)}, \qquad x \pm \frac{\delta}{2}\varphi(x) \in [-1,1],$$

and for $g \in \mathbb{C}(-1,1)$, and $r \geq 1$, denote

$$\omega_{k,r}^{\varphi}(g,t) := \sup_{0 \leq h \leq t} \sup_{x} |\varphi_{kh}^{r}(x)\Delta_{h\varphi(x)}^{k}g(x)|, \quad t \geq 0,$$

where the inner supremum is taken over all x so that

$$x \pm \frac{k}{2}h\varphi(x) \in (-1,1).$$

Note that for $g \in \mathbb{C}[-1,1]$,

$$\omega_{k,0}^{\varphi}(g,t) := \omega_{k}^{\varphi}(g,t).$$

We will have estimates involving $\omega_{k,r}^{\varphi}(g^{(r)},t)$, $r > 0$, and in order to guarantee that $\omega_{k,r}^{\varphi}(g^{(r)},t) \to 0$, as $t \to 0$, we need to restrict our functions to the space $\mathbb{C}_{\varphi}^{r}$, namely, those functions in $\mathbb{C}^{r}(-1,1)$ for which $\lim_{x \to \pm 1} \varphi^{r}(x)g^{(r)}(x) = 0$. For completion we put $\mathbb{C}_{\varphi}^{0} := \mathbb{C}[-1,1]$.

For $g \in \mathbb{C}_{\varphi}^{r}$ and $0 \leq p < r$, it can be shown that

$$\omega_{k+r-p,p}^{\varphi}(g^{(p)},t) \leq C(k,r)t^{r-p}\omega_{k,r}^{\varphi}(g^{(r)},t), \quad t \geq 0,$$

and, if $g \in \mathbb{B}^{r}$ and $0 \leq p < r$, then $g \in \mathbb{C}_{\varphi}^{p}$, and

$$\omega_{r-p,p}^{\varphi}(g^{(p)},t) \leq C(r)t^{r-p}\|\varphi^{r}g^{(r)}\|, \quad t \geq 0.$$

While conversely, if $g \in \mathbb{C}_{\varphi}^{p}$ and $\omega_{r-p,p}^{\varphi}(g^{(p)},t) \leq t^{r-p}$, then $g \in \mathbb{B}^{r}$ and

$$\|\varphi^{r}g^{(r)}\| \leq C(r);$$

or if say $g \in \mathbb{C}[-1,1]$, $p < \alpha < k$ and $\omega_{k}^{\varphi}(g,t) \leq t^{\alpha}$, then $g \in \mathbb{C}_{\varphi}^{p}$ and

$$\omega_{k-p,p}^{\varphi}(g^{(p)},t) \leq C(\alpha,k)t^{\alpha-p}, \quad t \geq 0. \tag{6}$$

The results we have for these moduli are

Theorem 2.7 ([11]) *If $f \in \mathbb{C}_{\varphi}^{r}$, with $r > 2s+2$, then*

$$E_{n}^{(1)}(f,Y_{s}) \leq \frac{C(k,r,s)}{n^{r}}\omega_{k,r}^{\varphi}\left(f^{(r)},\frac{1}{n}\right), \quad n \geq k+r.$$

To the contrary we have,

Theorem 2.8 ([11]) *Let $r \leq 2s+2$, excluding the case $r = 0$ and $k = 1$. Then for any constant $A > 0$ and every $n \geq 1$, there is function $g := g_{k,r,s,n,A} \in \mathbb{C}_{\varphi}^{r}$ which changes monotonicity s times in $[-1,1]$, for which*

$$e_{n}^{(1,s)}(g) > A\omega_{k,r}^{\varphi}(g^{(r)},1).$$

Again we can salvage something, namely,

Theorem 2.9 ([11]) *If* $f \in C_\varphi^r$, *with* $r > 2$, *then*

$$E_n^{(1)}(f, Y_s) \leq \frac{C(k, r, Y_s)}{n^r} \omega_{k,r}^\varphi \left(f^{(r)}, \frac{1}{n} \right), \quad n \geq k + r,$$

and

$$E_n^{(1)}(f, Y_s) \leq \frac{C(k, r, s)}{n^r} \omega_{k,r}^\varphi \left(f^{(r)}, \frac{1}{n} \right), \quad n \geq N(k, r, Y_s).$$

We wish to point out that Theorem 2.9 is also valid for $r = 2$ and $k = 1$, and for $r = 1$ and $k = 2$, (see [13]). On the other hand, the following negative result holds true.

Theorem 2.10 ([13]) *There is a function* $g \in C_\varphi^2$, *which changes monotonicity at the* y_i *'s, satisfying*

$$\limsup_{n \to \infty} \frac{n^2 E_n^{(1)}(g, Y_s)}{\omega_{3,2}^\varphi(g'', 1/n)} = \infty.$$

Remark 2.11 *There is even a function* $g \in C^2[-1, 1]$, *which changes monotonicity at the* y_i *'s, for which Theorem 2.10 is true. Moreover, the proof in* [13] *provides a corresponding result for monotone approximation as well.*

We are now in a position to summarize in truth tables what is known about the estimates

$$E_n^{(1)}(g, Y_s) \leq \frac{C}{n^r} \omega_{k,r}^\varphi \left(g^{(r)}, \frac{1}{n} \right), \quad n \geq N, \tag{7}$$

for every $g \in \Delta^{(1)}(Y_s) \cap C_\varphi^r$. It is natural to combine these tables with the truth tables for the estimates

$$E_n^{(1)}(g, Y_s) \leq \frac{C}{n^r} \| \varphi^r g^{(r)} \|, \quad n \geq N, \tag{8}$$

for every $g \in \Delta^{(1)}(Y_s) \cap \mathbb{B}^r$. Thus, in the tables below the column $k = 0$ is related to (8), whereas the other columns are related to (7).

r	⋮	⋮	⋮	⋮	⋮	⋮	⋰
4	+	+	+	+	+	+	⋯
3	+	+	+	+	+	+	⋯
2	+	⊖	?	−	−	−	⋯
1	+	+	⊖	?	−	−	⋯
0		+	+	⊖	−	−	⋯
	0	1	2	3	4	5	k

Fig. 8, $s = 0$ (the monotone case).

r							
	$\vdots$	$\vdots$	$\vdots$	$\vdots$	$\vdots$	$\vdots$	$\cdot\,\cdot^{\cdot}$
6	+	+	+	+	+	+	$\cdots$
5	+	+	+	+	+	+	$\cdots$
4	⊕	⊕	⊕	⊕	⊕	⊕	$\cdots$
3	+	⊕	⊕	⊕	⊕	⊕	$\cdots$
2	⊕	⊕	?	−	−	−	$\cdots$
1	+	⊕	⊕	?	−	−	$\cdots$
0		+	⊕	−	−	−	$\cdots$
	0	1	2	3	4	5	k

Fig. 9, $s = 1$.

r							
	$\vdots$	$\vdots$	$\vdots$	$\vdots$	$\vdots$	$\vdots$	$\cdot\,\cdot^{\cdot}$
$2s+4$	+	+	+	+	+	+	$\cdots$
$2s+3$	+	+	+	+	+	+	$\cdots$
$2s+2$	⊕	⊕	⊕	⊕	⊕	⊕	$\cdots$
$\vdots$	$\vdots$	$\vdots$	$\vdots$	$\vdots$	$\vdots$	$\vdots$	$\vdots$
3	⊕	⊕	⊕	⊕	⊕	⊕	$\cdots$
2	⊕	⊕	?	−	−	−	$\cdots$
1	+	⊕	⊕	?	−	−	$\cdots$
0		+	⊕	−	−	−	$\cdots$
	0	1	2	3	4	5	k

Fig. 10, $s > 1$.

Remark 2.12 *As we aluded to in Remark* 2.1, *in Fig.* 8 *we have several cases of "⊖". That* (7) *is invalid for these cases with absolute constants C and N, is due to Kopotun and Listopad* [8]. *We* [12] *have shown that these are "⊖" cases. Note that in all Figs.* 8–10 *we still have outstanding cases for which it is not known if we should write "−" or "⊖" (in Figs.* 8–9) *and "−", "⊖" or "⊕" (in Fig.* 10). *That is, we do not know whether Theorem* 2.10 *is valid when we replace* $w^{\varphi}_{3,2}$ *by* $w^{\varphi}_{2,2}$.

If we denote, as usual,

$$E_n(f) := \inf_{p_{n-1}} \|f - p_{n-1}\|$$

the degree of the best uniform approximation of f, then obviously,

$$E_n(f) \le E_n^{(1)}(f, Y_s). \tag{9}$$

Cases "+" and "⊕" together with (6) and the Ditzian-Totik [3] inverse theorem

$$\omega_k^{\varphi}(f, t) \le C(k) t^k \sum_{n=k}^{[1/t]} n^{k-1} E_n(f), \quad 0 < t \le \frac{1}{k}, \tag{10}$$

provide a partial inverse to (9), analogous to Theorems 2.3 and 2.5, (except for the case $2s + 2 < \alpha \leq 2s + 3$ and the case $s = 1$ and $2 < \alpha < 3$ in Theorem 2.13 and $\alpha = 3$ in Theorem 2.14, which we have to prove separately), namely,

Theorem 2.13 ([11]) *Assume that either $0 < \alpha < 1$, or $s = 1$ and $2 < \alpha < 3$, or $\alpha > 2s + 2$. Then, if*

$$E_n(f) < \frac{1}{n^\alpha} , \quad \forall n > \alpha, \tag{11}$$

then

$$E_n^{(1)}(f, Y_s) < \frac{C(\alpha, s)}{n^\alpha} , \quad \forall n > \alpha.$$

and

Theorem 2.14 ([11]) *Assume that $\alpha > 0$. Then, if (11) holds, then*

$$E_n^{(1)}(f, Y_s) < \frac{C(\alpha, Y_s)}{n^\alpha} , \quad \forall n > \alpha,$$

and

$$E_n^{(1)}(f, Y_s) < \frac{C(\alpha, s)}{n^\alpha} , \quad \forall n > N(\alpha, Y_s).$$

We note that it is impossible to make these constants independent of Y_s. Indeed we have

Theorem 2.15 ([11]) *Let $1 \leq \alpha \leq 2s + 2$, excluding the case $s = 1$ and $2 < \alpha < 3$. Then for any constant $B > 0$ and each $n > \alpha$, there exists a function $g := g_{s,\alpha,n,B}$ which changes monotonicity s times in $[-1, 1]$ and for which we simultaneously have,*

$$E_m(g) \leq \frac{1}{m^\alpha}, \quad \forall m > \alpha$$

and

$$e_n^{(1,s)}(g) \geq B.$$

Remark 2.16 *A corresponding result to Theorem 2.13 in monotone approxima-tion is valid for all $\alpha > 0$, except for the exceptional case $\alpha = 2$. The latter is of the type "⊖" (see [8] and [12]).*

3 Nearly Comonotone Approximation

We are going to give up the monotonicity constraint in the neighborhood of the points Y_s and near the endpoints of the interval. (Recall that $f \in \Delta^{(1)}(Y_s)$.) This will result in improved order of approximation, but only up to a certain extent. To this end, we write the Chebyshev nodes $x_j := x_{n,j} := \cos \pi j / n$, $j = 0, \ldots, n$ and put $x_{-1} := x_0$, $x_{n+1} := x_n$. Given Y_s, if $x_{j+1} \leq y_k < x_j$, then we delete the neighborhood (x_{j+2}, x_{j-1}). In some cases we may have to delete also small neighborhoods of the end points. Namely, we set

$$O(n, Y_s) := \bigcup_{j:[x_{j+1},x_j) \cap Y_s \neq \emptyset} (x_{j+2}, x_{j-1}),$$

and

$$O^*(n, Y_s) := O(n, Y_s) \cup [-1, -1 + 1/n^2] \cup [1 - 1/n^2, 1].$$

First we have

Theorem 3.1 ([10]) *There are constants $c = c(s, k)$ and $C = C(s, k)$ for which, if $f \in C^1[-1, 1]$, then for each $n \geq k$, a polynomial p_n of degree not exceeding n, which is comonotone with f on $[-1, 1] \setminus O([n/c], Y)$ (i.e.,*

$$p_n'(x) \prod_{i=1}^{s} (x - y_i) \geq 0, \quad x \in [-1, 1] \setminus O([n/c], Y)),$$

exists such that

$$|f(x) - p_n(x)| \leq C(s, k) \, \rho_n(x) \omega_k(f', \rho_n(x)), \quad x \in [-1, 1].$$

Also,

Theorem 3.2 ([10]) *There are constants $c = c(s, k)$ and $C(s, k)$ such that for every $n \geq 2$, a polynomial p_n of degree not exceeding n, which is comonotone with f on $[-1, 1] \setminus O^*([n/c], Y_s)$ exists, such that*

$$|f(x) - p_n(x)| \leq C(s) \, \omega_3(f, \rho_n(x)).$$

Finally, the party is over when we try to get ω_4. Indeed, in [2] we prove that this is impossible. In order to state our theorem we need some notation. Given $\epsilon > 0$ and a nondecreasing function $g \in C[-1, 1]$, we denote

$$E_n^{(1)}(g; \epsilon) := \inf_{p_{n-1}} \|g - p_{n-1}\|,$$

where the infimum is taken over all polynomials p_{n-1} of degree not exceeding $n - 1$ satisfying

$$\text{meas}(\{x : p_{n-1}'(x) \geq 0\} \cap [-1, 1]) \geq 2 - \epsilon.$$

Then

Theorem 3.3 ([2]) *For each sequence $\bar{\epsilon} = \{\epsilon_n\}_{n=1}^{\infty}$, of nonnegative numbers tending to 0, there exists a monotone function $g := g_{\bar{\epsilon}} \in C[-1, 1]$, such that*

$$\limsup_{n \to \infty} \frac{E_n^{(1)}(g; \epsilon_n)}{\omega_4(g, 1/n)} = \infty.$$

This result is extended to the comonotone case in [14]. Thus we can summarize these results in a truth table, which is the same for all $s \geq 0$, namely,

$$
\begin{array}{c|cccccc}
r & \vdots & \vdots & \vdots & \vdots & \vdots & \cdot^{\cdot^{\cdot}} \\
2 & + & + & + & + & + & \cdots \\
1 & + & + & + & + & + & \cdots \\
0 & + & + & + & - & - & \cdots \\
\hline
& 1 & 2 & 3 & 4 & 5 & k
\end{array}
$$

Fig. 11, All s and all types of estimates.

Remark 3.4 *Theorems 3.1 and 3.2 give pointwise estimates for nearly comonotone approximation, involving the ordinary moduli of smoothness. However, we have the same truth tables for estimates involving all other types of moduli considered in this paper if we relax the comonotonicity requirements near the points of monotonicity change and near the end points. One should emphasize though, that sometimes in the analogs of Theorem 3.1 we have to replace O with O^*. It should also be noted, that the constant c in Theorems 3.1 and 3.2 cannot be too small (see [11]), whereas in their analogs this is not essential [14].*

References

[1] R. K. Beatson and D. Leviatan, *On comonotone approximation*, Canadian Math. Bull. **26** (1983), 220–224.

[2] R. A. DeVore, D. Leviatan and I. A. Shevchuk, *Approximation of monotone functions: a counter example*, In Curves and Surfaces with Applications in CAGD, Proceedings of the Chamonix Conference, 1996, A. Le Méhauté, C. Rabut and L. L. Schumaker (eds.), Vanderbilt University Press, Nashville TN, 1997, pages 95-102.

[3] Z. Ditzian and V. Totik, *Moduli of smoothness*, Springer series in Computational Mathematics, Springer-Verlag, New York, 1987.

[4] G. A. Dzyubenko, J. Gilewicz and I. A. Shevchuk, *Piecewise monotone pointwise approximation*, Constr. Approx. **14** (1998), 311–348.

[5] J. Gilewicz and I. A. Shevchuk, *Comonotone approximation*, Fundamentalnaya i Prikladnaya Mathematica, **2** (1996), 319–363 (in Russian).

[6] G. L. Iliev, *Exact estimates for partially monotone approximation*, Analysis Math. **4** (1978), 181–197.

[7] K. Kopotun and D. Leviatan, *Comonotone polynomial approximation in $L_p[-1,1]$, $0 < p \leq \infty$*, Acta Math. Hungarica, **77** (1997), 279–288.

[8] Kopotun, K. A. and V. V. Listopad, *Remarks on monotone and convex approximation by algebraic polynomials*, Ukrainian Math. J., **46** (1994), 1266–1270.

[9] D. Leviatan and I. A. Shevchuk, *Some positive results and counterexamples in comonotone approximation*, J. Approx. Theory, **89** (1997), 195–206.

[10] D. Leviatan and I. A. Shevchuk, *Nearly comonotone approximation*, J. Approx. Theory **95** (1998), 53–81.

[11] D. Leviatan and I. A. Shevchuk, *Some positive results and counterexamples in comonotone approximation II*, J. Approx. Theory, to appear.

[12] D. Leviatan and I. A. Shevchuk, *Monotone approximation estimates involving the third modulus of smoothness*, Proceedings of the Ninth International Conference on Approx. Theory, Nashville TN 1998, Charles Chui and Larry Schumaker (eds.), Vanderbilt Univ. Press, to appear.

[13] D. Leviatan and I. A. Shevchuk, *More on comonotone polynomial approximation*, Constr. Approx., to appear.

[14] D. Leviatan and I. A. Shevchuk, *Nearly comonotone approximation II*, Acta Sci. Math. (Szeged), to appear.

[15] D. J. Newman, *Efficient comonotone approximation*, J. Approx. Theory, **25** (1979), 189–192.

[16] A. S. Shvedov, *Orders of coapproximation of functions by algebraic polynomials*, Mat. Zametki, **29** (1981), 117-130, English transl. in Math. Notes **29** (1981), 63–70.

[17] X. Wu and S. P. Zhou, *A counterexample in comonotone approximation in L^p space*, Colloq. Math., **114** (1993), 265–274.

D. Leviatan
School of Mathematical Sciences
Sackler Faculty of Exact Sciences
Tel Aviv University
Tel Aviv 69978, Israel
Email address: leviatan@math.tau.ac.il

I. A. Shevchuk
Institute of Mathematics
National Academy of Sciences of Ukraine
Kyiv 252601
Ukraine
Email address: shevchuk@imath.kiev.ua

International Series of Numerical Mathematics
Vol. 132, © 1999 Birkhäuser Verlag Basel/Switzerland

Will Ramanujan kill Baker-Gammel-Wills?
(A Selective Survey of Padé Approximation)

D.S. Lubinsky

Abstract

Every formal power series

$$f(z) = \sum_{j=0}^{\infty} a_j z^j$$

admits a formal continued fraction expansion

$$f(z) = c_0 + \frac{c_1 z^{\ell_1}|}{|1} + \frac{c_2 z^{\ell_2}|}{|1} + \frac{c_3 z^{\ell_3}|}{|1} + \cdots$$

where each $c_j \in \mathbf{C}$ and ℓ_j is a positive integer. The truncations (or *convergents*) of this continued fraction are also Padé approximants. A 1961 conjecture of Baker–Gammel–Wills asserts that if f is analytic in $|z| < 1$ except for poles, then a subsequence of the convergents to its continued fraction converges uniformly in compact subsets of $|z| < 1$ omitting poles. We discuss the status of this conjecture and briefly review Padé convergence theory. Moreover, we present evidence that Ramanujan's continued fraction

$$1 + \frac{qz|}{|1} + \frac{q^2 z|}{|1} + \frac{q^3 z|}{|1} + \cdots$$

with suitable q on the unit circle, provides a counterexample.

1 Introduction

Let

$$f(z) = \sum_{j=0}^{\infty} a_j z^j \tag{1}$$

be a formal power series with complex coefficients $\{a_j\}_{j=0}^{\infty}$. We may expand f in a formal continued or C–fraction, in the following way: write

$$f(z) = c_0 + \frac{c_1 z}{1 + z g(z)},$$

where g is also a formal power series. By multiplying both sides by $1 + zg(z)$ and equating coefficients of powers of z, we obtain infinitely many linear equations involving the power series coefficients of f and g. It turns out that this system is triangular, and we can successively solve for the constant coefficient of g, then the coefficient of z, and so on. Of course, this is almost the same as finding the reciprocal power series of a given formal power series.

Well, what we did to f, we can also do to g, and so we obtain

$$f(z) = c_0 + \cfrac{c_1 z}{1 + \cfrac{c_2 z}{1 + c_3 h(z)}},$$

where h is also a formal power series. Continuing in this way, we obtain an infinite formal continued fraction expansion

$$f(z) = c_0 + \cfrac{c_1 z}{1 + \cfrac{c_2 z}{1 + \cfrac{c_3 z}{1 + \ddots}}}. \qquad (2)$$

Actually, we have simplified issues a little: in general this process may yield higher powers of z, so that we really should write

$$f(z) = c_0 + \cfrac{c_1 z^{\ell_1}}{1 + \cfrac{c_2 z^{\ell_2}}{1 + \cfrac{c_3 z^{\ell_3}}{1 + \ddots}}} \qquad (3)$$

where $\ell_1, \ell_2, \ell_3, \ldots$ are positive integers. While this is the general case, (2) is the so–called *normal* case. It takes quite a lot of space to write (3); so we introduce the notation

$$f(z) = c_0 + \frac{c_1 z^{\ell_1}|}{|1} + \frac{c_2 z^{\ell_2}|}{|1} + \frac{c_3 z^{\ell_3}|}{|1} + \cdots. \qquad (4)$$

To give analytic meaning to the continued fraction, we recall that a series is defined as the limit of its partial sums, when that limit exists. In much the same way, we may truncate the continued fraction (3) to obtain the *nth convergent*

$$\frac{\mu_n}{\nu_n}(z) := c_0 + \frac{c_1 z^{\ell_1}|}{|1} + \frac{c_2 z^{\ell_2}|}{|1} + \frac{c_3 z^{\ell_3}|}{|1} + \cdots + \frac{c_n z^{\ell_n}|}{|1}.$$

We define the value of the continued fraction to be

$$\lim_{n \to \infty} \frac{\mu_n}{\nu_n}(z)$$

when that limit exists.

At first the notation $\frac{\mu_n}{\nu_n}$ for the nth convergent may seem curious, but it is suggestive of the fact that it is a rational function of z. We thus may assume that μ_n and ν_n are the numerator and denominator polynomials in this rational function, suitably normalized. In the normal case (2), one may show that μ_n and ν_n have degree approximately $n/2$ (the error in approximation is at most 1). In general the degrees will obviously depend on the size of ℓ_j, $1 \leq j \leq n$.

The convergents are also *Padé approximants* to the power series f:

$$(f\nu_n - \mu_n)(z) = O\left(z^{\deg(\mu_n)+\deg(\nu_n)+1}\right).$$

Here deg obviously indicates the degree of the polynomials, and the order relation indicates that the coefficients of $1, z, z^2, \ldots, z^{\deg(\mu_n)+\deg(\nu_n)}$ on the left–hand side vanish. More generally, given a power series (1), and integers $m, n \geq 0$, the $[m/n]$ *Padé approximant* to f is a rational function $[m/n] = P/Q$, where P, Q have degree $\leq m, n$ respectively, Q is not identically 0, and

$$(fQ - P)(z) = O(z^{m+n+1}). \tag{5}$$

It turns out that this last relation generates a system of linear equations for the coefficients of P and Q. This system can be solved, and while P and Q are not separately unique, the ratio P/Q is unique.

Henri Eugene Padé, a student of C. Hermite, arranged the approximants into a table

$$
\begin{array}{cccccc}
[0/0] & [0/1] & [0/2] & [0/3] & [0/4] & \cdots \\
[1/0] & [1/1] & [1/2] & [1/3] & [1/4] & \cdots \\
[2/0] & [2/1] & [2/2] & [2/3] & [2/4] & \cdots \\
[3/0] & [3/1] & [3/2] & [3/3] & [3/4] & \cdots \quad \cdot \\
[4/0] & [4/1] & [4/2] & [4/3] & [4/4] & \cdots \\
\vdots & \vdots & \vdots & \vdots & \vdots & \ddots
\end{array}
$$

The sequence of convergents $\{\mu_n/\nu_n\}_{n=1}^{\infty}$ to a normal continued fraction (2) occupies the stair step sequence

$$[0/0], [1/0], [1/1], [2/1], [2/2], [3/2], [3/3], [4/3], \ldots$$

in the above table (it is instructive to draw in arrows on the above table to verify the stair step). For a general continued fraction (3), the convergents still "stair step" down the table, but the steps may have size larger than 1. These bigger steps correspond to square blocks of equal approximants in the table. The structure of the Padé table has been thoroughly investigated [5]. However, there is still a nice unsolved problem as to what patterns of square blocks, or "square tilings of the plane" may arise for arbitrary f [33].

2   Some Convergence Theorems

In this brief survey, we shall focus on the convergence properties of the sequence $\{\mu_n/\nu_n\}_{n=1}^{\infty}$, but viewed as Padé approximants. There is, roughly speaking, a distinction between the philosophy of those that regard themselves as researchers into continued fractions, and those that view themselves as Padé approximators. The former tend to place hypotheses on the continued fraction coefficients $\{c_j\}$ and then investigate convergence; the latter have tended to place hypotheses on the analytic nature of the underlying function f. Both approaches have yielded important and beautiful results. In adopting the latter approach here, we are not underestimating the value of the former – indeed for a vast array of special functions, the $\{c_j\}$ are known explicitly, and investigation of the convergence of continued fractions under hypotheses on the $\{c_j\}$ has enriched function theory, special functions, numerical analysis, ... [17], [19].

From a Padé view, another serious omission is that we consider only the stair–step sequence $\{\mu_n/\nu_n\}_{n=1}^{\infty}$. For convergence results along other paths in the Padé table, the reader may consult [2], [5]. We shall also not discuss the important "converse" results obtained by the Russian school led by A.A. Goncar.

The complexity of the convergence theory is best illustrated by two examples:

Example 2.1: A good example For $z \in \mathbf{C} \backslash \left(-\infty, -\frac{1}{4}\right]$,

$$1 + \frac{z|}{|1} + \frac{z|}{|1} + \frac{z|}{|1} + \ldots = \frac{1}{2} + \sqrt{z + \frac{1}{4}}. \tag{6}$$

We see that the continued fraction on the left has an exceptionally simple form, with all $c_j = 1$. Indeed, if it converges to a value $f(z)$, then the repeating (or periodic) nature of the terms indicates that

$$1 + \frac{z}{f(z)} = f(z).$$

This gives a quadratic equation for $f(z)$, whose solution is the right–hand side of (6). There are several interesting features:

- The continued fraction converges throughout the cut plane $\mathbf{C} \backslash \left(-\infty, -\frac{1}{4}\right]$, whereas the Maclaurin series of the function converges only in the ball centre 0, radius $\frac{1}{4}$. This shows that continued fractions/Padé approximants can be used for analytic continuation.

- There are infinitely many ways to cut the plane from $-\frac{1}{4}$ to ∞ to obtain a single valued square root, but the continued fraction chooses the "most natural" branch cut, along the negative real axis. In this example, which is a so–called Stieltjes series [2], [5], all the zeros and poles of the approximants even lie along this cut. This feature of choosing the "best" branchcut, occurs very generally [31].

- The convergents μ_n/ν_n converge with geometric rate: given any $z \in \mathbf{C} \backslash \left(-\infty, -\frac{1}{4}\right]$, the error $f - \frac{\mu_n}{\nu_n}$ decays to 0 like $\rho(z)^n$ for some $\rho(z) \in (0, 1)$.

This simple example illustrates why Padé approximants became so popular in problems involving series expansions in the 1960's. Even where the underlying Maclaurin series has zero radius of convergence, one could form Padé approximants/continued fractions and these would converge in a suitably cut plane, for example, $\mathbf{C}\backslash(-\infty, 0]$. Moreover, the approximants converge rapidly when they converge. Their poles may also be used to predict the location of the singularities of the underlying function.

Now for the bad news:

Example 2.2: A bad example In the early 1970's Hans Wallin of Umeå University, constructed [34] an entire function, such that when we form the sequence $\{\mu_n/\nu_n\}_{n=1}^{\infty}$, it diverges at each point of the plane other than 0:

$$\limsup_{n\to\infty} |\mu_n/\nu_n|(z) = \infty, \ z \in \mathbf{C}\backslash\{0\}. \tag{7}$$

Thus despite the Maclaurin series converging everywhere, the continued fraction diverges except at 0.

While there were earlier examples of pathological behaviour due to Dumas, Perron and others, Wallin's example is the most impressive. The divergence in Wallin's example occurs because the convergents have *spurious poles*: poles that do not reflect the analytic properties of the underlying function. Thus although f is analytic at $z = 1$, Wallin's construction allows μ_n/ν_n to have a pole at $z = 1$ for infinitely many n. Indeed, one may replace 1 by any countable set, and in addition the denominators ν_n will be small on a set determined in advance.

The contrast between Examples 2.1 and 2.2 portrays a little of the complexity of the convergence theory of continued fractions/Padé approximants. As a very rough rule, it seems that the latter converge well, when the power series coefficients are "smooth" in some sense. Most special functions are "smooth" in this sense, and hence the convergence theory for their continued fractions is well understood, and has impressive positive results. In Wallin's example, the Maclaurin series coefficients do not behave "smoothly". Nevertheless, note that there was a lim sup in (7), and not a limit. It turns out that although some subsequence of $\{\mu_n/\nu_n\}_{n=1}^{\infty}$ diverges, some other subsequence converges uniformly in compact subsets of the plane.

This helps to motivate the *Baker–Gammel–Wills, or Padé, conjecture*, published in 1961 [3], and still unsolved. The prime mover in this conjecture was George Baker of Los–Alamos Scientific Laboratory, who in the 1960's and 1970's, applied Padé approximants to solve numerous problems in physics, while also contributing heavily to the theoretical development of the approximants. His team would compute, for example, the first 50 convergents μ_n/ν_n from a given series, and would observe that, say, 45 of the approximants are "good" and 5 are "bad". However the 5 bad ones would not necessarily be the first 5: they can appear anywhere in the 50. Once these were omitted, there would be a definite convergence trend, and physicists, being braver than numerical analysts, let alone pure mathematicians, were prepared to chuck out the bad ones!

Based on their numerical experience, they formulated several forms of the famous Baker–Gammel–Wills Conjecture [3], [5]. (Incidentally Wills was a vacation student, doing the programming. How many researchers would be as generous as George Baker in including the programmer amongst the co–authors on an important paper?) Here we present only one form:

Baker–Gammel–Wills Conjecture *Let $f(z)$ be analytic in $|z| < 1$, except for poles, none at 0. There exists an infinite subsequence $\{\mu_n/\nu_n\}_{n\in S}$ of $\{\mu_n/\nu_n\}_{n=1}^{\infty}$ that converges away from the poles of f. More precisely, uniformly in compact subsets of $|z| < 1$ omitting poles of f, we have*

$$\lim_{\substack{n\to\infty \\ n\in S}} \mu_n/\nu_n(z) = f(z).$$

Note that we require f to be analytic at 0 so that we may form a Maclaurin series and continued fraction. We shall use the abbreviation BGW for the conjecture.

The conjecture is widely believed to be false in the above form, though possibly true for functions that are meromorphic in the whole plane, that is analytic except for poles. Credence for this modification is provided by a related easier conjecture for the columns of the Padé table, the so–called Baker–Graves–Morris Conjecture [4]. It turned out to be true for functions meromorphic in the whole plane, but false for functions meromorphic in the unit ball [6].

The main difficulty in proving the conjecture is to construct a subsequence $\{\mu_n/\nu_n\}_{n\in S}$ without spurious poles: thus one needs to ensure that if f is analytic in some closed ball B, then for large $n \in S$, μ_n/ν_n has no poles in the interior of B.

Even if the conjecture is proved in full generality, it is somewhat depressing that its conclusion refers only to subsequences. Can anything at all be said about the convergence of the full sequence $\{\mu_n/\nu_n\}_{n=1}^{\infty}$? Well, there are numerous types of convergence concepts available, several of which one would learn in a first course on measure theory: pointwise convergence, convergence almost everywhere, convergence in norm, and *convergence in measure*. It was the Canadian mathematical physicist John Nuttall who realized [27] in the late 1960's that the latter concept is the correct one: for a very large class of functions, continued fractions/Padé approximants converge in measure:

Theorem 2.1 Nuttall's Theorem, 1970 *Let f be meromorphic in $\mathbf{C}$ and analytic at 0. Fix $r, \varepsilon > 0$. Then*

$$meas\left\{z: \ |z| \leq r \ and \ \left|f - \frac{\mu_n}{\nu_n}\right|(z) > \varepsilon\right\} \to 0, \ n \to \infty. \tag{8}$$

Here meas denotes planar Lebesgue measure, that is area.

Thus the area of the set on which the error in approximation by μ_n/ν_n is larger than ε, shrinks to 0, as $n \to \infty$. At last, a positive statement that applies very generally! It was not long before the theorem was generalized and sharpened in several directions, the most important one due to the reknowned function theorist C. Pommerenke. Following is a special case of Pommerenke's Theorem [28]:

Theorem 2.2 Pommerenke's Theorem, 1973 *Let f be analytic at 0 and in $\mathbf{C}\backslash\mathcal{E}$, where $cap\,(\mathcal{E}) = 0$. Fix $r, \varepsilon > 0$. Then*

$$cap\left\{z : |z| \le r \text{ and } \left|f - \frac{\mu_n}{\nu_n}\right|(z) > \varepsilon^n\right\} \to 0, \ n \to \infty. \tag{9}$$

Here cap denotes logarithmic capacity.

Before discussing logarithmic capacity, it is instructive to point out the main differences between the theorems of Nuttall and Pommerenke: Pommerenke's theorem allows any countable set of singularities, including poles or essential singularities, while Nuttall's theorem allows only the former. Thus, the latter treats $\exp\left(\frac{1}{1-z}\right)$ but Nuttall's does not. In addition, there is *geometric* convergence in capacity in Pommerenke's theorem – we have ε^n rather than ε. It also turns out that *cap* is the "natural" set function in theorems of this type.

Logarithmic capacity is a remarkable set function that admits at least three strikingly different definitions [15]. The simplest definition for compact $E \subset \mathbf{C}$ is via the Chebyshev constant:

$$cap(E) := \lim_{n\to\infty} \left(\min_{P(z)=z^n+\cdots} \|P\|_{L_\infty(E)}\right)^{1/n}. \tag{10}$$

Here the minimum is taken over all monic polynomials P of degree n, and the minimizing polynomial is called the nth Chebyshev polynomial for E. In the case $E = [-1, 1]$, the minimizing polynomial is indeed $T_n(x)/2^{n-1}$, the classical Chebyshev polynomials, normalized to be monic. The proof that the limit in (10) exists is fairly straightforward. One may extend the definition to general sets F as an inner capacity, in much the same way as one defines an inner measure:

$$cap(F) := \sup\{cap(E) : E \subset F, E \text{ compact}\}.$$

Note however that *cap* is not even finitely additive (or subadditive), so is not a measure in the usual sense.

Some examples:

$$
\begin{aligned}
cap\,[a, b] &= \frac{1}{4}(b - a)\,; \\
cap\,\{z : |z| \le r\} &= cap\,\{z : |z| = r\} = r\,; \\
meas(E) &\le \pi\,(cap(E))^2\,; \\
cap(E) &= 0 \Rightarrow meas(E) = 0.
\end{aligned}
$$

In fact, the zero sets of *cap* are thin indeed: not only do they have area 0, but their intersection with every line has linear measure 0, and they have Hausdorff dimension zero. While any countable set has *cap* 0, there are uncountable sets with this property.

In complex function theory, *cap* is more natural than any measure. One illustration of this is in removability of singularities. Recall that if f is bounded and analytic in $\{z : |z| < 1\}\backslash\{0\}$, then 0 is a removable singularity for f, that is, f may be analytically continued to 0. What happens if we replace $\{0\}$ by an arbitrary set $E \subset \{z : |z| < 1\}$? It turns out [12, pp.318–320] that every function f bounded and harmonic in $\{z : |z| < 1\}\backslash E$ has an harmonic continuation to all of the unit ball iff $cap(E) = 0$.

There are several reasons why *cap* is the correct set function to use in the context of Padé approximation. If we replace *cap* by some set function σ in (9), and insist that (9) holds only for entire f, then it follows [20] for compact E that

$$cap(E) = 0 \Rightarrow \sigma(E) = 0.$$

Thus *cap* is the set function with the "thinnest" zero sets for which (9) holds. Secondly if f has singularities of positive capacity, then the conclusion of Theorem 1.4 need not hold: the author and E.A. Rahmanov [20], [21], [29] constructed functions analytic in $|z| < 1$ but not on the unit circle, for which the sequence $\{\mu_n/\nu_n\}$ does not converge in measure in any open ball in $|z| < 1$. Of course, for these functions, BGW is still true, as the pathological behaviour occurs only in a subsequence.

The Nuttall–Pommerenke theorem has been generalized in many directions: Padé approximants may be replaced by rational functions that interpolate at an array of points; the function f may more generally lie in the Goncar–Walsh class – see [13], [18]. However all the latter results admit only functions that are single valued in their Weierstrass domain of analytic continuation. In particular, functions with branchpoints such as $\log(1+z)$, or the $\frac{1}{2}+\sqrt{\frac{1}{4}+z}$ of Example 2.1 are omitted.

A thorough treatment of the latter has been given by H. Stahl [30], [31], building to some extent on earlier work of Nuttall. One of the initially surprising conclusions is that from the viewpoint of approximation by rational functions, branchpoints are very strong singularities: the Padé approximants/convergents will pack most of their poles along a very definite path determined by the locations of the branch points; the form of this path minimizes the *cap* of a certain set outside which f is analytic. The domain in which we obtain convergence in capacity is thus extremal in a certain sense:

Theorem 2.3 Stahl's Theorem 1980's   *Let f be analytic at 0 and analytic (but not necessarily single valued) in $\mathbf{C}\backslash\mathcal{E}$, where $cap(\mathcal{E}) = 0$. There is an extremal domain $\mathcal{D}$ determined by the set $\mathcal{E}$ and with the following property: Let $\mathcal{K} \subset \mathcal{D}$ be* *compact. There exists $\eta \in (0,1)$ such that*

$$cap\left\{z : z \in \mathcal{K} \text{ and } \left|f - \frac{\mu_n}{\nu_n}\right|(z) > \eta^n\right\} \to 0, \ n \to \infty. \tag{11}$$

Outside the extremal domain Stahl established divergence with geometric rate; the size of η is determined by the values of generalized Green's functions [31]. As an example, for $f(z) = \log(1+z)$, we may take $\mathcal{E} = \{-1\}$ and $\mathcal{D} = \mathbf{C}\backslash(-\infty, 0]$.

If a sequence of functions converges in measure with respect to some measure, then a subsequence converges almost everywhere with respect to that measure. A similar conclusion holds for capacity. So a consequence of the Nuttall/Pommerenke/Stahl theorems, is that a subsequence of $\{\mu_n/\nu_n\}_{n=1}^\infty$ converges in the relevant domain $\mathcal{D}$ outside a set of cap 0 and hence also of planar measure 0. This may be viewed as a weak form of BGW.

However the conjecture itself remains unresolved. For functions whose Maclaurin series coefficients decay very rapidly to 0, the conjecture was proved in 1996 in [24]:

Theorem 2.4   (a) *Assume that the Maclaurin series coefficients $\{a_j\}$ of f satisfy*

$$\limsup_{j\to\infty} |a_j|^{1/j^2} < 1. \tag{12}$$

Then there exists a subsequence $\{\mu_n/\nu_n\}_{n\in S}$ of $\{\mu_n/\nu_n\}_{n=1}^\infty$ that converges to f, uniformly in compact subsets of $\mathbf{C}$.

(b) *More generally, if f is analytic at 0 and meromorphic in $\mathbf{C}$, and if for some $r > 0$, the errors of rational approximation,*

$$E_{jj}(r) := \inf\left\{\left\|f - \frac{p}{q}\right\|_{L_\infty(|z|\leq r)} : \deg(p), \ \deg(q) \leq j\right\}$$

satisfy

$$\limsup_{j\to\infty} E_{jj}^{1/j^2}(r) < 1 \tag{13}$$

then there exists a subsequence $\{\mu_n/\nu_n\}_{n\in S}$ of $\{\mu_n/\nu_n\}_{n=1}^\infty$ that converges uniformly in compact subsets of $\mathbf{C}$ omitting poles of f.

The conditions (12), (13) are very severe: they imply for example that the functions are respectively entire or meromorphic of order 0. Although this is a small class of functions, it is the only class restricted only by growth conditions for which the conjecture has been established. It provides some evidence that BGW may well be true for functions meromorphic in $\mathbf{C}$.

The idea of the above proof is to show that in the presence of spurious poles, we can "remove" the latter to obtain rapid rates of decay of $E_{jj}(r)$ as $j \to \infty$.

While this is possible for a subsequence that has spurious poles, this leads to a contradiction if every μ_n/ν_n has spurious poles.

Under the more general conditions of Pommerenke's theorem, it seems extremely difficult to construct a subsequence without spurious poles, that is for which, the poles eventually leave regions in which f is analytic. One weak result in this direction is due to the author [22]: under the conditions of Pommerenke's theorem, we showed that there is a subsequence $\{\mu_n/\nu_n\}_{n\in S}$ of $\{\mu_n/\nu_n\}_{n=1}^{\infty}$ such that in any compact set K in which f is analytic, μ_n/ν_n has for $n \in S$, $o(n)$ poles in K. Of course, BGW would require $o(1)$!

Notice that even this result does not apply when all we know of f is that it is meromorphic or analytic in the unit ball. Can anything at all be currently said for this situation, which after all is the hypothesis of BGW on the function f? We remarked earlier that the full sequence $\{\mu_n/\nu_n\}_{n=1}^{\infty}$ need not converge in measure or capacity in this situation [20], [29]. It was Goncar who first posed the question as to whether still a subsequence of $\{\mu_n/\nu_n\}_{n=1}^{\infty}$ converges in measure or capacity. This conjecture was repeated in [32], where there are several interesting weaker conjectures around BGW. The only positive result in this direction seems to be [23], where the author showed that for a subsequence $\{\mu_n/\nu_n\}_{n\in S}$, the point 0 attracts a zero proportion of poles and as a consequence, there is a weak convergence in capacity property in neighbourhoods of 0.

3 Ramanujan's Continued Fraction

In searching for examples that may resolve BGW, it is natural to look at special functions, where explicit formulae are available for their convergents. The "older" hypergeometric special functions all have convergent fractions. With the growth of the *q–disease* (to quote Richard Askey), it seemed natural to look at basic hypergeometric series, or *q*–series. The latter are typically studied for $|q| < 1$ or $|q| > 1$ [11], where there is uniform convergence of their continued fractions, but consideration of q on the unit circle has proved just as interesting.

The author and Ed Saff [26] considered the partial theta function

$$f(z) = \sum_{j=0}^{\infty} q^{j^2} z^j$$

for q on the unit circle; these turned out to contradict the Baker–Graves–Morris Conjecture for q not a root of unity, but they still satisfy BGW. Subsequently, Kathy Driver in her thesis investigated a class of power series whose continued fraction coefficients were derived by Peter Wynn, including

$$f(z) = 1 + \sum_{j=1}^{\infty} z^j \left(\prod_{k=1}^{j} \frac{A - q^{k+\alpha}}{B - q^{k+\beta}} \right).$$

For q not a root of unity, and suitable restrictions on A, B, α, β, these have natural boundaries, and their continued fractions exhibit some fascinating convergence/divergence phenomena, but they still satisfy BGW [7–10].

One continued fraction that escaped this earlier scrutiny is an especially simple one due to Ramanujan:

$$H_q(z) = 1 + \frac{qz|}{|1} + \frac{q^2 z|}{|1} + \frac{q^3 z|}{|1} + \frac{q^4 z|}{|1} + \dots \tag{14}$$

In 1920, Ramanujan noted that for $|q| < 1$, and $z \in \mathbf{C}$, there is the identity

$$H_q(z) = \frac{G_q(z)}{G_q(qz)} \tag{15}$$

where

$$G_q(z) = 1 + \sum_{j=1}^{\infty} \frac{q^{j^2}}{(1-q)(1-q^2)(1-q^3)\dots(1-q^j)} z^j \tag{16}$$

(See [1]). For $|q| < 1$, it is easily seen that G_q is entire of order 0, and because the coefficients q^j of the continued fraction decay to 0 as $j \to \infty$, the continued fraction converges in the plane to a meromorphic function H_q.

For $|q| = 1$, it follows from Worpitzky's theorem, perhaps the oldest and simplest convergence theorem on continued fractions, that the continued fraction converges for $|z| < \frac{1}{4}$. Example 2.1, for which $q = 1$, shows that $\frac{1}{4}$ is sharp. But what happens beyond this disc? We shall use H_q to denote the analytic continuation of this continued fraction to its Weierstrass domain of analytic continuation, even where the continued fraction diverges. In recent years, the author has become convinced that H_q with q not a root of unity, may resolve BGW in the negative for the unit ball.

The first question that arises is what is the nature of H_q for such q? Also does the identity (15) persist? We see that for q not a root of unity, that is $q^j \neq 1$, $j = 1, 2, 3, \dots$, at least the Maclaurin series coefficients of G_q are well defined. But one can say more [25]:

Theorem 3.1 *For almost all $\theta \in [-\pi, \pi]$ (with respect to linear Lebesgue measure) and $q = e^{i\theta}$,*

  (I) G_q is analytic in the unit ball and has a natural boundary on the unit circle;

  (II) H_q is meromorphic in the unit ball with a natural boundary on the unit circle;

  (III) The identity (15) holds in $|z| < 1$ away from the poles of H_q.

We note that as $q \to 1$, H_q will have a growing number of poles near $-\frac{1}{4}$ since H_q attempts to imitate the branchcut of H_1 along $\left(-1, -\frac{1}{4}\right]$. Thus there really will be poles of H_q inside the unit ball.

In the thin set of q (actually it is still thinner than linear measure zero, it has Hausdorff dimension 0 and logarithmic dimension 2) in which the conclusion of Theorem 3.1 fails, there are some fascinating features. For q a root of unity, H_q will have square root branchpoints, while for q not a root of unity, G_q may have a natural boundary on $|z| = r$, for any $r \in (0,1)$. In particular, G_q may have a natural boundary on $|z| = \frac{1}{8}$ and the identity (15) holds inside that circle, but H_q continues analytically to at least $|z| < \frac{1}{4}$.

Of course what we really want to know is if $\{\mu_n/\nu_n\}_{n=1}^\infty$ converges or diverges. Analysis is facilitated by explicit formulae that were first established by the Australian mathematician M.D. Hirschhorn [16]:

$$\mu_n(z) = \sum_{k=0}^{\left[\frac{n+1}{2}\right]} z^k q^{k^2} \left[\begin{array}{c} n+1-k \\ k \end{array} \right]; \tag{17}$$

$$\nu_n(z) = \mu_{n-1}(qz) = \sum_{k=0}^{\left[\frac{n}{2}\right]} z^k q^{k(k+1)} \left[\begin{array}{c} n-k \\ k \end{array} \right]. \tag{18}$$

Here $[x]$ denotes the greatest integer$\leq x$, and

$$\left[\begin{array}{c} n \\ k \end{array} \right] := \frac{(1-q^n)(1-q^{n-1})\dots(1-q^{n+1-k})}{(1-q^k)(1-q^{k-1})\dots(1-q)}$$

is the Gaussian binomial coefficient (as $q \to 1$, it reduces to the ordinary binomial coefficient). We note that μ_n and ν_n have no common factors, so that every zero of ν_n is a pole of μ_n/ν_n.

In describing the limiting behaviour of $\{\mu_n\}_{n=1}^\infty$, $\{\nu_n\}_{n=1}^\infty$, one needs the fact that for q not a root of unity, or equivalently, $\theta/(2\pi)$ irrational, $\{q^n\}_{n=1}^\infty = \{e^{in\theta}\}_{n=1}^\infty$ is dense on the unit circle. Thus given β with $|\beta| = 1$, we can find an infinite sequence of positive integers S with

$$\lim_{\substack{n \to \infty \\ n \in S}} q^n = \beta. \tag{19}$$

Using the explicit formulae (17), (18) and identities from the theory of q–series one may prove [25]:

Theorem 3.2 *Let q satisfy the conclusions of Theorem 3.1. Let $|\beta| = 1$ and let S be an infinite sequence of positive integers for which (19) holds. Then uniformly in compact subsets of $|z| < 1$,*

$$\lim_{\substack{n \to \infty \\ n \in S}} \mu_n(z) = G_q(z)\overline{G_q(\bar{\beta}qz)}; \tag{20}$$

$$\lim_{\substack{n \to \infty \\ n \in S}} \nu_n(z) = G_q(qz)\overline{G_q(\bar{\beta}qz)}. \tag{21}$$

Thus the numerator $\mu_n(z)$ in the nth convergent converges to the numerator $G_q(z)$ in $H_q(z)$, but multiplied by an extra factor that depends on β. Likewise, the denominator $\nu_n(z)$ in the nth convergent converges to the denominator $G_q(qz)$ in $H_q(z)$, but multiplied by the same extra factor.

At first sight, this seems perfectly consistent with uniform convergence of μ_n/ν_n to H_q: after the division, the extra factor $\overline{G_q(\overline{\beta q z})}$ cancels. However, this is not the case. Because of Hurwitz's theorem, each zero of the right–hand side of (21) attracts zeros of ν_n as $n \to \infty$, $n \in S$. The zeros of $G_q(qz)$ are indeed poles of $H_q(z)$, but the zeros of $\overline{G_q(\overline{\beta q z})}$ may not be. Thus the zeros of the latter function will attract zeros of ν_n that will be spurious poles of μ_n/ν_n.

What is the relation between the zeros of $G_q(z)$ and $\overline{G_q(\overline{\beta q z})}$? We see that the latter function involves multiplication by β which rotates the variable z and the conjugations reflect about the real axis. These operations preserve circles centre 0. Thus if $H_q(z)$ has, say, 3 poles on $|z| = r$, that is $G_q(qz)$ has 3 zeros there, then the right–hand side of (21) will have 6 zeros on that circle (counting multiplicity). Then for large n, ν_n will have 6 zeros near that circle, and hence μ_n/ν_n will have 6 poles near that circle. Although we have pursued this argument through a subsequence S satisfying (19), the fact that the unit circle is compact implies that we can always extract S with (19) holding for some β. We may thus deduce:

Corollary 3.3 *Let $0 < r < 1$. If H_q has ℓ poles counting multiplicity on $|z| = r$, then for large enough n, μ_n/ν_n has 2ℓ poles in any small enough neighbourhood of that circle.*

This is the first known example in which every convergent of large enough order to a function meromorphic in the unit ball has spurious poles, and even double as many poles, as the function from which it is formed. In every other example that this author knows, some subsequence of the convergents behaves pathologically, but another subsequence has the correct number of poles and satisfies BGW.

This brings us to the main question: does this contradict BGW? Well, almost! For most β, the set of zeros of $G_q(qz)$ will not be the same as the set of zeros of $\overline{G_q(\overline{\beta q z})}$. But we have not ruled out the possibility that there is a single β for which the sets of zeros of the two functions is the same. Then for the corresponding subsequence, μ_n/ν_n would still converge pointwise and even uniformly away from the poles of H_q and BGW would be true. However I do not believe that this is the case. Taking account of the conjugations and rotations in $\overline{G_q(\overline{\beta q z})}$, we can say:

Corollary 3.4 *If the zeros of $G_q(z)$ are not symmetric about any line through the origin, then H_q contradicts the Baker–Gammel–Wills Conjecture.*

The best way to assimilate this is to recall that Maclaurin series that have real coefficients have zeros that occur in conjugate pairs, that is are symmetric about the real axis. Effectively, we do not want the zeros of $G_q(\gamma z)$ to occur in conjugate pairs for any γ with $|\gamma| = 1$. Looking at just how oscillatory are the

Maclaurin series coefficients of G_q, there is certainly no reason to expect such a symmetry!

Having failed for sometime to disprove the symmetry, I turned to my colleague Prof. A. Knopfmacher, for help with Mathematica calculations. He has computed and plotted the zeros of the partial sums of G_q of high order for many values of q. From theoretical considerations, one expects the zeros of the partial sums to provide good approximations to the zeros of G_q in, for example, $|z| < \frac{3}{4}$. Increasing the order of the partial sums and careful calculation indicates that the zeros are indeed good approximations. In every case, the approximate zeros in $|z| < \frac{3}{4}$ are not symmetric about any line through 0. Thus there is convincing evidence that H_q does indeed contradict BGW. We shall present these numerical results elsewhere. Moreover, spurred on by this numerical evidence, I have a new idea, but still no proof.

Will BGW turn 40 in 2001?

Acknowledgements

I would like to thank the organisers for the kind invitation to the Dortmund conference, for their hard work, and the enjoyable and stimulating conference. I would also like to thank Prof. A. Knopfmacher for the calculations that have given new stimulus to investigating H_q.

References

[1] C. Adiga, B.C. Berndt, S. Bhargava, G.N. Watson, Chapter 16 of Ramanujan's Second Notebook: Theta Functions and q–Series, Memoirs of the American Mathematical Society, 53(1985), no. 315.

[2] G.A. Baker, Essentials of Padé Approximants, Academic Press, 1975.

[3] G.A. Baker, J.L. Gammel and J.G. Wills, An Investigation of the Applicability of the Padé Approximant Method, J. Math. Anal. Applns., 2(1961), 405–418.

[4] G.A. Baker and P.R. Graves–Morris, Convergence of Rows of the Padé Table, J. Math. Anal. Applns., 57(1977), 323–339.

[5] G.A. Baker and P.R. Graves–Morris, Padé Approximants, 2nd Edn., Encyclopaedia of Mathematics and its Applications, Vol. 59, Cambridge University Press, Cambridge, 1996.

[6] V.I. Buslaev, A.A. Goncar, S.P. Suetin, On the Convergence of Subsequences of the mth Row of the Padé Table, Math. USSR. Sbornik, 48(1984), 535–540.

[7] K.A. Driver, Convergence of Padé Approximants for Some q-hypergeometric Series (Wynn's Power Series I,II and III), Ph. D Thesis, Witwatersrand University, Johannesburg, 1991.

[8] K.A. Driver and D.S. Lubinsky, Convergence of Padé Approximants for a q–hypergeometric Series (Wynn's Power Series I), Aequationes Mathematicae, 42(1991), 85–106.

[9] K.A. Driver and D.S. Lubinsky, Convergence of Padé Approximants for a q–hypergeometric Series (Wynn's Power Series II), Colloquia Mathematica (Janos Bolyai Society), 58(1990), 221–239.

[10] K.A. Driver and D.S. Lubinsky, Convergence of Padé Approximants for a q–hypergeometric Series (Wynn's Power Series III), Aequationes Mathematicae, 45(1993), 1–23.

[11] G. Gasper and M. Rahman, Basic Hypergeometric Series, Cambridge University Press, Cambridge, 1990.

[12] G.M. Goluzin, Geometric Theory of Functions of a Complex Variable, AMS Translations of Math. Monographs, Vol. 26, American Mathematical Society, Providence, Rhode Island, 1969.

[13] A.A. Goncar, A Local Condition of Single–Valuedness of Analytic Functions, Math. USSR. Sbornik, 18(1972), 151–167.

[14] A.A. Goncar and E.A. Rahmanov, Equilibrium Distributions and Degree of Rational Apoproximation of Analytic Functions, Math. USSR. Sbornik, 62(1989), 305–348.

[15] E.M. Hille, Analytic Function Theory, Vol. 2, Chelsea, New York, 1987.

[16] M.D. Hirschhorn, A Continued Fraction of Ramanujan, J. Austral. Math. Soc. (Series A), 29(1980), 80–86.

[17] W.B. Jones and W.J. Thron, Continued Fractions: Theory and Applications, Encyclopaedia of Mathematics and its Applications, Vol. 11, Addison Wesley, London, 1980.

[18] J. Karlsson, Rational Interpolation and Best Rational Approximation, J. Math. Anal. Applns., 53(1976), 38–52.

[19] L. Lorentzen and H. Waadeland, Continued Fractions with Applications, North Holland, 1992.

[20] D.S. Lubinsky, Diagonal Padé Approximants and Capacity, J. Math. Anal. Applns., 78(1980), 58–67.

[21] D.S. Lubinsky, Divergence of Complex Rational Approximations, Pacific Journal of Mathematics, 108(1983), 141–153.

[22] D.S. Lubinsky, Distribution of Poles of Diagonal Rational Approximants to Functions of Fast Rational Approximability, Constr. Approx., 7(1991), 501–519.

[23] D.S. Lubinsky, Convergence of Diagonal Padé Approximants for Functions Analytic Near 0, Trans. Amer. Math. Soc., 347(1995), 3149–3157.

[24] D.S. Lubinsky, On the Diagonal Padé Approximants of Meromorphic Functions, Indagationes Mathematicae, 7(1)(1996), 97–110.

[25] D.S. Lubinsky, manuscript.

[26] D.S. Lubinsky and E.B. Saff, Convergence of Padé Approximants of Partial Theta Functions and the Rogers–Szegö Polynomials, Constr. Approx., 3(1987), 331–361.

[27] J. Nuttall, Convergence of Padé Approximants of Meromorphic Functions, J. Math. Anal. Applns., 31(1970), 147–153.

[28] C. Pommerenke, Padé Approximants and Convergence in Capacity, J. Math. Anal. Applns., 41(1973), 775–780.

[29] E.A. Rahmanov, On the Convergence of Padé Approximants in Classes of Holomorphic Functions, Math. USSR. Sbornik, 40(1981), 149–155.

[30] H. Stahl, General Convergence Results for Padé Approximants, (in) Approximation Theory VI, (eds. C.K. Chui, L.L. Schumaker, J.D. Ward), Academic Press, San Diego, 1989, pp.605–634.

[31] H. Stahl, The Convergence of Padé Approximants to Functions with Branch Points, J. Approx. Theory, 91(1997), 139–204.

[32] H. Stahl, Conjectures Around the Baker–Gammel–Wills Conjecture: Research Problems 97-2, Constr. Approx., 13(1997), 287–292.

[33] L.N. Trefethen, Square Blocks and Equioscillation in the Padé, Walsh, and CF Tables, (in) Rational Approximation and Interpolation, (eds. P.R. Graves–Morris, E.B. Saff, R.S. Varga), Springer Lecture Notes in Maths, Vol. 1105, 1984, pp. 170–181.

[34] H. Wallin, The Convergence of Padé Approximants and the Size of the Power Series Coefficients, Applicable Analysis, 4(1974), 235–251.

Centre for Applicable Analysis and Number Theory
Department of Mathematics
Witwatersrand University
Wits 2050
South Africa
Email address: 036dsl@cosmos.wits.ac.za

International Series of Numerical Mathematics
Vol. 132, © 1999 Birkhäuser Verlag Basel/Switzerland

Approximation Operators of Binomial Type

Alexandru Lupaş *

Abstract

Our objective is to present a unified theory of the approximation operators of binomial type by exploiting the main technique of the so- called "umbral calculus" or "finite operator calculus" (see [18], [20]-[22]). Let us consider the basic sequence $(b_n)_{n \geq 0}$ associated to a certain delta operator Q. By supposing that $b_n(x) \geq 0$, $x \in [0, \infty)$, our purpose is to put in evidence some approximation properties of the linear positive operators $(L_n^Q)_{n \geq 1}$ which are defined on $C[0, 1]$ by

$$L_n^Q f = \sum_{k=0}^{n} \beta_{n,k}^Q f\left(\frac{k}{n}\right) , \ \beta_{n,k}^Q(x) := \frac{1}{b_n(n)} \binom{n}{k} b_k(nx) b_{n-k}(n - nx).$$

This paper is first of all concerned with the construction of such operators. The construction method is performed by means of umbral calculus. The Bernstein operators are in fact $(L_n^D)_{n \geq 1}$, $D-$ being the derivative. It is shown that such operators leave invariant the cone of the convex functions of higher order and there are establishhed some identities between the elements of the "Q-basis" $\{\beta_{n,0}^Q, ..., \beta_{n,n}^Q\}$. Such representations turn out to be useful since they may be used to investigate qualitative properties of the operators.

AMS Subject Classification : 41A36

Key Words : Bernstein-type operators, binomial polynomials , delta operators.

1   Introduction

This paper is a survey of the role of binomial polynomials in Approximation Theory, which includes discrete linear positive operators, convexity preservation, and so on, and possibly, if we interpret it broadly, various applications of "Umbral Calculus". As the role of polynomial sequences in Approximation is extremely broad and diverse, I will not try to cover all the connections; rather I will focus on the topics with which I am familiar and in which I am interested. Therefore, this survey is biased by my mathematical interest.

The paper is aimed at nonspecialist in Umbral Calculus, so some material well known is included in the first section.

*Supported by Alexander von Humboldt Foundation

1.1   Binomial Sequences, Delta Operators

This section contains some basic facts needed in the subsequent analysis. The reader is referred to Gian-Carlo Rota and Steven Roman (1978) for the details about delta operators, as well as for historical remarks on the study of binomial sequences.

A sequence $(p_n)_{n \geq 0}$ of polynomials where for all n, p_n is exactly of degree n is called a *polynomial sequence*. Examples of such sequences are $(p_0(x) := 1)$:

1. $e_n(x) = x^n$, (the monomials)
2. $a_n(x) = x(x - na)^{n-1}$, (Abel)
3. $(x)_n = x(x + 1) \ldots (x + n - 1)$, (upper-factorials)
4. $< x >_n = x(x - 1) \ldots (x - n + 1)$, (lower-factorials)
5. $t_n(x) = e^{-x} \sum\limits_{k=0}^{\infty} \frac{x^k k^n}{k!} = \sum\limits_{k=0}^{n} [0, 1, 2, \ldots, k; e_n] x^k$

    (Touchard or exponential polynomials)
6. $G_n(x) = G_n(x; a, b) = \frac{x}{x - an} \left\langle \frac{x - an}{b} \right\rangle_n$ (Gould)

A polynomial sequence $(b_n)_{n \geq 0}$ is called *binomial* , iff for all x, y

$$b_n(x + y) = \sum_{k=0}^{n} \binom{n}{k} b_k(x) b_{n-k}(y) \ , \ \ n = 0, 1, 2, \ldots .$$

All above mentioned sequences are of binomial type. These binomial sequences occur in analysis and in combinatorics. A polynomial sequence $\pi = (b_n)_{n \geq 0}$ which is binomial has a generating function of the form

$$e^{x \Phi(t)} = \sum_{k=0}^{\infty} b_k(x) \frac{t^k}{k!} \ , \tag{1}$$

where $\Phi(t)$ is a formal power series

$$\Phi(t) = \sum_{k=0}^{\infty} c_k \frac{t^k}{k!} \ ,$$

with $c_0 = 0$, $c_1 \neq 0$, $c_k \in \mathbf{C}$.

At the same time, if $\pi = (b_n)_{n \geq 0}$ is generated by (1), then $(b_n)_{n \geq 0}$ is binomial.

Further, let us denote by Π the (real) linear space of all polynomials with real coefficients. The symbol Π_n is devoted to denote the linear space of all polynomials of degree at most n . If $h \in \Pi_n$ and $degree(h) = n$ we shall use the notation $h \in \Pi_n^*$. Let us put in evidence some operators $\Pi \to \Pi$. For instance, I is the identity, D is the derivative, E^a is *the shift-operator* ,i.e. $(E^a f)(x) = f(x + a)$. We shall use the notation

$$\Sigma = \{U : \Pi \to \Pi \ , \ U \text{ linear}\}, \ \ \Sigma_t = \{U \in \Sigma \ ; \ E^a U = U E^a \ , \ \forall a\},$$

$$\Sigma_t^* = \{U \in \Sigma_t \ ; \ U^{-1} \text{ exists .}\}$$

Let us remind that if $U_1, U_2 \in \Sigma_t$, then $U_1 U_2 = U_2 U_1$.

Definition 1.1 *(i) An operator Q is called a **delta operator**, iff :*

$$Q \in \Sigma_t \text{ and } Qe_1 = \text{const.} \neq 0 .$$

(ii) Σ_δ denotes the set of all delta operators.

*(iii) Let K , $K \in \Sigma$, be defined as $(Kh)(t) = th(t)$, $h \in \Pi$. If U is a shift-invariant operator, i.e. $U \in \Sigma_t$, then the linear operator U' defined by $U' = UK - KU$, is the so-called **Pincherle derivative of U**.*

Examples of delta operators :
1) $Q_1 = D(I - D)$
2) $\Delta_a = E^a - I = e^{aD} - I$, $a \neq 0$,
3) $\nabla = I - E^{-1}$,
4) $A = DE^a$ – Abel operator ,
5) $L = \frac{D}{D-I}$, - Laguerre operator
6) $G = E^{a+b} - E^a = \Delta_b E^a$ – Gould operator
7) $T = \ln(I + D)$ – Touchard operator .

Supposing $Q \in \Sigma_t$, then $Q \in \Sigma_\delta$ if and only if $Q = DP$, where $P \in \Sigma_t$, $Pe_0 \neq 0$,i.e. $P \in \Sigma_t^*$. Likewise, $Q \in \Sigma_\delta$ iff $Q = \phi(D)$, $\phi(t)$ being a formal power series with $\phi(0) = 0$, $\phi'(0) \neq 0$. At the same time is well-known that if Q is a delta operator, then $Q(\Pi_n^*) \subseteq \Pi_{n-1}^*$.
It is known that

$$U \in \Sigma_t \implies U' \in \Sigma_t , \quad Q \in \Sigma_\delta \implies Q'^{-1} \text{ exists} .$$

Definition 1.2 *Let $Q \in \sum_\delta$. If*

$$b_0(x) = 1 , b_n(0) = 0 , Qb_n = nb_{n-1} , n \geq 1 ,$$

*then (b_n) is a **basic sequence associated** to Q.*

Theorem 1.3 ([18],[20]-[22])

i) *Every delta operator has an unique sequence of basic polynomials.*

ii) *If (b_n) is a basic sequence for some delta operator Q, then it is binomial.*

iii) *If (b_n) is a binomial sequence, then it is a basic sequence for some delta operator.*

iv) *Let Q be a delta operator with basic sequence (b_n), and $Q = \phi(D)$. Let $\phi^{-1}(t)$ be the inverse formal power series of $\phi(t)$. Then*

$$\sum_{k=0}^{\infty} \frac{b_k(x)}{k!} t^k = e^{x\Phi(t)} , \quad \Phi(t) = \phi^{-1}(t) , \tag{2}$$

and

$$b_n(x) = \left(KQ'^{-1}b_{n-1}\right)(x) . \tag{3}$$

1.2   Two Representations of the Binomial Polynomials

One of the simplest delta operator is the *forward difference* $\Delta = E - I$, $(\Delta f)(x) = f(x+1) - f(x)$. In this case $\Delta = h(D)$ with $h(t) = e^t - 1$. The basic sequence for Δ is the "lower-factorials" or the *falling factorial sequence* $w = (w_n)_{n \geq 0}$, where

$$w_n(x) = <x>_n = x(x-1)(x-2)\ldots(x-n+1) \ . \tag{4}$$

Let $\pi = (b_n)_{n \geq 0}$ be a polynomial sequence of binomial type associated for the delta operator $Q = \phi(D)$,

$$b_n(x) = \sum_{k=0}^{n} c_{k,n} x^k.$$

It follows that the sequence $b^\Delta = (b_n^\Delta)_{n \geq 0}$, where

$$b_n^\Delta(x) = (\tau_w p_n)(x) = \sum_{k=0}^{n} c_{k,n} w_k(x),$$

is again a binomial sequence relative to $Q_0 = \phi(\Delta)$ (see [22]). The sequence $b^\Delta = (b_n^\Delta)_{n \geq 0}$ is called the *difference analogue* of the sequence $(b_n)_{n \geq 0}$.

The *exponential polynomials*, $(t_n)_{n \geq 0}$, introduced by Steffensen and studied by Touchard, are basic polynomials for the delta operator $T = \ln(I + D)$. These polynomials are $t_n(x) = \sum_{k=0}^{n} S(n,k) x^k$, $n = 0, 1, \ldots$, where $S(n,k)$, $k = 0, 1, \ldots$, denote the Stirling numbers of the second kind, defined by $x^n = \sum_{k=0}^{n} S(n,k) w_k(x)$, where we have used the notation (4). Let us observe that $S(n,k)$ is a divided difference. More precisely $S(n,k) = [0, 1, \ldots, k \ ; \ e_n]$, where

$$[x_0, x_1, \ldots, x_k \ ; \ f] = \sum_{j=0}^{k} \frac{f(x_j)}{w'(x_j)} \ , \quad w(x) = (x - x_0)\ldots(x - x_k).$$

Therefore

$$t_n(x) = \sum_{k=0}^{n} [0, 1, \ldots, k; e_n] x^k \ . \tag{5}$$

The following remarkable identity is known in the literature [18] as *Dobinski formula*

$$t_n(x) = e^{-x} \sum_{k=0}^{\infty} \frac{k^n}{k!} x^k \ . \tag{6}$$

A generalization of (6) may be obtained as in [11], [20]-[22] .

Theorem 1.4 *Let* $\pi = (b_n)_{n \geq 0}$ *be a polynomial sequence of binomial type. If* $b^\Delta = (b_n^\Delta)_{n \geq 0}$ *is its difference analogue of the sequence , then*

$$b_n(x) = e^{-x} \sum_{k=0}^{\infty} \frac{b_n^\Delta(k)}{k!} x^k = \sum_{k=0}^{n} [0, 1, 2, \ldots, k \ ; b_n^\Delta] x^k. \tag{7}$$

As above let us consider that $\pi = (b_n)_{n \geq 0}$ is binomial with respect to a delta operator Q. It is of interest to find the representation of b_n in the terms of the Bernstein basis defined by

$$\mathcal{B}_D := \{b_{n,0}, b_{n,1}, ..., b_{n,n}\} \quad, \quad b_{n,k}(x) = \binom{n}{k} x^k (1 - x)^{n-k} . \tag{8}$$

Theorem 1.5 *If* $Q = DP$, $P \in \Sigma_t^*$, *and*

$$q_n(x) := \sum_{j=0}^{n} \binom{nx}{j} \left(P^{-j} b_{n-j} \right)(0) \quad,$$

then

$$b_n(x) = \sum_{k=0}^{n} b_{n,k}(x) q_n \left(\frac{k}{n} \right) . \tag{9}$$

Proof. If $h \in \Pi_n$ and $Z_{n,k} : \Pi \to \Pi$, $k = 0, 1, ..., n$, are the linear operators

$$Z_{n,k} = {}_1F_1(-k; -n; D) \quad,$$

that is

$$(Z_{n,k} h)(x) = \sum_{\nu=0}^{k} \frac{(-k)_\nu}{(-n)_\nu} \frac{h^{(\nu)}(x)}{\nu!} \quad,$$

then it is known that (see [12] ,p.205)

$$h(x) = \sum_{k=0}^{n} b_{n,k}(x) \left(Z_{n,k} h \right)(0) \quad.$$

Now the proof is complete taking into account that

$$D^\nu b_n = P^{-\nu} Q^\nu b_n = \nu! \binom{n}{\nu} P^{-\nu} b_{n-\nu} \quad, \quad (Z_{n,k} b_n)(0) = q_n \left(\frac{k}{n} \right) .$$

$\blacksquare$

2   The Operator Class $\mathcal{B}$

Further we denote

$$\beta_{n,k}^Q(x) = \frac{1}{b_n(n)} \binom{n}{k} b_k(nx) b_{n-k}(n - nx) \quad, \quad k = 0, 1, ... \tag{10}$$

where $(b_n)_{n \geq 0}$ is the **basic sequence associated to** $Q \in \Sigma_\delta$, $b_n(n) \neq 0$.
Then $\beta^D_{n,k}(x) = \binom{n}{k} x^k (1-x)^{n-k} = b_{n,k}(x)$ and

$$\beta^Q_{n,k}(x) = \sum_{j=0}^{n} A_{j,n}(k) x^{n-j} \quad ,$$

with

$$A_{0,n}(k) = \frac{(nc_1)^n}{b_n(n)} \binom{n}{k} \ , \quad c_1 = b'_1(0) \neq 0 \ .$$

Definition 2.1 *The system of polynomials*

$$\mathcal{B}_Q \ = \ \left\{ \ \beta^Q_{n,0}(x) \ , \ \beta^Q_{n,1}(x) \ , \ ... , \ \beta^Q_{n,n}(x) \ \right\}$$

is called the Q *-binomial basis of the linear space* Π_n .

Our aim is to construct (by means of the "umbral calculus" - [18],
[20]-[22]) sequences of approximation operators $L^Q_n : C(\mathcal{I}) \to C(\mathcal{I})$, $n \geq 1$,
$\mathcal{I} = [0,1]$, of the following form

$$(L^Q_n f)(x) = \sum_{k=0}^{n} \binom{n}{k} \beta^Q_{n,k}(x) f\left(\frac{k}{n}\right) \ . \tag{11}$$

Definition 2.2 *(i) The class* $\mathcal{B}$ *includes all sequences of linear* **positive** *operators* $\left(L^Q_n\right)_{n \geq 1}$ *defined as in (11).*
(ii) $\mathcal{F}$ *is the set of all formal power series* $\sigma(t)$ *of the form*

$$\sigma(t) = \sum_{k=1}^{\infty} d_k \frac{t^k}{k!} \quad , \quad with \ d_1 > 0 \ , \ d_k \geq 0 \ (k \geq 2) \ .$$

Theorem 2.3 (*T.Popoviciu [19]-1932*) *Let* $Q \in \Sigma_\delta$, $Q = \phi(D)$, *and* L^Q_n *as
in (11). Then* $(L^Q_n) \in \mathcal{B}$ *if and only if*

$$\phi^{-1}(t) \ := \ \sum_{k=1}^{\infty} c_k \frac{t^k}{k!} \ \in \mathcal{F} \ . \tag{12}$$

Moreover, if $\phi^{-1}(t)$ *belongs to* $\mathcal{F}$, *then* $b_n(x)$, $n = 1, 2, ...,$ *generated by (2)
has only non-negative coefficients.*

Remark: see also: H.Brass [1] -1971, P.Sablonniere [23] -1995 .
Remark: The Bernstein polynomial

$$(B_n f)(x) \ := \ \sum_{k=0}^{n} b_{n,k}(x) f\left(\frac{k}{n}\right) \tag{13}$$

admits the representation

$$(B_n f)(x) = \frac{1}{e_n(n)} \sum_{k=0}^{n} \binom{n}{k} e_k(nx) e_{n-k}(n - nx) f\left(\frac{k}{n}\right);$$

that is $B_n = L_n^D$ and $(B_n)_{n \geq 1} \in \mathcal{B}$.

2.1 The Behaviour on the Cone of Convex Functions

Let $\mathbf{K_s}$ be the set of all f , $f \in C(\mathcal{I})$, which are non-concave of s-th order on $\mathcal{I}$. This means that for all systems $x_1, x_2, ..., x_{s+2}$ of distinct points from $\mathcal{I}$,

$$[x_1, x_2, ..., x_{s+2} \; ; \; f] \geq 0 \quad , \quad f \in \mathbf{K_s} .$$

Suppose further that $(r_n)_{n \geq 0}$ is a sequence of polynomials generated by

$$e^{x\sigma(t)} = \sum_{k=0}^{\infty} r_k(x) \frac{t^k}{k!} \quad , \quad \sigma(t) \in \mathcal{F} , \tag{14}$$

and put

$$\vartheta_{m,k}^{<a_n>}(x) = \frac{1}{r_m(a_n)} \binom{m}{k} r_k(a_n x) r_{m-k}(a_n - a_n x) \quad , \quad n \geq 1, \tag{15}$$

where $k \in \{0, 1, ..., m\}$, and $(a_n)_{n \geq 1}$ is a sequence of positive numbers. By means of Theorem 2.3 we observe that

$$\vartheta_{m,k}^{<a_n>}(x) \geq 0 \quad , \quad \forall x \in \mathcal{I} .$$

Let us define the linear positive operators $\mathcal{L}_m^{<a_n>} : C(\mathcal{I}) \to C(\mathcal{I})$ $m \geq 1$, $n \geq 1$, with the images

$$\mathcal{L}_m^{<a_n>} f = \sum_{k=0}^{m} \vartheta_{m,k}^{<a_n>} f\left(\frac{k}{m}\right) . \tag{16}$$

In the particular case

$$a_n = m = n \; , \; \sigma(t) = \phi^{-1}(t) \; , \; \phi(t) \in \mathcal{F} ,$$

the operator $\mathcal{L}_m^{<a_n>}$ is the same with L_n^Q , $Q = \phi(D) \in \Sigma_\delta$, $\left(L_n^Q\right)_{n \geq 1} \in \mathcal{B}$.

Theorem 2.4 *Let $\mathcal{L}_m^{<a_n>}$ be defined as in (15)-(16) . Then*

(a) $\mathcal{L}_m^{<a_n>}\left(\mathbf{K_s}\right) \subseteq \mathbf{K_s}$, $s \in \{-1, 0, 1, 2, ...\}$;

(b)   *If $B_m : C(\mathcal{I}) \to C(\mathcal{I})$, $m \in \mathbf{N}$, are the Bernstein operators and $(m, n) \in \mathbf{N} \times \mathbf{N}$, then*

$$f \leq B_m f \leq \mathcal{L}_m^{<a_n>} f \; , \quad f \in \mathbf{K_1} \quad , \quad on \; \mathcal{I} .$$

Proof. According to [1] if $(p_n)_{n\geq0}$ is a polynomial sequence having a generating function of the form

$$e^{x\nu(t)} = \sum_{k=0}^{\infty} p_k(x)\frac{t^k}{k!} \quad , \quad \nu(t) \in \mathcal{F} , \tag{17}$$

then with notation

$$(\mathcal{M}_m f)(x) = \frac{1}{p_m(1)} \sum_{k=0}^{m} \binom{m}{k} p_k(x)p_{m-k}(1-x)f\left(\frac{k}{m}\right)$$

$$(x \in \mathcal{I} , f \in C(\mathcal{I})),$$

one has

$$f \in \mathbf{K_s} \implies \mathcal{M}_m f \in \mathbf{K_s} , \tag{18}$$

as well as for $x \in \mathcal{I}$

$$|(\mathcal{M}_m f)(x) - f(x)| \geq |(B_m f)(x) - f(x)| , \quad \forall f \in \mathbf{K_1} .$$

Next, we observe that , by means of Jensen inequality, the above inequality implies

$$f(x) \leq (B_m f)(x) \leq (\mathcal{M}_m f)(x) , \quad f \in \mathbf{K_1} , x \in \mathcal{I}. \tag{19}$$

Let us denote $p_m^{<a_n>}(x) = r_m(a_n x)$, $a_n > 0$, where a_n is considered as a parameter. Since

$$e^{x\sigma^{<a_n>}(t)} = \sum_{k=0}^{\infty} p_k^{<a_n>}(x)\frac{t^k}{k!} \quad , \quad \sigma^{<a_n>}(t) = a_n\sigma(t) ,$$

it follows that $\sigma^{<a_n>}(t) \in \mathcal{F}$, and moreover the linear positive operators $\mathcal{M}_m^{<a_n>}$, with

$$(\mathcal{M}_m^{<a_n>} f)(x) = \frac{1}{p_m^{<a_n>}(1)} \sum_{k=0}^{m} \binom{m}{k} p_k^{<a_n>}(x)p_{m-k}^{<a_n>}(1-x)f\left(\frac{k}{m}\right) \tag{20}$$

$$(x \in \mathcal{I} , f \in C(\mathcal{I})),$$

satisfy (18) and (19). Now it is easy to observe that $\mathcal{M}_m^{<a_n>} = \mathcal{L}_m^{<a_n>}$ which completes the proof. ∎

We note that a proof of the fact that the linear positive operators $\mathcal{L}_m^{<a_n>}$ preserve the convexity of any order may be performed by means of S.Karlin theory on total positivity (see [6]-[7]).

In the case $m = n$ the operators defined by (20) may be considered as a generalization of D.D.Stancu operators (see [24] - [25]).

By considering in the above theorem $m = n$, $a_n = n$, we give the following theorem.

Theorem 2.5 *Suppose that* L_n^Q *is defined by (11) and satisfies* $L_n^Q \in \mathcal{B}$. *Then for* $n \geq 1$

(a) $L_n^Q(\mathbf{K_s}) \subseteq \mathbf{K_s}$, $s \in \{-1, 0, 1, 2, ...\}$;

(b) *for* $f \in \mathbf{K_1}$

$$\min_{L_n^Q \in \mathcal{B}} \| f - L_n^Q f \| = \| f - B_n f \| \ ,$$

pace*1.5cm*where* $B_n = L_n^D$ *is the Bernstein operator* ;

(c) *with the notation* $\{z\} = z - [z]$, *for* $f \in \mathbf{K_1}$

$$f(x) \leq \left(1 - \{nx\}\right) f\left(\frac{[nx]}{n}\right) + \{nx\} f\left(\frac{1 + [nx]}{n}\right) \leq (B_n f)(x) \leq$$

$$\leq (L_n^Q f)(x) \ , \ x \in [0, 1) \ .$$

Proof. The assertions (a) and (b) follow from Theorem 2.4. Now let us consider the linear positive mapping $f \to V_n f$, with $(V_n f)(1) = f(1)$ and

$$(V_n f)(x) = \left(1 - \{nx\}\right) f\left(\frac{[nx]}{n}\right) + \{nx\} f\left(\frac{1 + [nx]}{n}\right) , x \in [0, 1) \ .$$

It is easy to observe that $(V_n f)\left(\frac{k}{n}\right) = f\left(\frac{k}{n}\right)$, $k = 0, 1, ..., n$, and moreover

$$V_n e_j = e_j \text{ for } j = 0, 1, \quad V_n(\mathbf{K_1}) \subseteq \mathbf{K_1} \ , \quad L_n^Q V_n f = L_n^Q f \ .$$

Therefore the inequalities $f \leq V_n f \leq L_n^Q V_n f = L_n^Q f$ are valid on the interval $\mathcal{I}$. ∎

2.2 Some Auxiliary Identities

Let $Q \in \Sigma_\delta$, $(Kf)(t) = tf(t)$, and $\Theta = KQ'^{-1}$. From (3)

$$(\Theta b_n)(x) = b_{n+1}(x) \ ,$$

and therefore the linear operator Θ is called the *shift operator* for the sequence $\pi = (b_n)_{n \geq 0}$. A representation of the shift-operator Θ is

$$(\Theta h)(x) = x \sum_{k=0}^{\infty} \frac{c_{k+1}}{k!} (Q^k h)(x) \ , \quad h \in \Pi \ , \tag{21}$$

where (see (12)) $c_k = b_k'(0)$. For instance, for $h = b_n$ we obtain

$$b_{n+1}(x) = x \sum_{k=0}^{n} \binom{n}{k} b_{n-k}(x) b_{k+1}'(0) \quad .$$

The following result extends many classical identities.

Theorem 2.6 *Let* $T_{k,y} : \Pi \to \Pi$ *be defined as*

$$T_{k,y} = \frac{1}{k!} \sum_{j=0}^{\infty} \frac{(-1)^j}{j!} \Theta^{j+k} Q^{j+k} E^y \ , \quad k = 0, 1, ..., \ y \in \mathbf{R} \ .$$

If $\pi = (b_n)_{n \geq 0}$ *is the basic sequence of* Q, *then*

$$\binom{n}{k} b_k(x) b_{n-k}(y) = \left(T_{k,y} b_n\right)(x)$$

(22)

$$\sum_{k=0}^{n} \binom{n}{k} b_k(x) b_{n-k}(y) f\left(\frac{k}{n}\right) =$$

$$= \sum_{k=0}^{n} \frac{k!}{n^k} \binom{n}{k} \left[0, \frac{1}{n}, ..., \frac{k}{n} \ ; \ f\right] \left(\Theta^k E^y b_{n-k}\right)(x) \ .$$

Proof. We have

$$b_k(x) b_{n-k}(y) = \sum_{j=0}^{n-k} \binom{n-k}{j} b_{n-k-j}(y) b_{j+k}(x) \sum_{\mu=0}^{j} (-1)^{\mu} \binom{j}{\mu} =$$

$$= \sum_{j=0}^{n-k} (-1)^{n-k-j} \binom{n-k}{j} \sum_{\mu=0}^{j} \binom{j}{\mu} b_{\mu+n-j}(x) b_{j-\mu}(y) =$$

$$= \sum_{j=0}^{n-k} (-1)^j \binom{n-k}{j} \left(\Theta^{j+k} E^y b_{n-k-j}\right)(x) \ .$$

But

$$Q^j b_{n-k} = j! \binom{n-k}{j} b_{n-k-j} \text{ if } j \leq n-k \ , \quad Q^j b_{n-k} = 0 \text{ for } j > n-k \ ,$$

enables us to write

$$\binom{n}{k} b_k(x) b_{n-k}(y) = \binom{n}{k} \sum_{j=0}^{n-k} \frac{(-1)^j}{j!} \left(\Theta^{j+k} Q^j E^y b_{n-k}\right)(x) =$$

$$= \frac{1}{k!} \sum_{j=0}^{\infty} \frac{(-1)^j}{j!} \left(\Theta^{j+k} Q^{j+k} E^y b_n\right)(x) = \left(T_{k,y} b_n\right)(x) \ .$$

Let us observe that if

$$Af := \sum_{k=0}^{n} a_{k,n} f\left(\frac{k}{n}\right) \ , \quad a_{k,n} \in \mathbf{C} \ ,$$

then according to Newton interpolation formula

$$f\left(\frac{k}{n}\right) = \sum_{j=0}^{k} \frac{j!}{n^j} \binom{k}{j} \left[0, \frac{1}{n}, ..., \frac{j}{n} \; ; \; f\right] \quad,$$

and therefore

$$Af = \sum_{k=0}^{n} \frac{k!}{n^k} \left[0, \frac{1}{n}, ..., \frac{k}{n} \; ; \; f\right] \Lambda_{k,n} \quad, \tag{23}$$

with

$$\Lambda_{k,n} = \frac{n^k}{k!} A h_k = \sum_{j=k}^{n} a_{j,n} \binom{j}{k} \quad, \quad h_k(t) = t(t - \frac{1}{n})...(t - \frac{k-1}{n}) \quad.$$

Using (23) with $a_{k,n} = a_{k,n}(x,y) = \binom{n}{k} b_k(x) b_{n-k}(y)$, we see that

$$\Lambda_{k,n} = \binom{n}{k} \left(\Theta^k E^y b_{n-k}\right)(x) \quad.$$

■

Theorem 2.7 *If* $L_n^Q \in \mathcal{B}$ *and*

$$d_{k,n}(x) = \frac{1}{b_n(n)} \left(\Theta^k E^{n-nx} b_{n-k}\right)(nx) \quad, \tag{24}$$

then $d_{k,n} \in \Pi_k^*$. *Moreover*

$$(L_n^Q f)(x) = \sum_{k=0}^{n} \frac{k!}{n^k} \binom{n}{k} \left[0, \frac{1}{n}, ..., \frac{k}{n} \; ; \; f\right] d_{k,n}(x) \quad.$$

Also $L_n^Q\left(\Pi_s\right) \subseteq \Pi_s$.

Proof. It is necessary to justify the last assertion, i.e. the implication that $L_n^Q\left(\Pi_s\right) \subseteq \Pi_s$. Suppose that $h \in \Pi_s$. Then

$$\left[0, \frac{1}{n}, ..., \frac{k}{n} \; ; \; h\right] = 0 \quad, \quad k \geq s+1 \quad.$$

At the same time, using the method of generating function from (20) we see that for any natural numbers m, k and arbitrary a

$$\Upsilon(k, m, a; x) := \sum_{j=0}^{k} \binom{k}{j} b_{j+m}(ax) b_{k-j}(a - ax)$$

may be written as

$$\Upsilon(k, m, a; x) = \sum_{\mu=0}^{m-1} (ax)^{m-\mu} H_{\mu,k}(a) \ ,$$

$(H_{0,k}(a))_{k \geq 0}$ being a polynomial sequence in the variable a and where for $\mu \geq 1$

$$H_{\mu,k}(a) \in \Pi_k \ .$$

Therefore, $\Upsilon(k, m, a; x)$ is a polynomial of degree m in x. Finally, the remark that $d_{k,n}(x) = \frac{1}{b_n(n)} \Upsilon(n - k, k, n; x)$ implies $d_{k,n} \in \Pi_k$. Further

$$(L_n^Q h)(x) = \sum_{k=0}^{s} \frac{k!}{n^k} \left[0, \frac{1}{n}, ..., \frac{k}{n} \ ; \ h\right] d_{k,n}(x)$$

completes the proof.   ∎

Let us observe that

$$d_{0,n}(x) = 1 \ , \ d_{1,n}(x) = x \ , \ d_{2,n}(x) = n^2 x(x - 1)\left(Q'^{-2} b_{n-2}\right)(n) + x. \qquad (25)$$

2.3   The Behaviour on Polynomials

If p is a natural number, let us denote $\mathcal{E} = \Theta Q$ and

$$S_p(x, y, n) = \sum_{k=0}^{n} \binom{n}{k} b_k(x) b_{n-k}(y) \left(\frac{k}{n}\right)^p \ , \qquad (26)$$

where it is supposed that $\pi = (b_n)_{n \geq 0}$ is the basic sequence associated to the delta operator Q, Θ being the shift of $\pi = (b_n)_{n \geq 0}$.

Lemma 2.8 *If $S_p(x, y, n)$ is defined as in (26), then*

$$S_p(x, y, n) = \frac{1}{n^p} \sum_{j=0}^{p} j! \binom{n}{j} [0, 1, ..., j \ ; e_p] \left(\Theta^j E^y b_{n-j}\right)(x) \ , \qquad (27)$$

that is $S_p(x, y, n) = \frac{1}{n^p} \left(\mathcal{E}^p E^y b_n\right)(x)$.

Proof. Because $\mathcal{E} b_k = k b_k$, it follows that $\mathcal{E}^p b_k = k^p b_k$, which enables us to write that $S_p(x, y, n) = \frac{1}{n^p} \left(\mathcal{E}^p E^y b_n\right)(x)$. By means of Newton interpolation formula

$$k^p = \sum_{j=0}^{p} j! \binom{k}{j} [0, 1, ..., j \ ; e_p] \ ,$$

and in this manner

$$S_p(x,y,n) = \frac{1}{n^p} \sum_{j=0}^{p} j![0,1,...,j\;;e_p] \sum_{k=0}^{n-j} \binom{n}{k+j}\binom{k+j}{j} b_{k+j}(x)b_{n-j-k}(y) =$$

$$= \frac{1}{n^p} \sum_{j=0}^{p} j!\binom{n}{j}[0,1,...,j\;;e_p] \sum_{k=0}^{n-j} \binom{n-j}{k}(\Theta^j b_k)(x)b_{n-j-k}(y) =$$

$$= \frac{1}{n^p} \sum_{j=0}^{p} j!\binom{n}{j}[0,1,...,j\;;e_p](\Theta^j E^y b_{n-j})(x)\;.$$

■

2.4 The Order of Approximation

In the following we need a certain inequality between the terms of a binomial sequence.

Theorem 2.9 *If* $Q = \phi(D)$, $Q \in \Sigma_\delta$ *and the positivity condition (12) is satisfied, then for* $x > 0$

$$0 < c_1 \frac{b_{n-1}(x)}{x} \le \left(Q'^{-2}b_{n-2}\right)(x) \le \frac{b_n(x)}{x^2}\;,\quad n \ge 2\;.$$

Proof. According to (21) we have

$$\frac{b_{n-1}(x)}{x} = \left(Q'^{-1}b_{n-2}\right)(x) = \sum_{k=0}^{n-2} \binom{n-2}{k}c_{k+1}b_{n-2-k}(x)\;,$$

which implies (see (3))

$$\begin{aligned}
\left(Q'^{-2}b_{n-2}\right)(x) &= \frac{1}{x}\sum_{k=0}^{n-2}\binom{n-2}{k}c_{k+1}b_{n-1-k}(x) = \\
&= \frac{1}{x}\sum_{k=0}^{n-2}\binom{n-1}{k}\left(1-\frac{k}{n-1}\right)c_{k+1}b_{n-1-k}(x) \le \\
&\le \frac{1}{x}\sum_{k=0}^{n-2}\binom{n-1}{k}c_{k+1}b_{n-1-k}(x) = \\
&= \frac{1}{x}\left(\frac{b_n(x)}{x}-c_n\right) \le \frac{b_n(x)}{x^2}\;,\quad x>0\;,
\end{aligned}$$

with equality cases iff $c_2 = c_3 = ... = c_n = 0$. At the same time, one observes that

$$\left(Q'^{-2}b_{n-2}\right)(x) \ge \frac{1}{x}c_1 b_{n-1}(x)\;,\quad x > 0\;.$$

■

Theorem 2.10 *Suppose that* $Q \in \Sigma_\delta$, $Q = \phi(D)$ *and* L_n^Q *is defined as in (11). Further let us assume* $L_n^Q \in \mathcal{B}$, *i.e. the positivity condition (12) is verified, and denote*

$$\rho_n(Q) = 1 - \frac{n(n-1)}{b_n(n)}(Q'^{-2}b_{n-2})(n) . \tag{28}$$

Then

1. *The sequence* $(\rho_n(Q))_{n \geq 1}$ *is bounded. More precisely*

$$\frac{1}{n} \leq \rho_n(Q) < 1 . \tag{29}$$

2. *For* $e_j(t) = t^j$ *one has* $L_n^Q e_j = e_j$, $j = 0, 1$, *and*

$$(L_n^Q e_2)(x) = e_2(x) + \rho_n(Q)x(1-x) . \tag{30}$$

Proof. From Theorem 2.9 we find

$$\frac{1}{n} \leq \rho_n(Q) \leq 1 - c_1(n-1)\frac{b_{n-1}(n)}{b_n(n)} < 1 .$$

The equalities (30) are obtained by means of (27) or by using Theorem 2.7 together with (25). ∎

As usually, for $f \in C(\mathcal{I})$, let $\omega_k(f; .)$ its k-th order modulus of smoothness. Further

$$\Omega_{j,x}(t) = |t - x|^j , \quad j > 0 . \tag{31}$$

It is easy to see that

$$(L_n^Q \Omega_{2,x})(x) = \rho_n(Q)x(1-x) ,$$

where $\rho_n(Q)$ is as in (28).
We have derived the following estimates, obtained as particular cases of some general results (for instance ,see [3]- [5]).

Theorem 2.11 *Suppose that* L_n^Q *is defined as as in (11). If* $(L_n^Q) \in \mathcal{B}$ *then for* $f \in C(\mathcal{I})$, $x \in \mathcal{I}$, $n \geq 2$, *we have*
 (i) $|f(x) - (L_n^Q f)(x)| \leq 2\omega_1(f; \sqrt{\rho_n(Q)x(1-x)})$;
 (ii) $\|f - L_n^Q f\| \leq \frac{5}{4}\omega_1(f; \sqrt{\rho_n(Q)})$;
 (iii) $|f(x) - (L_n^Q f)(x)| \leq \frac{3}{2}\omega_2(f; \sqrt{\rho_n(Q)x(1-x)})$;
 (iv) $\|f - L_n^Q f\| \leq \frac{9}{8}\omega_2(f; \sqrt{\rho_n(Q)})$.

An interesting problem is to find how large is $|(L_n^Q f)(x)|$, when x is outside the interval $\mathcal{I}$. We start with the following simple result.

Lemma 2.12 *Suppose* $(L_n^Q) \in \mathcal{B}$. *Let* $h \in \Pi_n$ *and* $T_n(z) = \cos n \arccos z$ *for* $|z| \le 1$. *If* x_0 *is arbitrary in* $[a, b]$, $-\infty < a < b < \infty$, *and for all* $t \in [a, b]$ *one has* $|h(t)| \le 1$, *then*

$$|h(x)| \le T_n\left(1 + \frac{2|x - x_0|}{b - a}\right) \quad \text{for all real } x .$$

Proof. Under our assumptions, a classical inequality of V.A.Markov asserts that $|(D^k h)(x_0)| \le \left(\frac{2}{b-a}\right)^k (D^k T_n)(1)$, $k = 0, 1, ..., n$. Therefore

$$|h(x)| = |\sum_{k=0}^{n} \frac{(x - x_0)^k}{k!} (D^k h)(x_0)| \le$$

$$\le \sum_{k=0}^{n} \frac{(D^k T_n)(1)}{k!} \left(\frac{2|x - x_0|}{b - a}\right)^k = T_n\left(1 + \frac{2|x - x_0|}{b - a}\right) .$$

∎

If we consider $a = 0$, $b = 1$, $f \in C(\mathcal{I})$, $\|f\| \ne 0$, $h = \frac{1}{\|f\|} L_n^Q f$, one finds

Corollary 2.13 *Suppose that* $x \le 0$, $y \ge 1$, $f \in C(\mathcal{I})$. *Then*

$$|(L_n^Q f)(x)| \le T_n(1 - 2x)\|f\| \quad , \quad |(L_n^Q f)(y)| \le T_n(2y - 1)\|f\| \quad ,$$

where $\|.\|$ *is the uniform norm on the interval* $\mathcal{I}$.

Another interesting result is the following.

Lemma 2.14 *If* $(L_n^Q) \in \mathcal{B}$, $1 \le a < b$, *then for all* $x \in \mathcal{I}$

$$\sqrt{x(1 - x)}|(DL_n^Q f)(x)| \le n\sqrt{\frac{a - x}{b - x}}\|f\|_{\mathcal{J}}$$

where $\mathcal{J} = \mathcal{I} \cup [a, b]$ *and* $\|.\|_{\mathcal{J}}$ *is the uniform norm on* $\mathcal{J}$.

Further it is our intention to find the asymptotic behaviour of the remainder term on certain subspace of $C(\mathcal{I})$.

Theorem 2.15 *Suppose that* $x_0 \in \mathcal{I}$, $f \in C(\mathcal{I})$ *and* $f''(x_0)$ *exists. Let* $L_n^Q \in \mathcal{B}$ *and* $d_{k,n}$, $\rho_n(Q)$ *defined by* (24), (28). *If*

$$\lim_{n \to \infty} n\rho_n(Q) = \mathcal{K} \quad , \quad \mathcal{K} > 0 \quad ,$$

and the condition

$$\lim_{n \to \infty} \frac{1}{n^3} \sum_{k=0}^{4} (k - nx)^4 \sum_{j=k}^{4} (-1)^{j-k} \binom{j}{k} d_{j,n}(x) = 0$$

is satisfied , then

$$\lim_{n\to\infty} n\big(f(x_0) - (L_n^Q f)(x_0)\big) = -\frac{\mathcal{K}x_0(1-x_0)}{2} f''(x_0) \quad . \tag{32}$$

Proof. In our case,

$$\lim_{n\to\infty} n\big(e_k(x_0) - (L_n^Q f)(x_0)\big) = r_k(x_0) \quad , \quad k = 1, 2,$$

where $r_1(x_0) = 0$, $r_2(x_0) = -\mathcal{K}x_0(1-x_0)$. Let us observe that

$$\lim_{n\to\infty} n\big(L_n^Q \Omega_{4,x_0}\big)(x_0) = 0 .$$

Using Theorem 4.II and Lemma 4.8 from [8], we find

$$\lim_{n\to\infty} n\big(f(x_0) - (L_n^Q f)(x_0)\big) =$$

$$= \big(f'(x_0) - x_0 f''(x_0)\big) r_1(x_0) + \frac{r_2(x_0)}{2} f''(x_0)$$

which in fact is our asymptotic formula (32).   ∎

3   Examples

Some examples, each of which includes, we would like to hope, a little novelty, are given in the following, both as an illustration of the theory and to show how much of the past literature on linear positive operators of Bernstein type is the iteration of a few basic principles.

It is clear that $\mathcal{B}_D$ is in fact the *Bernstein basis* of the linear space Π_n . This means that L_n^D is the Bernstein operator.

If

$$l_{n,k}(x) = \prod_{j=0,j\neq k}^{n} \frac{nx - j}{k - j} = (-1)^{n-k} \binom{nx}{k} \binom{nx - k - 1}{n - k} \quad ,$$

then for $k \in \{0, 1, ..., n\}$ we have

$$\beta_{n,k}^Q(x) = (L_n^Q l_{n,k})(x) = \frac{1}{b_n(n)} \Big(T_{k,n-nx} b_n\Big)(nx) ,$$

the operators $T_{k,y}$ being defined as in Theorem 2.6 .

3.1   The Sequence $(L_n^T)_{n \geq 1}$

In this case $T = \ln(I + D)$ and

$$\left(L_n^T f\right)(x) = \frac{1}{t_n(n)} \sum_{k=0}^{n} \binom{n}{k} t_k(nx) t_{n-k}(n - nx) f\left(\frac{k}{n}\right) \quad ,$$

where $t_n(x)$ is as in (5)-(6).

Firstly, let us remark that if $(S_n)_{n \geq 1}$ is the sequence of Favard-Szász operators, i.e.

$$\left(S_n g\right)(x) = e^{-nx} \sum_{k=0}^{\infty} \frac{(nx)^k}{k!} g\left(\frac{k}{n}\right) \quad ,$$

$$\left(g : [0, \infty) \to \mathbf{R} \quad , \quad |g(x)| \leq A e^{Bx} \right)$$

then $t_n(x)$ admits the representation

$$t_n(x) = n^n \left(S_n e_n\right)\left(\frac{x}{n}\right) \quad .$$

Theorem 3.1 *If B_n is the Bernstein operator defined as in (13), then*

$$L_n^T = \mathcal{T}_n B_n \quad ,$$

where $\mathcal{T}_n : \Pi \to \Pi$, $n \geq 1$, are the linear positive operators

$$\mathcal{T}_n = \frac{e^{-n}}{t_n(n)} \sum_{j=1}^{\infty} \frac{n^j j^n}{j!} B_j \quad .$$

Proof. With the notation (8), using Dobinski formula we obtain

$$\binom{n}{k} t_k(nx) t_{n-k}(n - nx) = e^{-n} \sum_{j=1}^{\infty} \frac{n^j j^n}{j!} \left(B_j b_{n,k}\right)(x)$$

which enables us to write

$$\left(L_n^T f\right)(x) = \frac{e^{-n}}{t_n(n)} \sum_{j=1}^{\infty} \frac{n^j j^n}{j!} \left(B_j B_n f\right)(x) = \left(\mathcal{T}_n B_n f\right)(x) \, ,$$

and the proof is complete.   ∎

Further

$$T'^{-1} = I + D \quad , \quad t_{n+1}(x) = x\left(t_n(x) + t_n'(x)\right) \quad .$$

The positivity of the Stirling numbers $S(n,k)$, $k = 1, ..., n$, provides us

$$0 < \frac{t_{n-1}(x)}{t_n(x)} \le \frac{1}{x} \quad , \quad x > 0, \ n \ge 1 .$$

Further

$$\left(T'^{-2} t_{n-2}\right)(x) = \frac{t_n(x) - t_{n-1}(x)}{x^2} \quad , \quad n \ge 2 ,$$

and

$$\rho_n(T) = \frac{1}{n} + \frac{n-1}{n} \frac{t_{n-1}(n)}{t_n(n)} .$$

Therefore

$$\frac{1}{n} < \rho_n(T) \le \frac{2}{n} .$$

3.2   The Operator L_n^∇

Let $\nabla = I - E^{-1}$ be the backward-difference operator having the basic sequence
$\left((x)_n \right)_{n \ge 0}$, $(x)_n = x(x+1)...(x+n-1)$.
In this case $L_n^\nabla : C[0,1] \to \Pi_n$, $n = 1, 2, ...,$ is defined as

$$(L_n^\nabla f)(x) = \sum_{k=0}^{n} \beta_{n,k}^\nabla(x) f\left(\frac{k}{n}\right) \quad , \tag{33}$$

where $\mathcal{B}_\nabla = \{ \beta_{n,0}^\nabla(x), ..., \beta_{n,n}^\nabla(x) \}$ is the "nabla basis" , i.e.

$$\beta_{n,k}^\nabla(x) = \frac{1}{(n)_n} \binom{n}{k} (nx)_k (n - nx)_{n-k} . \tag{34}$$

Some properties of the $\nabla-$ basis are the following:

Theorem 3.2 i). *The following equalities are satisfied*

$$\beta_{n,k}^\nabla(x) = \frac{1}{k!} \sum_{j=k}^{n} \frac{(-n)_j (-j)_k}{(n)_j} \frac{(nx)_j}{j!} \quad ;$$

$$_2F_1(-n, nx; n; 1 - t) = \sum_{k=0}^{n} \beta_{n,k}^\nabla(x) t^k \quad ; \tag{35}$$

$$x^n = \sum_{k=0}^{n} c(n,k) \beta_{n,k}^\nabla(x)$$

with

$$c(n,k) = \frac{1}{n^n} \sum_{j=0}^{k} (-1)^{n-j} \frac{(-k)_j (n)_j}{(-n)_j} S(n,j) \quad ,$$

$S(n,j)$ *being the Stirling number of the second kind.*

ii). *If* $f \in \Pi_n$ *and*

$$\beta(k, n; f) := \sum_{j=0}^{k} \frac{(-k)_j (n)_j}{(-n)_j n^j} \left[0, -\frac{1}{n}, -\frac{2}{n}, ..., -\frac{j}{n}; f\right],$$

then

$$f(x) = \sum_{k=0}^{n} \beta(k, n; f) \beta_{n,k}^{\nabla}(x) \ .$$

iii). *For* $k \in \{0, 1, ..., n-1\}$ *the elements of the " $\nabla-$ basis " satisfy the recurrence relation*

$$\left(x - \frac{k}{n}\right) \beta_{n,k}^{\nabla}(x) =$$

$$= (x-2)\left(\frac{k}{n} \beta_{n,k}^{\nabla}(x) - \frac{k+1}{n} \beta_{n,k+1}^{\nabla}(x)\right) + \frac{k^2}{n^2} \beta_{n,k}^{\nabla}(x) - \frac{(k+1)^2}{n^2} \beta_{n,k+1}^{\nabla}(x).$$

iv). *The mean deviation of the discrete distribution* $d(k) := \beta_{n,k}^{\nabla}(x)$ *is given by*

$$\sum_{k=0}^{n} \beta_{n,k}^{\nabla}(x)|k - nx| = \frac{(n - [nx])(nx + [nx])}{n} \beta_{n,[nx]}^{\nabla}(x),$$

where $x \in \mathcal{I}$ *and* $[.]$ *denotes the integral part.*

In the following we list some properties of the sequence $\left(L_n^{\nabla}\right)_{n \geq 1}$. The next result gives a representation of the inverse operator.

Theorem 3.3 $L_n^{\nabla}(\Pi_n) = \Pi_n$. *More precisely, if* $h \in \Pi_n$ *and* $\mathcal{H}_n : \Pi_n \rightarrow \Pi_n$ *is the linear operator*

$$\left(\mathcal{H}_n h\right)(x) = \sum_{j=0}^{n} \frac{(-nx)_j (n)_j}{(-n)_j \ j!} \left(\nabla_{\frac{1}{n}}^j h\right)(0) \ , \quad \nabla_{\frac{1}{n}} = I - E^{-\frac{1}{n}} \ ,$$

then

$$L_n^{\nabla} \mathcal{H}_n = I \quad on \ \Pi_n \ .$$

Lemma 3.4 *The following equalities are satisfied*

1) $L_n^{\nabla} e_j = e_j , j = 0, 1 \ , \quad (L_n^{\nabla} e_2)(x) = e_2(x) + \frac{2x(1-x)}{n+1} \ ,$

$$\left(L_n^{\nabla} e_3\right)(x) = e_3(x) + \frac{6x(1-x) + 6nx^2(1-x)}{(n+1)(n+2)} \ ,$$

$$\left(L_n^{\nabla} e_4\right)(x) = e_4(x) + \frac{2x(1-x)}{(n+1)(n+2)(n+3)} \left(6n(n+3)x - \right.$$

$$\left. -6(n^2+1)x(1-x) + \frac{13n-1}{n}\right) ;$$

2) *with the notation from (28) and (31), we have*

$$\rho_n(\nabla) = \tfrac{2}{n+1}$$

$$(L_n^\nabla \Omega_{1,x})(x) = \frac{\left(n-[nx]\right)\left(nx+[nx]\right)}{n^2}\beta_{n,[nx]}^\nabla(x)\,,$$

$$(L_n^\nabla \Omega_{2,x})(x) = \frac{2x(1-x)}{n+1}\,,$$

$$(L_n^\nabla \Omega_{4,x})(x) = \frac{2x(1-x)}{(n)_4}\left(6n(n-7)x(1-x)+13n-1\right)$$

$$\max_{x\in[0,1]}(L_n^\nabla \Omega_{4,x})(x) = \frac{(3n-1)}{4n(n+1)(n+3)} < \frac{3}{4n^2}\,,\quad n>7\,.$$

Let us introduce the polynomials

$$\lambda_{i,j}(n,x) = \frac{n!}{(n)_n}\frac{(nx)_i(n-nx)_j}{i!\,j!}\,.$$

It is of interest to find the derivatives of the images $L_n^\nabla f$, $f \in C(\mathcal{I})$.

Lemma 3.5 *If D is the derivative, then*

$$DL_n^\nabla f = \sum_{k=0}^{n-1}\sum_{j=0}^{k}\left[\frac{j}{n},\frac{j+n-k}{n}\,;f\right]\lambda_{j,k-j}\,,$$

$$D^2 L_n^\nabla f = \sum_{k=0}^{n-2}\sum_{j=0}^{k}\sum_{\mu=0}^{j}\left(\mathcal{D}_{\mu,j}(n-k-1;f)+\mathcal{D}_{\mu,j}(k-j+1;f)\right)\lambda_{\mu,j-\mu}\,,$$

where $\mathcal{D}_{\mu,j}(i;f) = \left[\frac{\mu}{n},\frac{\mu+i}{n},\frac{\mu+n-j}{n}\,;f\right]$.

Let $s(m,k)$-be the Stirling number of the first kind, and

$$c(m,k) = (-1)^{m-k}s(m,k) > 0\quad.$$

Theorem 3.6 *Let $A \in \Sigma$, $\Pi_p \subseteq Ker(A)$ and*

$$Q_{k,j}(t) := \left|t-\frac{j}{n}\right|_+^k\,,\quad W_{k,j} := A(L_n^\nabla Q_{k,j})\,.$$

Then for $f \in C(\mathcal{I})$

$$A(L_n^\nabla f) = (p+1)\sum_{k=1}^{p}\frac{c(p,k)}{n^{p+1-k}}\sum_{j=p}^{n-1}W_{k,j}\left[\frac{j-p}{n},\frac{j-p+1}{n},...,\frac{j+1}{n}\,;f\right].$$

It is possible to put in evidence a "discrete" equality which involves the Bernstein operator and the operator L_n^∇.

Theorem 3.7 *Let* $d_{j,n}(t) = |t - \frac{i}{n}|_+$, $\xi_{j,n} = L_n^\nabla d_{j,n} - B_n d_{j,n}$. *Then* $\xi_{j,n}(x) \geq 0$, $x \in \mathcal{I}$, $j = 1, 2, ..., n-1$, *and*

$$(L_n^\nabla f)(x) = (B_n f)(x) + \frac{2}{n} \sum_{j=1}^{n-1} \xi_{j,n}(x) \left[\frac{j-1}{n}, \frac{j}{n}, \frac{j+1}{n} ; f \right] \quad . \tag{36}$$

Moreover, for $f \in C^2(\mathcal{I})$, $1 < r \leq \infty$,

$$\|f''\|_r = \left(\int_0^1 |f''(t)|^r \, dt \right)^{\frac{1}{r}} \quad , \quad 1 < r < \infty \quad , \quad \|f''\|_\infty = \max_{\mathcal{I}} |f''| \, ,$$

then

$$\left| \, (L_n^\nabla f)(x) - (B_n f)(x) \, \right| \leq c_r \frac{x(1-x)}{n^{1-\frac{1}{r}}} \|f''\|_r \quad , \quad x \in \mathcal{I} \, , \tag{37}$$

with

$$c_r = \frac{1}{2^{\frac{1}{r}}} \left(\frac{r-1}{2r-1} \right)^{1-\frac{1}{r}} \quad , \quad 1 < r < \infty \quad , \quad c_\infty = \frac{1}{2} \, .$$

Proof. In order to prove (37) it is used (36) together with the following inequality (see [10]): *if*

$$f \in C^2(\mathcal{I}) \, , \, 1 < r \leq \infty \, , \, \zeta_j(n) \geq 0 \, , \, \sum_{j=1}^{n-1} \zeta_j(n) = 1 \, ,$$

then for $h \in (0, \frac{1}{2}]$, $h \leq x_j \leq 1 - h$

$$\left| \, \sum_{j=1}^{n-1} \zeta_j(n) [x_j - h, x_j, x_j + h \, ; f] \, \right| \leq \frac{c_r}{h^{\frac{1}{r}}} \|f''\|_r \quad .$$

◼

Other approximation properties of the sequence $(L_n^\nabla)_{n \geq 1}$ are the following.

Theorem 3.8 a). *For* $f \in C(\mathcal{I})$

$$\|f - L_n^\nabla f\|_\infty \leq \frac{11}{8} \omega_2 \left(f; \frac{1}{\sqrt{n}} \right) \, ;$$

b). *In the case* $f \in C^1(\mathcal{I})$ *the following estimate holds*

$$|f(x) - (L_n^\nabla f)(x)| \leq \sqrt{\frac{x(1-x)}{2(n+1)}} \omega_1 \left(f'; \sqrt{\frac{8x(1-x)}{n+1}} \right) \quad , \quad x \in \mathcal{I} \, ;$$

c). *Suppose that* $f \in C(\mathcal{I})$ *is such that* $f''(x_0)$, $x_0 \in \mathcal{I}$, *exists. Then*

$$\lim_{n \to \infty} n \left(f(x_0) - (L_n^\nabla f)(x_0) \right) = -x_0(1 - x_0) f''(x_0) \, ;$$

The proofs of the above results concerning the sequence $(L_n^\nabla)_{n \geq 1}$ are given in [14] and [15].

3.3   The Sequence $(L_n^L)_{n \geq 1}$

In this case the delta operator is $L = \phi(D) = \frac{D}{I+D}$. From $\phi^{-1}(t) = \frac{t}{1-t}$ we observe that the positivity condition (12) is satisfied. The basic sequence for L is $(l_n(x))_{n \geq 0}$, where

$$l_0(x) = 1 \quad , \quad l_n(x) = \sum_{k=1}^{n} \frac{n!}{k!} \binom{n-1}{k-1} x^k \quad , \quad n \geq 1 .$$

It may be shown that (see [15])

$$\rho_n(L) = -\frac{3n+2}{n^2} + \frac{2}{n^2} \frac{l_{n+1}(n)}{l_n(n)}$$

and $\frac{1}{n} < \rho_n(L) \leq \frac{3}{n}$, where $\rho_n(L)$ is prescribed in (28). Therefore , the sequence $(L_n^L)_{n \geq 1}$ converges pointwise to the identity on the whole space $C(\mathcal{I})$.

4   A conjecture

Conjecture 4.1 *If* $(L_n^Q) \in \mathcal{B}$, *then there exists a sequence of linear operators* $S_n^Q : C(\mathcal{I}) \to C(\mathcal{I}), n = 1, 2, ...,$ *with*

$$f \in \mathbf{K_s} \implies S_n^Q f \in \mathbf{K_s} \quad , \quad s = -1, 0, 1, ...,$$

such that

$$L_n^Q = S_n^Q B_n \quad , \qquad on \ C(\mathcal{I}) ,$$

where B_n *is the Bernstein operator.*

The reasons of this statement are :

1.   The extremal property from Theorem 2.5 (see (b)) ,

2.   There are some particular cases in which such representations are valid. For instance

a). Let $Q = T = \ln (I + D)$. In this case , according to Theorem 3.1 we have $L_n^T = S_n^T B_n$ with

$$S_n^T = \frac{e^{-n}}{t_n(n)} \sum_{k=0}^{\infty} \frac{k^n n^k}{k!} B_k \quad ;$$

b). $Q = \nabla$. Then $L_n^\nabla = S_n^\nabla B_n$ where

$$\left(S_n^\nabla f\right)(x) = \int_0^1 \frac{t^{nx-1}(1-t)^{n-nx-1}}{B(nx, n-nx)} f(t) \, dt \quad , \quad x \in (0,1) ,$$

$$\left(S_n{}^\nabla f\right)(x_0) \; := f(x_0) \; , \; x_0 \in \{0,1\}.$$

Other approximation operators constructed by means of some binomial sequences are given in [2], [16] , [17] .

Acknowledgment. The author thanks Dr.P.Sablonniere for his helpful comments and careful reading of the manuscript.

References

[1] H.Brass, *Eine Verallgemeinerung der Bernsteinschen Operatoren,* Abhandl.Math.Sem. Hamburg **36** (1971), 111-222.

[2] E.W.Cheney and A.Sharma, *On a generalization of Bernstein polynomials,* Riv.Mat.Univ.Parma (2) **5** (1964), 77-82.

[3] I.Gavrea,H.H.Gonska,D.P.Kacsó , *Positive linear operators with equidistant nodes* , Comput.Math.Appl. **8** (1996), 23-32.

[4] I.Gavrea, H.H.Gonska, D.P.Kacsó , *On discretely defined positive linear polynomial operators giving optimal degrees of approximation* , Schriftenreihe des Fachbereichs Mathematik, Univ.Duisburg/Germany, SM-DU-364 (1996) (submitted for publication).

[5] D.P.Kacsó , *Approximation in* $C^r[a,b]$, Ph.D. Thesis,Cluj-Napoca **1997**.

[6] S.Karlin, *Total positivity and convexity preserving transformations* , Proc.Symposia in Pure Mathematics vol.**VII**,Convexity, AMS. (1963), 329-347.

[7] S.Karlin, *Total Positivity*, Stanford University Press,**1968** .

[8] A. Lupaş , *Contribuţii la teoria aproximării prin operatori liniari*, Ph.D.Thesis, Cluj-Napoca, **1976** .

[9] A. Lupaş, *A generalization of Hadamard inequalities for convex functions,* Univ. Beograd Publ. Elektrotehn. Fak. Ser. Mat. Fiz. **No. 544-576**, (1976), 115–121.

[10] A.Lupaş , *Inequalities for divided differences* , Univ.Beograd Publ.Elektrotehn.Fak.Ser.Mat.Fiz.**No.678-715**, (1980), 24-28.

[11] A. Lupaş , *Dobinski-Type Formula for Binomial Polynomials* , Studia Univ.Babeş-Bolyai, Mathematica, **XXXIII** , 2, (1988), 40-44.

[12] A. Lupaş , *The approximation by means of some linear positive operators* , Approximation Theory, Proc.IDoMAT 95 ; Edited by M.W.Müller, M.Felten, D.H.Mache ,Mathematical Research, Vol.86,pp.201-229,Akademie Verlag,Berlin **1995** .

[13] A.Lupaş , *Classical Polynomials and Approximation Theory* , Colloquiumvortrag, Angewandte Analysis, Uni-Duisburg, Dec. **1996**.

[14] A.Lupaş ,*Approximation properties of the operators* $(L_n^\nabla)_{n\geq 1}$, (submitted for publication) .

[15] L.Lupaş and A.Lupaş, *Polynomials of binomial type and approximation operators* , Studia Univ.Babeş-Bolyai, Mathematica, **XXXII** ,4 , (1987) , 61-69.

[16] C.Manole, *Dezvoltări in serii de polinoame Appell generalizate cu aplicaţii la aproximarea funcţiilor* , Ph.D.Thesis , Cluj-Napoca, **1984**.

[17] Gr. Moldovan , *Generalizări ale polinoamelor lui S.N.Bernstein* , Ph.D.Thesis ,Cluj-Napoca, **1971** .

[18] R.Mullin and G.C.Rota , *On the foundations of combinatorial theory* (III), Theory of binomial enumeration, Graph Theory and its Applications, Academic Press, New York, **1970** , 167-213 .

[19] T.Popoviciu , *Remarques sur les polynômes binomiaux* , Mathematica **6** (1932), 8-10.

[20] S.Roman, *The Umbral Calculus* , Academic Press,New York, **1984** .

[21] S.M.Roman and G.C.Rota , *The Umbral Calculus* , Advances in Mathematics **27** ,2, (1978), 95-188.

[22] G.C.Rota, D.Kahaner and A.Odlyzko, *Finite Operator Calculus* , J.Math.Anal.Appl. **42** (1973), 685-760.

[23] P. Sablonniere, *Positive Bernstein-Sheffer Operators,* J.Approx.Theory **83** (1995), 330-341.

[24] D.D.Stancu, *Approximation of functions by a new class of linear positive operators,* Rev.Roum.Math.Pures et Appl. **13** (1968), 1173-1194.

[25] D.D.Stancu, *Approximation of functions by means of some new classes of positive linear operators,* " Numerische Methoden der Approximationstheorie", Proc.Conf.Oberwolfach 1971 ISNM vol.16, Birkhäuser- Verlag,Basel, **1972**, 187-203.

Department of Mathematics
Faculty of Sciences
University ,,Lucian Blaga" of Sibiu
Str.I.Ratiu nr.7 , 2400 Sibiu , Romania
Email address: lupas@cs.sibiu.ro

International Series of Numerical Mathematics
Vol. 132, © 1999 Birkhäuser Verlag Basel/Switzerland

Certain Results Involving Gammaoperators

A. Lupaş, D.H. Mache, V. Maier, M.W. Müller

Dedicated to Professor Werner Meyer - König

Abstract

A systematic and general investigation of the moments of Gammaoperators and the higher derivatives of its kernel leads in a very natural way to Laguerre polynomials and to a sequence (φ_k) of polynomials with degree $[\frac{k}{2}]$. The paper is closed with a computational algorithm for a special linear combination $L_{n,r}$ of Gammaoperators.

Key words and phrases: Approximation by positive linear operators, Gammaoperators in L_p - spaces, linear combination, Laguerre polynomials.

1   Introduction

Gammaoperators G_n, $n \geq 2$ are defined for functions $f \in L_{1,loc}(0,\infty)$ by

$$G_n(f;x) := (G_n f)(x) := \int_0^\infty g_n(x,t) f(\frac{n}{t}) dt, \ x > 0, \tag{1}$$

with the kernel $g_n(x,t) = \frac{x^{n+1}}{n!} e^{-xt} t^n$. A change of variables yields the second representation

$$(G_n f)(x) = \frac{1}{n!} \int_0^\infty e^{-t} t^n f(\frac{nx}{t}) dt, \quad x > 0. \tag{2}$$

These operators have been introduced in [5] and investigated in subsequent papers [3],[6],[7],[9]. The global rate of weighted simultaneous approximation in the L_p - metric was characterized in [1]. V. Totik [9] has proved that the method $(G_n)_{n \geq 2}$ is globally saturated of order $\mathcal{O}(n^{-1})$.

For a special linear combination of Gammaoperators in L_p - spaces having a much better degree of approximation than Gammaoperators themselves direct, converse and an equivalence theorem have been proved in [2]. All these investigations require very careful estimates of the local decay of the moments of Gammaoperators for large n and a suitable representation for the derivatives of the kernel. In the earlier papers these basic tools have been obtained in each case by ad hoc methods. For the proofs in our last paper [2] a systematic and general treatment of these tools was necessary. This was one of the reasons to write this parallel paper. Its results are interesting in themselves.

2 The sequence (φ_k)

Let $(\varphi_k)_{k\in\mathbb{N}_0}$ denote the sequence of functions defined by means of the generating function

$$e^{xH(t)} = \sum_{k=0}^{\infty} \varphi_k(x)\frac{t^k}{k!}, \tag{3}$$

where

$$H(t) := -(t + \ln(1-t)) = \sum_{k=2}^{\infty} \frac{t^k}{k}, \quad |t| < 1.$$

From combinatorial theory (see e.g. [8]) it is well known that (φ_k) is a sequence of polynomials. The problem is now to give an explicit representation for φ_k, a recurrence relation and to determine the exact degree of φ_k. For $a \in \mathbb{C}$ we use the notation

$$(a)_k := a(a+1)\cdots(a+k-1) = \frac{\Gamma(a+k)}{\Gamma(a)}, \quad k \in \mathbb{N}_0. \tag{4}$$

According to this

$$(1-t)^{-x} = \sum_{k=0}^{\infty}(x)_k\frac{t^k}{k!}, \quad |t| < 1, x \in \mathbb{R}. \tag{5}$$

Now we obtain immediately from (3) and

$$
\begin{aligned}
e^{xH(t)} &= e^{-xt-x\ln(1-t)} = e^{-xt}(1-t)^{-x} \\[2mm]
&= \left(\sum_{k=0}^{\infty}(-x)^k\frac{t^k}{k!}\right)\left(\sum_{k=0}^{\infty}(x)_k\frac{t^k}{k!}\right) \\[2mm]
&= \sum_{k=0}^{\infty}\left(\sum_{l=0}^{k}(-1)^{k-l}\binom{k}{l}x^{k-l}(x)_l\right)\frac{t^k}{k!}
\end{aligned}
$$

the representation

$$\varphi_k(x) = \sum_{l=0}^{k}(-1)^{k-l}\binom{k}{l}x^{k-l}(x)_l, \quad k \in \mathbb{N}_0. \tag{6}$$

The first polynomials are $\varphi_0(x) = 1$, $\varphi_1(x) = 0$, $\varphi_2(x) = x$, $\varphi_3(x) = 2x$ and $\varphi_4(x) = 3x^2 + 6x$. From (6) and (4) it is clear that

$$\varphi_k(0) = 0, \quad k \in \mathbb{N}_0. \tag{7}$$

Differentiating (3) with respect to x one obtains by similar considerations as above easily the following recurrence relation for the derivatives

$$\frac{d}{dx}\varphi_{k+2}(x) = \sum_{l=0}^{k}\binom{k+2}{l}(k+1-l)!\varphi_l(x), \qquad k \in \mathbb{N}_0. \tag{8}$$

By differentiating (3) with respect to t one has, using the fact that $\varphi_1(x) = 1$ and $H'(t) = \frac{t}{1-t}$ for $|t| < 1$,

$$\frac{t}{1-t}\sum_{k=0}^{\infty}x\varphi_k(x)\frac{t^k}{k!} = \sum_{k=0}^{\infty}x\varphi_{k+1}(x)\frac{t^k}{k!}$$

and from this

$$t\sum_{k=0}^{\infty}x\varphi_k(x)\frac{t^k}{k!} = t\sum_{k=0}^{\infty}\frac{\varphi_{k+2}(x)}{k+1}\frac{t^k}{k!} - t\sum_{k=0}^{\infty}\varphi_{k+1}(x)\frac{t^k}{k!}.$$

A comparison of coefficients gives the three term recurrence relation

$$\varphi_{k+2}(x) = (k+1)\left(\varphi_{k+1}(x) + x\varphi_k(x)\right), \qquad k \in \mathbb{N}_0 \tag{9}$$

with $\varphi_0(x) = 1$, $\varphi_1(x) = 0$.

Because of $e^{(x+y)H(t)} = e^{xH(t)}e^{yH(t)}$, $x, y \in \mathbb{R}$, $|t| < 1$ the sequence $(\varphi_k)_{k \in \mathbb{N}_0}$ satisfies the binomial identity

$$\varphi_k(x+y) = \sum_{l=0}^{k}\binom{k}{l}\varphi_l(x)\varphi_{k-l}(y), \qquad x, y \in \mathbb{R}, \qquad k \in \mathbb{N}_0. \tag{10}$$

The auxiliary material at hand permits now to determine the exact degree of the polynomial φ_k :

Suppose that $m_j := deg(\varphi_j)$, $j \in \mathbb{N}_0$. $m_0 = m_1 = 0, m_2 = 1$ and (9) imply by induction that the sequence (m_j) is increasing. From (8) we see that $m_{j+2} - m_j = 1$, $j \in \mathbb{N}_0$. Summing up these equations for $j = 1, \cdots, k-1$ in case $k = 2j$ and for $j = 1, \cdots, k$ in case $k = 2j+1$ we obtain

$$deg(\varphi_k) = \left[\frac{k}{2}\right], \qquad k \in \mathbb{N}_0. \tag{11}$$

In later applications we will need the leading coefficient of the polynomial φ_k. For this purpose we insert into (9) the monomial representation of φ_k, $k \geq 2$

$$\varphi_k(x) = l_{0,k}x^{[\frac{k}{2}]} + l_{1,k}x^{[\frac{k}{2}]-1} + \cdots + l_{[\frac{k}{2}]-1,k}x. \tag{12}$$

A comparison of the leading coefficients gives then for $k = 2j$, $j \in \mathbb{N}$

$$l_{0,2j+2} = (2j+1)l_{0,2j}, \qquad l_{0,2} = 1$$

and thus

$$l_{0,2n} = \frac{(2n)!}{2^n n!}, \qquad n \in \mathbb{N}. \tag{13}$$

For $k = 2j - 1$, $j \geq 2$ we obtain in a similar manner with (13)

$$l_{0,2j+1} = 2j(l_{0,2j} + l_{0,2j-1}) = 2j\frac{(2j)!}{2^j j!} + 2jl_{0,2j-1},$$

which may be written in the form

$$\frac{l_{0,2j+1}}{2^j j!} - \frac{l_{0,2j-1}}{2^{j-1}(j-1)!} = 2\left(\frac{j(2j+1)!}{3 \cdot 4^j [j!]^2} - \frac{(j-1)(2j-1)!}{3 \cdot 4^{(j-1)}[(j-1)!]^2}\right) \qquad j \geq 2.$$

Summing these equations for $j = 2, \cdots, n$ gives

$$\frac{l_{0,2n+1}}{2^n n!} - \frac{l_{0,3}}{2} = \frac{2n(n+1)!}{3 \cdot 4^n [n!]^2} - 1$$

and thus on account of $l_{0,3} = 2$

$$l_{0,2n+1} = \frac{(2n+1)!}{3 \cdot 2^{n-1}(n-1)!}, \qquad n \in \mathbb{N}. \tag{14}$$

Making use of

$$(x)_l = \frac{\Gamma(x+l)}{\Gamma(x)} = \frac{1}{\Gamma(x)} \int_0^\infty e^{-t} t^{x-1+l} dt, \qquad x > 0, l \in \mathbb{N}_0$$

we obtain the integral representation

$$\varphi_k(x) = \frac{1}{\Gamma(x)} \int_0^\infty e^{-t} t^{x-1}(t-x)^k \, dt, \qquad x > 0, k \in \mathbb{N}_0 \tag{15}$$

for the polynomials φ_k on $(0, \infty)$.

3   The moments of the Gammaoperators

Let us define as usual the monomials $e_r(t) := t^r$, $r \in \mathbb{N}_0$. Note that $G_n e_0 = e_0$ and $G_n e_1 = e_1$ for $n \geq 2$

Lemma 3.1 *For* $n > r \geq 2$, $r \in \mathbb{N}$

$$G_n(e_r; x) = x^r \frac{n^r(n-r)!}{n!} = x^r \sum_{l=0}^{r-1} c_{r,l} \frac{n}{n-l}, \tag{16}$$

where

$$c_{r,l} = \frac{(-1)^{r-1-l}}{(r-1)!}\binom{r-1}{l} l^{r-1}, \qquad \sum_{l=0}^{r-1} c_{r,l} = 1.$$

Proof. In order to prove the right hand part of (16) we consider the Lagrange polynomial of degree $r-1$ interpolating the monomial e_{r-1} at the knots $0, 1, \cdots, r-1$. Evidently

$$x^{r-1} = \sum_{l=0}^{r-1} \frac{\omega(x)}{(x-l)\omega'(l)} l^{r-1}, \quad \omega(x) := x(x-1)\cdots(x-r+1).$$

Taking here $x = n$ gives the representation of the coefficients $c_{r,l}$. It should be noticed that $c_{r,0} = 0$. Moreover

$$\sum_{l=0}^{r-1} c_{r,l} = \frac{1}{(r-1)!} \sum_{l=0}^{r-1} (-1)^{r-1-l} \binom{r-1}{l} l^{r-1} = [0, 1, \cdots, r-1; e_{r-1}] = 1,$$

where $[x_0, \cdots, x_m; f]$ denotes the m - th divided difference of a function f at the knots $x_0, \cdots, x_m$. ∎

Remark 3.2 *The left hand part of (16) is true for $n > r \geq 0$.*

Lemma 3.3 *Let $n > r \geq 2$. Then*

$$\frac{n^r(n-r)!}{n!} = \sum_{l=0}^{r-1} c_{r,l} \frac{n}{n-l} = \sum_{k=0}^{\infty} \frac{S(k+r-1, r-1)}{n^k},$$

where $S(i, j) := [0, 1, \cdots, j; e_i]$ with $i, j \in \mathbb{N}$ denote the Stirling numbers of the second kind.

Proof. From (16), for $1 \leq l \leq r - 1 < n$ and because of $c_{r,0} = 0$

$$\sum_{l=0}^{r-1} c_{r,l} \frac{n}{n-l} = \sum_{l=1}^{r-1} \frac{(-1)^{r-1-l}}{(r-1)!} \binom{r-1}{l} l^{r+1} \left(\sum_{k=0}^{\infty} \frac{l^k}{n^k} \right)$$

$$= \sum_{k=0}^{\infty} \left(\sum_{l=0}^{r-1} \frac{(-1)^{r-1-l}}{(r-1)!} \binom{r-1}{l} l^{k+r-1} \right) \frac{1}{n^k}$$

$$= \sum_{k=0}^{\infty} [0, 1, \cdots, r-1; e_{k+r-1}] \frac{1}{n^k}.$$

∎

For deeper investigations of the so-called moments of the Gammaoperators, i.e.

$$T_{m,n}(x) := G_n((t-x)^m; x), \quad m, n \in \mathbb{N}_0, \, n > m, \, x > 0 \tag{17}$$

Laguerre polynomials are unavoidable.

Let us remind that the k-th degree Laguerre polynomial, $k \in \mathbb{N}_0$, is defined in the following way:

$$L_k^{(\alpha)}(z) := \frac{(\alpha+1)_k}{k!} \sum_{j=0}^k (-1)^j \binom{k}{j} \frac{z^j}{(\alpha+1)_j}, \qquad \alpha \in \mathbb{C}, z \in \mathbb{C}, \qquad (18)$$

where the symbol $(a)_k$ has been explained by (4). For α real, $\alpha > -1$ the Laguerre polynomials are orthogonal on $(0, \infty)$ with regard to the weight function $e^{-t} t^\alpha$.

Theorem 3.4 *Let* $n > m \geq 1$ *and* $x > 0$. *Then*

$$T_{m,n}(x) = \frac{m!(-x)^m}{(-n)_m} L_m^{(-n-1)}(-n). \qquad (19)$$

Proof. Using (3.2) and (16), we have

$$G_n((t-x)^m; x) = \sum_{j=0}^m (-1)^{m-j} x^{m-j} \binom{m}{j} G_n(e_j; x)$$

$$= x^m \sum_{j=0}^m (-1)^{m-j} \binom{m}{j} \frac{n^j (n-j)!}{n!}$$

and with $\frac{n^j(n-j)!}{n!} = \frac{(-n)^j}{(-n)_j}$, $j = 0, 1, \cdots, m$

$$G_n((t-x)^m; x) = \frac{m!(-1)^m x^m}{(-n)_m} \frac{(-n)_m}{m!} \sum_{j=0}^m (-1)^j \binom{m}{j} \frac{(-n)^j}{(-n)_j}$$

$$= \frac{m!(-x)^m}{(-n)_m} L_m^{(-n-1)}(-n). \qquad \blacksquare$$

Taking into account the recurrence relation for Laguerre polynomials [4, p. 241]

$$m L_m^{(\alpha)}(x) = (2m - 1 + \alpha - x) L_{m-1}^{(\alpha)}(x) - (m - 1 + \alpha) L_{m-2}^{(\alpha)}(x),$$

(19) gives immediately

Corollary 3.5 *The moments of the Gammaoperators satisfy for* $n > m \geq 2$ *and* $x > 0$ *the recurrence relation*

$$T_{m,n}(x) = \frac{(m-1)x}{n-m+1} \left(2 T_{m-1,n}(x) + x T_{m-2,n}(x)\right) \qquad (20)$$

with $T_{0,n}(x) = 1, T_{1,n}(x) = 0$.

Theorem 3.6 *Fix $m \geq 2$ and let $n > m \geq 2$, $x > 0$. Then*

$$\lim_{n \to \infty} n^{\left[\frac{m+1}{2}\right]} T_{m,n}(x) = D_{0,m} x^m, \tag{21}$$

where $D_{0,m} = l_{0,m}$ for m even, $D_{0,m} = 2l_{0,m}$ for m odd and $l_{0,m}$ is defined by (13) and (14) respectively.

Proof. With $n = m - 1 + l$, $l \geq 2$ and (15) we obtain

$$
\begin{aligned}
T_{m,n}(x) &= \frac{1}{n!} \int_0^\infty e^{-t} t^n \left(\frac{nx}{t} - x\right)^m dt \\[2mm]
&= \frac{x^m}{n!} \int_0^\infty e^{-t} t^{l-1} ((m-1) - (t-l))^m \, dt \\[2mm]
&= \frac{x^m (n-m)!}{n!} \sum_{k=0}^m (-1)^k \binom{m}{k} (m-1)^{m-k} \varphi_k(n-m+1).
\end{aligned}
$$

From the binomial identity (10) we have

$$\varphi_k(n - m + 1) = \sum_{j=0}^k \binom{k}{j} \varphi_j(n) \varphi_{k-j}(1 - m)$$

and this implies

$$T_{m,n}(x) = \frac{x^m (n-m)!}{n!} \sum_{k=0}^m d_{k,m} \varphi_k(n) \tag{22}$$

with

$$d_{k,m} := (-1)^k \binom{m}{k} \sum_{j=0}^{m-k} (-1)^j \binom{m-k}{j} (m-1)^{m-k-j} \varphi_j(1-m).$$

The $d_{k,m}$ are independent of n. We have especially

$$d_{m,m} = (-1)^m, \quad d_{m-1,m} = (-1)^{m-1} m(m-1).$$

Inserting (12) into (22) we obtain easily

$$T_{m,n}(x) = \frac{x^m (n-m)!}{n!} \left(d_{0,m} + \sum_{k=2}^m d_{k,m} \sum_{j=1}^{\left[\frac{k}{2}\right]} l_{\left[\frac{k}{2}\right]-j,k} n^j \right) \tag{23}$$

$$= \frac{x^m (n-m)!}{n!} \left(D_{0,m} n^{\left[\frac{m}{2}\right]} + R(n) \right),$$

where $R(n)$ is a polynomial of degree at most $[\frac{m}{2}] - 1$ in the variable n and

$$
\begin{aligned}
D_{0,m} &= d_{m,m}l_{0,m} + \frac{1 - (-1)^m}{2} d_{m-1,m}l_{0,m-1} \\[2mm]
&= \begin{cases} l_{0,m} & \text{for} \quad m \text{ even} \\[4mm] 2l_{0,m} & \text{for} \quad m \text{ odd} . \end{cases}
\end{aligned}
$$

(21) is now a consequence of the fact that $\frac{(n-m)!}{n!}$ behaves asymptotically like n^{-m}.   ∎

A combination of (23) with Lemma 3.3 yields easily the important

Corollary 3.7 *Let $n > m \geq 2$ and $x > 0$. Then*

$$
T_{m,n}(x) = \frac{x^m}{n^{[\frac{m+1}{2}]}} \sum_{k=0}^{\infty} \frac{\beta_k}{n^k} , \tag{24}
$$

where the $\beta_k = \beta_k(m)$ are independent of n and x;

$$
0 \leq T_{m,n}(x) \leq C \frac{x^m}{n^{[\frac{m+1}{2}]}} \tag{25}
$$

with a suitable $C = C(m)$.

In order to justify in (25) the positivity of $T_{m,n}(x)$ it is sufficient to use (20).

4   Derivatives of the kernel of Gammaoperators

Here again Laguerre polynomials and the polynomials φ_k will play the most important role.
Using the definition (18) of Laguerre polynomials and the representation (6) of the polynomials φ_k we obtain after short calculations the very useful relation

$$
L_k^{(n+1-k)}(n+1) = \frac{(-1)^k}{k!} \varphi_k(-n-1), \quad k \in \mathbb{N}, n \geq k - 1. \tag{26}
$$

The following two lemmata are linking the partial derivatives of the kernel $g_n(x,t)$ to Laguerre polynomials and Laguerre polynomials to the polynomials φ_k respectively.

Lemma 4.1 *Let $k \in \mathbb{N}, n \geq k - 1, x > 0$ and $t \geq 0$. Then*

$$
\frac{\partial^k g_n(x,t)}{\partial x^k} = \frac{k!}{x^k} g_n(x,t) L_k^{(n+1-k)}(xt) .
$$

Proof. Applying Leibniz's rule we obtain in the first instance

$$\frac{\partial^k g_n(x,t)}{\partial x^k} = \frac{t^n}{n!}\frac{\partial^k}{\partial x^k}(x^{n+1}e^{-xt})$$

$$= g_n(x,t)\frac{k!}{x^k}\sum_{j=0}^{k}\frac{(-1)^j}{k!}\binom{k}{j}\frac{(n+1)!}{(n+1-k+j)!}(xt)^j .$$

Now

$$\frac{(n+1)!}{k!(n+1-k+j)!} = \binom{n+1}{k}\frac{1}{(n+2-k)_j} = \frac{(n+2-k)_k}{k!(n+2-k)_j}$$

and (26) implies

$$\frac{\partial^k g_n(x,t)}{\partial x^k} = g_n(x,t)\frac{k!}{x^k}\frac{(n+2-k)_k}{k!}\sum_{j=0}^{k}(-1)^j\binom{k}{j}\frac{(xt)^j}{(n+2-k)_j}$$

$$= g_n(x,t)\frac{k!}{x^k}L_k^{(n+1-k)}(xt) .$$

∎

Lemma 4.2 *Let $k \in \mathbb{N}$, $n \geq k - 1$, $x > 0$ and $t \geq 0$. Then*

$$L_k^{(n+1-k)}(xt) = \frac{x^k}{k!}\sum_{j=0}^{k}(-1)^j\binom{k}{j}(\frac{n+1}{x} - t)^{k-j}\frac{\varphi_j(-n-1)}{x^j} .$$

Proof. By Taylor's formula

$$L_k^{(n+1-k)}(xt) = \sum_{j=0}^{k}\frac{(xt - n - 1)^j}{j!}\frac{d^j}{dy^j}L_k^{(n+1-k)}(y)|_{y=n+1}$$

and by induction

$$\frac{d^j}{dy^j}L_m^{(\alpha)}(y) = (-1)^j L_{m-j}^{(\alpha+j)}(y), \quad \alpha > -1, \ 0 \leq j \leq m .$$

Thus

$$\frac{d^j}{dy^j}L_k^{(n+1-k)}(y)|_{y=n+1} = (-1)^j L_{k-j}^{(n+1-k+j)}(n+1), \quad k \in \mathbb{N}, \ n \geq k - 1, \ 0 \leq j \leq k$$

and with (26)

$$
L_k^{(n+1-k)}(xt) = \sum_{j=0}^{k} \frac{(n+1-xt)^j}{j!} L_{k-j}^{(n+1-k+j)}(n+1)
$$

$$
= \sum_{j=0}^{k} \frac{(n+1-xt)^j}{(k-j)!} L_j^{(n+1-j)}(n+1)
$$

$$
= \frac{x^k}{k!} \sum_{j=0}^{k} (-1)^j \binom{k}{j} (\frac{n+1}{x} - t)^{k-j} \frac{\varphi_j(-n-1)}{x^j} .
$$

∎

We are able now to describe the structure of partial derivatives of the kernel in a very clear way.

Theorem 4.3 *Let $k \in \mathbb{N}$, $n \geq k-1$, $x > 0$ and $t \geq 0$. Then*

$$
\frac{\partial^k g_n(x,t)}{\partial x^k} = (\frac{n+1}{x} - t)^k g_n(x,t) + \sum_{j=2}^{k} \sum_{\mu=1}^{[\frac{j}{2}]} c_{j\mu} \frac{(n+1)^\mu}{x^j} (\frac{n+1}{x} - t)^{k-j} g_n(x,t),
$$

(27)

where $c_{j\mu} := c_{j\mu}(k) := (-1)^{j+\mu} \binom{k}{j} l_{[\frac{j}{2}]-\mu,j}$.

Remark 4.4 *For $k = 1$ (27) reduces to $\frac{\partial g_n(x,t)}{\partial x} = (\frac{n+1}{x} - t) g_n(x,t)$.*

Remark 4.5 *(27) says roughly that the k-th partial derivative of the kernel contains at most terms of the form*

$$
\frac{(n+1)^b}{x^a} (\frac{n+1}{x} - t)^{k-a} g_n(x,t), \quad 0 \leq a \leq k, 1 \leq b \leq [\frac{a}{2}].
$$

Proof of Thereom 4.3. Combining the above two lemmata and bearing in mind that $\varphi_0(x) = 1, \varphi_1(x) = 0$ and by (12)

$$
\varphi_j(x) = \sum_{\mu=0}^{[\frac{j}{2}]-1} l_{\mu,j} x^{[\frac{j}{2}]-\mu} = \sum_{\mu=1}^{[\frac{j}{2}]} l_{[\frac{j}{2}]-\mu,j} x^\mu \quad j \geq 2
$$

we obtain

$$
\frac{\partial^k g_n(x,t)}{\partial x^k} = (\frac{n+1}{x} - t)^k g_n(x,t) + \sum_{j=2}^{k} (-1)^j \binom{k}{j} \frac{\varphi_j(-n-1)}{x^j} (\frac{n+1}{x} - t)^{k-j} g_n(x,t),
$$

where the last sum becomes equal to

$$\sum_{j=2}^{k}\sum_{\mu=1}^{[\frac{j}{2}]}(-1)^j\binom{k}{j}l_{[\frac{j}{2}]-\mu,j}(-1)^\mu\frac{(n+1)^\mu}{x^j}(\frac{n+1}{x}-t)^{k-j}g_n(x,t).$$

With $c_{j\mu} := c_{j\mu}(k) := (-1)^{j+\mu}\binom{k}{j}l_{[\frac{j}{2}]-\mu,j}$ the theorem is proved. ∎

By means of Lemma 4.1 and the theory of confluent hypergeometric functions (see e.g.[8]) the following recurrence relation can be established for $2 \le k \le n+1$, $x > 0$:

$$\frac{\partial^k g_n(x,t)}{\partial x^k} = (\frac{n-k+2}{x}-t)\frac{\partial^{k-1}g_n(x,t)}{\partial x^{k-1}} - \frac{(k-1)t}{x}\frac{\partial^{k-2}g_n(x,t)}{\partial x^{k-2}}. \tag{28}$$

5 A sequence of linear combinations of Gammaoperators

We fix $r \in \mathbb{N}$, $r \ge 2$ and consider for $n \ge r+1$ special linear combinations of Gammaoperators of the form

$$L_{n,r} := \sum_{j=1}^{r}\alpha_j(n,r)G_{jn} \tag{29}$$

with the coefficients

$$\alpha_j(n,r) = \frac{(-1)^{r-j}}{n^r}\binom{r}{j}\binom{jn}{r} = \frac{(-1)^{r-j}}{(r-1)!}\binom{r-1}{j-1}(jn-1)\cdots(jn-r+1).$$

These new operators are acting e.g. on the space $L_{1,loc}(0,\infty)$, they are no longer positive, but composed of positive operators.
For instance

$$L_{n,2} = (-1+\frac{1}{n})G_n + (2-\frac{1}{n})G_{2n},$$

$$L_{n,3} = (\frac{1}{2}-\frac{3}{2n}+\frac{1}{n^2})G_n + (-4+\frac{6}{n}-\frac{2}{n^2})G_{2n} + (\frac{9}{2}-\frac{9}{2n}+\frac{1}{n^2})G_{3n}.$$

Lemma 5.1 *For $r \in \mathbb{N}$, $n \ge r+1$ the coefficients satisfy the conditions*

$$\sum_{j=1}^{r}\alpha_j(n,r) = 1, \tag{30}$$

$$\sum_{j=1}^{r}|\alpha_j(n,r)| \le C \tag{31}$$

with $C = C(r)$.

Proof. Evidently

$$\sum_{j=1}^{r} \alpha_j(n,r) = [n, 2n, \ldots, rn; (t-1)(t-2)\cdots(t-r+1)]$$

$$= [n, 2n, \ldots, rn; e_{r-1} + q] \quad \text{with} \quad q \in \Pi_{r-2}$$

$$= 1,$$

which proves (30). From

$$|\alpha_j(n,r)| = \frac{1}{(r-1)!}\binom{r-1}{j-1}(j-\frac{1}{n})\cdots(j-\frac{r-1}{n})$$

$$\leq \frac{1}{(r-1)!}\binom{r-1}{j-1}r^{r-1}, \, 1 \leq j \leq r$$

we obtain

$$\sum_{j=1}^{r}|\alpha_j(n,r)| \leq \frac{r^{r-1}}{(r-1)!}\sum_{j=1}^{r}\binom{r-1}{j-1} = \frac{r^r}{r!}2^{r-1} =: C,$$

which proves (31).   ∎

Remark 5.2 *(30) implies formally $L_{n,1} = G_n$.*

The following very useful identity can easily be proved:

$$\sum_{j=1}^{r}\alpha_j(n,r)f(jn) = [n, 2n, \ldots, rn; (t-1)\cdots(t-r+1)f(t)], \qquad (32)$$

for every function $f : [n, \infty) \to \mathbb{R}$.

Lemma 5.3 *For $r \geq 2$, $n \geq r+1$*

$$L_{n,r}h = h, \qquad \text{for all} \quad h \in \Pi_r.$$

Proof. $L_{n,r}e_0 = e_0$ by (30). With (16) and (32) we obtain for e_s, $1 \leq s \leq r$

$$L_{n,r}e_s = e_s\sum_{j=1}^{r}\alpha_j(n,r)\frac{(jn)^s(jn-s)!}{(jn)!} = e_s[n, 2n, \ldots, rn; h],$$

where $h(t) = t^{s-1}(t-s)(t-s-1)\cdots(t-r+1)$, which admits the representation $h = e_{r-1} + g$, $g \in \Pi_{r-2}$. Therefore $L_{n,r}e_s = e_s$, $1 \leq s \leq r$, which proves the lemma.   ∎

Remark 5.4 *In the same way as in the above proof we get*

$$L_{n,r}e_{r+1} = e_{r+1}[n, 2n, \ldots, rn; \frac{t^r}{t-r}],$$

which is unequal to e_{r+1}. Therefore the degree of exactness of the linear combinations $L_{n,r}$ is exactly r.

Remark 5.5 *The sequence $(G_n)_{n \geq 2}$ of Gammaoperators is known to be a positive linear approximation method on the spaces $L_p(0, \infty)$, $1 \leq p < \infty$ with respect to the L_p-metric (see 11), i.e. $\lim_{n \to \infty} \|G_n f - f\|_p = 0$, for all $f \in L_p(0, \infty)$, $1 \leq p < \infty$. Due to (29) - (31) this implies for $r \geq 2$, $n \geq r + 1$*

$$0 \leq \|L_{n,r}f - f\|_p \leq \sum_{j=1}^{r} |\alpha_j(n, r)| \cdot \|G_{jn}f - f\|_p$$

$$\leq C \sum_{j=1}^{r} \|G_{jn}f - f\|_p,$$

where the upper bound tends to zero for $n \to \infty$, i.e

$$\lim_{n \to \infty} \|L_{n,r}f - f\|_p = 0, \qquad \text{for all} \qquad f \in L_p(0, \infty), 1 \leq p < \infty. \tag{33}$$

In other words the sequence $(L_{n,r})_{n \geq r+1}$ is a linear approximation method on the spaces $L_p(0, \infty)$, $1 \leq p < \infty$ with respect to the L_p- metric.

The approximation properties (direct and converse results) will be studied in a forthcoming paper. We close with

Theorem 5.6 **(Computational Algorithms for $L_{n,r}$)**
Let $r \geq 2$, $n \geq r + 1$ and put

$$U_n(s, s) := (r-1)! \binom{sn-1}{r-1} G_{sn}, \, s = 1, \ldots, r. \tag{34}$$

If the linear operators $U_n(j, k)$ on $L_{1,loc}(0, \infty)$ are recursively given by

$$U_n(j, k) = \frac{U_n(j+1, k) - U_n(j, k-1)}{(k-j)n}, \tag{35}$$

for $k = 2, 3, \ldots, r$ and $j = 1, 2, \ldots, k - 1$, then

$$U_n(1, r) = L_{n,r}. \tag{36}$$

Proof. We consider the Gamma-type operator

$$G_f(t,x) := \frac{1}{\Gamma(t+1)} \int_0^\infty e^{-y} y^t f(\frac{tx}{y})\, dy, \quad t > 0,\ x > 0,\ f \in L_{1,loc}(0,\infty).$$

Since

$$G_f((j+l)n, x) = G_{(j+l)n}(f;x), \quad l \in \mathbb{N}_0$$

we have according to (35) and with (32)

$$(U_n(j,k)f)(x) = [jn, (j+1)n, \ldots, kn; (t-1)(t-2)\cdots(t-r+1)G_f(t,x)]_t.$$

This implies for $j = 1$ and $k = r$

$$U_n(1,r) \quad = \quad \frac{1}{(r-1)!n^{r-1}} \sum_{k=1}^r (-1)^{r-k} \binom{r-1}{k-1}(kn-1)(kn-2)\cdots(kn-r+1)G_{kn}$$

$$= \quad \sum_{k=1}^r \alpha_k(n,r)G_{kn} \qquad \text{by (29)}$$

$$= \quad L_{n,r}. \qquad\qquad\qquad \blacksquare$$

Example $(r = 3)$.

In the same way as for divided differences we have to compute the elements from the following array:

$$U_n(1,1)$$
$$U_n(1,2)$$
$$U_n(2,2) \qquad\qquad U_n(1,3) = L_{n,3}$$
$$U_n(2,3)$$
$$U_n(3,3)$$

The first column contains the "initial values"

$$U_n(s,s) \quad = \quad (sn-1)(sn-2)G_{sn} \qquad \text{for} \quad s = 1,2,3\,, \text{i.e}$$

$$U_n(1,1) \quad = \quad (n^2 - 3n + 2)G_n, \qquad U_n(2,2) = (4n^2 - 6n + 2)G_{2n}\,,$$

$$U_n(3,3) \quad = \quad (9n^2 - 9n + 2)G_{3n}\,.$$

By means of (35) the elements of the second column are calculated, namely

$$U_n(1,2) \quad = \quad (4n - 6 + \frac{2}{n})G_{2n} - (n - 3 + \frac{2}{n})G_n$$

$$U_n(2,3) \quad = \quad (9n - 9 + \frac{2}{n})G_{3n} - (4n - 6 + \frac{2}{n})G_{2n}\,.$$

Finally

$$
\begin{aligned}
L_{n,3} &= U_n(1,3) = \frac{U_n(2,3) - U_n(1,2)}{2n} \\
&= (\frac{1}{2} - \frac{3}{2n} + \frac{1}{n^2})G_n + (-4 + \frac{6}{n} - \frac{2}{n^2})G_{2n} + (\frac{9}{2} - \frac{9}{2n} + \frac{1}{n^2})G_{3n} ,
\end{aligned}
$$

which has been obtained at the beginning of this chapter directly from the definition.

Remark 5.7 *In [2] the authors considered a linear combination of $2r - 1$ instead of r Gammaoperators, called $M_{n,r}$. At first sight the $L_{n,r}$ seem to have a much simpler form than the $M_{n,r}$ in [2]. However, in order to construct $M_{n,r}$, we need [2, Chapter 4] only one initial value $D_{n,0}$ and $2r-2$ "iterations" $D_{n,1}, \ldots, D_{n,2r-2}$. At the same time the algorithm for $L_{n,r}$ needs r initial values $U_n(1,1), \ldots, U_n(r,r)$ and $\frac{r(r-1)}{2}$ "iterations" $U_n(j,k)$.*

References

[1] A. LUPAŞ, D.H. MACHE and M.W. MÜLLER, *Weighted L_p - approximation of derivatives by the method of Gammaoperators*, Results in Mathematics, VOL. 28, No. 3/4 (1995), 277 – 286.

[2] A. LUPAŞ, D.H. MACHE, V. MAIER and M.W. MÜLLER, *Linear combinations of Gammaoperators in L_p- spaces*, Results in Mathematics, VOL. 34, (1998), 156 – 168.

[3] A. LUPAŞ and M.W. MÜLLER, *Approximationseigenschaften der Gammaoperatoren*, Math. Zeitschr., 98 (1967), 208 – 226.

[4] W. MAGNUS, F. OBERHETTINGER and P.P. SONI, *Formulas and Theorems for the Special Functions of Mathematical Physics*, Springer - Verlag, Berlin - Heidelberg - New York (1966).

[5] M.W. MÜLLER, *Die Folge der Gammaoperatoren*, Dissertation, Stuttgart (1967).

[6] M.W. MÜLLER, *Punktweise und gleichmäßige Approximation durch Gammaoperatoren*, Math. Zeitschr., 103 (1968), 227 – 238.

[7] M.W. MÜLLER, *Einige Approximationseigenschaften der Gammaoperatoren*, Mathematica, 10 (33), (1968), 303 – 310.

[8] E.D. RAINVILLE, *Special Functions*, 4 th edition, Macmillan Company, New York (1967).

[9] V. TOTIK, *The gammaoperators in L^p spaces*, Publ. Math. Debrecen 32 (1985) 43 – 55.

Alexandru Lupaş
Universitatea din Sibiu
Facultatea de Stiinţe
Str. I.Ratiu nr.7
RO - 2400 Sibiu, Romania
Email address: lupas@vectra.sibiu.ro

Detlef H. Mache
Ludwig-Maximilians Universität München
Mathematisches Institut
Lehrstuhl für Numerische Analysis
Theresienstr. 39
D - 80333 München, Germany
Email address: mache@rz.mathematik.uni-muenchen.de
                (or: mache@math.uni-dortmund.de)

Volker Maier & Manfred W. Müller
Universität Dortmund
Lehrstuhl VIII für Mathematik
Vogelpothsweg 87
D - 44221 Dortmund, Germany
Email address: volker.maier@math.uni-dortmund.de
                manfred.mueller@math.uni-dortmund.de

International Series of Numerical Mathematics
Vol. 132, © 1999 Birkhäuser Verlag Basel/Switzerland

Recent research at Cambridge on radial basis functions

M.J.D. Powell

Abstract

Much of the research at Cambridge on radial basis functions during the last
four years has addressed the solution of the thin plate spline interpolation
equations in two dimensions when the number of interpolation points, n say,
is very large. It has provided some techniques that will be surveyed because
they allow values of n up to 10^5, even when the positions of the points are
general. A close relation between these techniques and Newton's interpolation method is explained. Another subject of current research is a new way
of calculating the global minimum of a function of several variables. It is
described briefly, because it employs a semi-norm of a large space of radial
basis functions. Further, it is shown that radial basis function interpolation
minimizes this semi-norm in a way that is a generalisation of the well-known
variational property of thin plate spline interpolation in two dimensions.
The final subject is the deterioration in accuracy of thin plate spline interpolation near the edges of finite grids. Several numerical experiments in the
one-dimensional case are reported that suggest some interesting conjectures
that are still under investigation.

1 Introduction

Let the values $f_i = f(\underline{x}_i)$, $i = 1, 2, \ldots, n$, of a real valued function $f(\underline{x})$, $\underline{x} \in \mathcal{R}^d$,
be given, where the points $\underline{x}_i \in \mathcal{R}^d$, $i = 1, 2, \ldots, n$, are all different. A radial basis
function interpolant to the data has the form

$$s(\underline{x}) = \sum_{j=1}^{n} \lambda_j \, \phi(\|\underline{x} - \underline{x}_j\|) + p(\underline{x}), \qquad \underline{x} \in \mathcal{R}^d, \tag{1.1}$$

for some function $\phi(r)$, $r \geq 0$, where the vector norm is Euclidean and where p is a
polynomial of degree at most m, say. Therefore m is a nonnegative integer, except
that we employ the notation $m = -1$ to denote the important case when p has to

be identically zero. For $m \geq 0$, we express p as the sum

$$p(\underline{x}) = \sum_{j=1}^{\hat{m}} c_j \pi_j(\underline{x}), \qquad \underline{x} \in \mathcal{R}^d, \tag{1.2}$$

where π_j, $j = 1, 2, \ldots \hat{m}$, is a basis of the linear space of polynomials of degree at most m from $\mathcal{R}^d$ to $\mathcal{R}$. After choosing ϕ and m, the coefficients of s are required to satisfy the $(n+\hat{m}) \times (n+\hat{m})$ system of linear equations

$$\left(\begin{array}{c|c} \Phi & P \\ \hline P^T & 0 \end{array} \right) \left(\begin{array}{c} \underline{\lambda} \\ \underline{c} \end{array} \right) = \left(\begin{array}{c} \underline{f} \\ 0 \end{array} \right), \tag{1.3}$$

where Φ and P are the $n \times n$ and $n \times \hat{m}$ matrices that have the elements $\Phi_{ij} = \phi(\|\underline{x}_i - \underline{x}_j\|)$ and $P_{ij} = \pi_j(\underline{x}_i)$, respectively. Further, $\underline{\lambda} \in \mathcal{R}^n$ and $\underline{c} \in \mathcal{R}^{\hat{m}}$ are the vectors of coefficients of s, and the components of $\underline{f}$ are the data f_i, $i = 1, 2, \ldots, n$. Thus the first n rows of expression (1.3) are the interpolation conditions $s(\underline{x}_i) = f_i$, $i = 1, 2, \ldots, n$.

We assume throughout this paper that, if $m \geq 1$, then the data points $\underline{x}_i$, $i = 1, 2, \ldots, n$, are such that the rank of P is $\hat{m}$, because otherwise the system (1.3) would be singular. In particular, if $m = 1$ and $d = 2$, the assumption is that the data points are not collinear. Then the crucial property of the radial basis function method is that several choices of $\phi(r)$, $r \geq 0$, are available such that the equations (1.3) have a unique solution, with no other restrictions on the data points (Micchelli, 1986), except that $n \geq 2$ is required occasionally for $m = -1$. Further, for each of these choices of ϕ there is an integer m_0 such that, if $m \geq m_0$ and if $\underline{v}$ is any nonzero vector that satisfies $P^T \underline{v} = 0$, then $\underline{v}^T \Phi \underline{v}$ is nonzero and its sign is opposite to that of $(-1)^{m_0}$, where $\underline{v}$ can be any nonzero vector in the case $m_0 = -1$. The most popular choices of ϕ are the Gaussian $\phi(r) = \exp(-c r^2)$, the multiquadric $\phi(r) = (r^2 + c^2)^{1/2}$ and the thin plate spline $\phi(r) = r^2 \log r$, c being a positive constant. The values of m_0 in these cases are -1, 0 and 1, respectively.

The nonsingularity of the system (1.3) when the data points are in general positions provides an excellent reason for studying the radial basis function method. Further, the approximation $s \approx f$ is highly useful in a wide range of practical applications (Hardy, 1990), because accuracy and smoothness properties are achieved that compare favourably with those of other methods (Franke, 1982). Many interesting theoretical questions occur too. Some stunning answers are derived from Fourier transforms by Buhmann (1990) when the data points form a regular infinite grid, but, for finite n, it is usual for numerical experiments to suggest orders of accuracy that have not been confirmed theoretically. Therefore the subject is still an active field of research at Cambridge. Three of the topics that are under investigation will be considered in the remaining three sections of this paper.

The first of them is an iterative procedure for solving the equations (1.3) when n is very large, in the important special case of thin plate spline interpolation in two dimensions. Therefore $d=2$ and $m=1$ hold, and ϕ is the function

$$\phi(r) = r^2 \log r, \qquad r \geq 0, \tag{1.4}$$

the value of $\phi(0)$ being zero. The work began about six years ago, as reported by Beatson and Powell (1994), and a breakthrough for data points in general positions is described by Powell (1997). Thus, by applying the Laurent expansion technique of Beatson and Newsam (1992) to generate the product $\Phi\underline{\lambda}$ for trial values of $\underline{\lambda}$, it is possible for n to exceed 10^4, although there is no useful sparsity in Φ. Another major advance was made by Goodsell last year, which reduces substantially the number of products that have to be calculated on each iteration. Thus a close relation between the procedure and Newton's interpolation formula was noticed, that assists a description of the new method. These findings are presented in Section 2.

Section 3 addresses an application of the radial basis function method to global optimization, the least value of $f(\underline{x})$, $\underline{x} \in \mathcal{X}$, being required, where $\mathcal{X}$ is a subset of $\mathcal{R}^d$. We take the view that it is expensive to calculate $f(\underline{x})$ for each choice of $\underline{x}$. Therefore we let the data $f_i = f(\underline{x}_i)$, $i = 1, 2, \ldots, n$, be all the values of the objective function that are known already, and we consider the crucial question of deciding on the position of $\underline{x}_{n+1}$, which is the vector of variables for the next calculation of f. A way of choosing this point is recommended. It depends on a measure of smoothness that is derived from the properties of Φ that are stated in the second paragraph of this section. Specifically, we find that the nonnegativity of $(-1)^{m_0+1}\underline{\lambda}^T\Phi\underline{\lambda}$ provides a useful semi-norm of the interpolant (1.1), and that there is an associated scalar product that assists the calculation of $\underline{x}_{n+1}$ by the recommended procedure. This semi-norm and scalar product belong to the large linear space that is studied by Schaback (1993).

The final topic includes some interesting examples of orders of accuracy that are suggested by numerical experiments. Specifically, we consider thin plate spline interpolation in only one dimension when the data points x_i, $i = 0, 1, \ldots, n$, are spaced uniformly throughout the interval $[0, 1]$. The numerical results indicate the accuracy

$$\max\{|f(x) - s(x)| : 0 \leq x \leq 1\} = \mathcal{O}(h^{3/2}) \tag{1.5}$$

when f is smooth, where h is the spacing between data points, although $\mathcal{O}(h^3)$ accuracy occurs away from the ends of the range (Bejancu, 1997). Therefore, as suggested by Beatson and Powell (1992) for multiquadrics, we investigate the replacement of the $P^T\underline{\lambda} = 0$ part of expression (1.3) by two other conditions that take up the degrees of freedom that remain in the coefficients of s after the interpolation equations are satisfied. Thus it is shown numerically in Section 4 that it is possible to reduce the right hand side of expression (1.5) from $\mathcal{O}(h^{3/2})$ to $\mathcal{O}(h^{5/2})$ when f is a general smooth function. Work has begun on trying to establish these curious orders of convergence analytically.

2   An iterative procedure for interpolation

The new procedure of this section is analogous to the following version of Newton's method for polynomial interpolation in one dimension. Let the interpolation conditions be $s(x_i) = f_i$, $i = 1, 2, \ldots, n$, where x_i, $i = 1, 2, \ldots, n$, are points of $\mathcal{R}$ that are all different, and where s is now required to be a polynomial of degree at most $n-1$ from $\mathcal{R}$ to $\mathcal{R}$. We introduce the Lagrange functions $\chi_n(x) = 1$, $x \in \mathcal{R}$, and

$$\chi_k(x) = \prod_{j=k+1}^{n} \frac{x - x_j}{x_k - x_j}, \qquad x \in \mathcal{R}, \qquad k = 1, 2, \ldots, n-1, \tag{2.1}$$

and we write s in the form

$$s(x) = \sum_{k=1}^{n} \mu_k \chi_k(x), \qquad x \in \mathcal{R}. \tag{2.2}$$

It follows that the partial sum

$$s_\ell(x) = \sum_{k=\ell}^{n} \mu_k \chi_k(x), \qquad x \in \mathcal{R}, \tag{2.3}$$

is the unique polynomial of degree at most $n - \ell$ that interpolates the last $n - \ell + 1$ data, where ℓ is any integer in $[1, n]$. Further, μ_ℓ is the coefficient of $x^{n-\ell}$ in s_ℓ divided by the coefficient of $x^{n-\ell}$ in χ_ℓ. Thus it is possible to calculate the parameters μ_k, $k = 1, 2, \ldots, n$, of formula (2.2). We are going to employ a construction of this kind.

Specifically, for thin plate spline interpolation in two dimensions to the data $f_i = f(\underline{x}_i)$, $i = 1, 2, \ldots, n$, the analogue of expression (2.1) is the Lagrange function

$$\chi_k(\underline{x}) = \sum_{j=k}^{n} \lambda_{kj} \, \phi(\|\underline{x} - \underline{x}_j\|) + p_k(\underline{x}), \qquad \underline{x} \in \mathcal{R}^2, \tag{2.4}$$

whose parameters are fixed by the conditions

$$\left. \begin{array}{ll} \chi_k(\underline{x}_j) = \delta_{jk}, & j = k, k+1, \ldots, n, \\ \sum_{j=k}^{n} \lambda_{kj} = 0 \quad \text{and} \quad \sum_{j=k}^{n} \lambda_{kj} \, \underline{x}_j = 0 \end{array} \right\}, \tag{2.5}$$

where ϕ is the function (1.4) and where p_k is a polynomial from $\mathcal{R}^2$ to $\mathcal{R}$ of degree at most one. Thus χ_k is defined uniquely for all integers k in $[1, \hat{n}]$, where $\hat{n}$ is an integer from $[1, n-3]$ such that the last $q = n - \hat{n}$ of the points $\underline{x}_i$, $i = 1, 2, \ldots, n$, are not collinear. Then, corresponding to equation (2.2), we write the required interpolant to all the data in the form

$$s(\underline{x}) = \sum_{k=1}^{\hat{n}} \mu_k \chi_k(\underline{x}) + \sigma(\underline{x}), \qquad \underline{x} \in \mathcal{R}^2, \tag{2.6}$$

where σ is the thin plate spline interpolant to the last q data. Thus the analogue of expression (2.3), namely the partial sum

$$s_\ell(\underline{x}) = \sum_{k=\ell}^{\hat{n}} \mu_k \chi_k(\underline{x}) + \sigma(\underline{x}), \qquad \underline{x} \in \mathcal{R}^2, \tag{2.7}$$

is the unique thin plate spline interpolant to the last $n-\ell+1$ data, the centres of the radial functions being at the interpolation points as usual, where ℓ is any integer from $[1, \hat{n}]$. It follows that, for each of these integers ℓ, the parameter μ_ℓ is the coefficient of $\phi(\|\underline{x}-\underline{x}_\ell\|)$ in s_ℓ divided by $\lambda_{\ell\ell}$. Fortunately, this remark provides a convenient way of calculating μ_ℓ, which will be explained in the case $\ell=1$.

Let A be the leading $n \times n$ submatrix of the inverse of the partitioned matrix of the system (1.3). It follows from this system that, for general right hand sides f_i, $i = 1, 2, \ldots, n$, the thin plate spline coefficients of expression (1.1) are the components of the product

$$\underline{\lambda} = A \underline{f}. \tag{2.8}$$

In particular, by choosing $\underline{f}$ to be the first coordinate vector in $\mathcal{R}^n$, we find that the elements of the first column of A are the coefficients λ_{1j}, $j = 1, 2, \ldots, n$, of the Lagrange function χ_1. Now A is a symmetric matrix and equation (2.8) gives $\lambda_1 = \sum_{j=1}^{n} A_{1j} f_j$. Further, the definition (2.7) with $\ell=1$ provides the formula

$$\mu_1 = \lambda_1 / \lambda_{11} = \sum_{j=1}^{n} \lambda_{1j} f_j / \lambda_{11}. \tag{2.9}$$

Thus it is straightforward to generate μ_1 if the coefficients of the Lagrange function χ_1 are available. Similarly, the conclusion of the previous paragraph gives the values

$$\mu_\ell = \sum_{j=\ell}^{n} \lambda_{\ell j} f_j / \lambda_{\ell\ell}, \qquad \ell=1, 2, \ldots, \hat{n}. \tag{2.10}$$

Therefore equation (2.6) with the coefficients (2.10) can be regarded as Newton's method for radial basis function interpolation.

The new procedure is similar to the one that has just been described, except that it makes approximations to the Lagrange functions that have been mentioned, and it is applied recursively, in order to correct the errors that arise from the approximations. Indeed, the value of q, introduced between expressions (2.5) and (2.6), is about 30, and the estimate of the function (2.4) has the form

$$\hat{\chi}_k(\underline{x}) = \sum_{j \in \mathcal{I}(k)} \hat{\lambda}_{kj} \phi(\|\underline{x}-\underline{x}_j\|) + \hat{p}_k(\underline{x}), \qquad \underline{x} \in \mathcal{R}^2, \tag{2.11}$$

where $\mathcal{I}(k)$ is a subset of the integers $\{k, k+1, \ldots, n\}$ that has only q elements. Specifically, $\hat{\chi}_k$ is the Lagrange function of thin plate spline interpolation at the points $\underline{x}_j$, $j \in \mathcal{I}(k)$, that has the properties $\hat{\chi}_k(\underline{x}_j) = \delta_{jk}$, $j \in \mathcal{I}(k)$. Further, it is usual to define the elements of $\mathcal{I}(k)$ by the condition that $\underline{x}_j$, $j \in \mathcal{I}(k)$, are the q points in the set $\{\underline{x}_j : j = k, k+1, \ldots, n\}$ that are closest to $\underline{x}_k$, including $\underline{x}_k$ itself. Then we deduce from the previous two paragraphs that the function

$$\hat{s}(\underline{x}) = \sum_{k=1}^{\hat{n}} \hat{\mu}_k \, \hat{\chi}_k(\underline{x}) + \hat{\sigma}(\underline{x}), \qquad \underline{x} \in \mathcal{R}^2, \tag{2.12}$$

may be a useful approximation to s, if we pick the coefficients

$$\hat{\mu}_\ell = \sum_{j \in \mathcal{I}(\ell)} \hat{\lambda}_{\ell j} \, f_j \, / \, \hat{\lambda}_{\ell\ell}, \qquad \ell = 1, 2, \ldots, \hat{n}, \tag{2.13}$$

where $\hat{\sigma}$ is the thin plate spline solution of the interpolation equations

$$\hat{\sigma}(\underline{x}_i) = f_i - \sum_{k=1}^{\hat{n}} \hat{\mu}_k \, \hat{\chi}_k(\underline{x}_i), \qquad i = \hat{n}+1, \hat{n}+2, \ldots, n, \tag{2.14}$$

in order that $\hat{s}$ interpolates the last $q = n - \hat{n}$ data.

In general, however, $\hat{s}$ does not interpolate all the data, because of the errors of the estimates $\hat{\chi}_k \approx \chi_k$, $k = 1, 2, \ldots, \hat{n}$. It happens typically in numerical experiments that $\hat{s}$ has the property

$$\max\{|f_i - \hat{s}(\underline{x}_i)| : i = 1, 2, \ldots, n\} \approx 0.1 \max\{|f_i| : i = 1, 2, \ldots, n\}. \tag{2.15}$$

Therefore the following iterative refinement method is recommended. We now let s be an estimate of the required interpolant that is set to zero initially. Then the data f_i, $i = 1, 2, \ldots, n$, are replaced by $f_i - s(\underline{x}_i)$, $i = 1, 2, \ldots, n$, throughout the description of the previous paragraph. In particular, equation (2.13) becomes the formula

$$\hat{\mu}_\ell = \sum_{j \in \mathcal{I}(\ell)} \hat{\lambda}_{\ell j} \, [f_j - s(\underline{x}_j)] \, / \, \hat{\lambda}_{\ell\ell}, \qquad \ell = 1, 2, \ldots, \hat{n}. \tag{2.16}$$

The resultant thin plate spline function $\hat{s}$ is added to s, which completes the iteration and which is expected to reduce $\max\{|f_i - s(\underline{x}_i)| : i = 1, 2, \ldots, n\}$ by about a factor of 10. Termination occurs when the maximum residual becomes acceptably small. All the coefficients of the approximate Lagrange functions are generated before the iterations are begun. Therefore most of the work of an iteration is the calculation of the residuals, which requires only $\mathcal{O}(n \log n)$ operations, because the Laurent expansion technique of Beatson and Newsam (1992) is applied. Thus interpolation equations with more than 10^4 data points can be solved, as claimed in Section 1.

The difference between this iterative procedure and the version that was in use a year ago is that the estimate s of the required interpolant was revised $\hat{n}+1$ times, instead of only once, on each iteration. Specifically, for each integer ℓ in

$[1, \hat{n}]$, the coefficient (2.16) was calculated, and then $\hat{\mu}_\ell \hat{\chi}_\ell$ was added to s before increasing ℓ for the next application of formula (2.16). After all of these changes to s, the thin plate spline $\hat{\sigma}$ was defined by the interpolation conditions

$$\hat{\sigma}(\underline{x}_i) = f_i - s(\underline{x}_i), \qquad i = \hat{n}+1, \hat{n}+2, \dots, n, \qquad (2.17)$$

which correspond to expression (2.14), and s was overwritten by $s+\hat{\sigma}$, which completed the iteration. Then the main task of each iteration was generating the residuals $f_j - s(\underline{x}_j)$, $j \in \mathcal{I}(\ell)$, for $\ell = 1, 2, \dots, \hat{n}$. Goodsell noticed, however, that the amount of work per iteration would be reduced by about a factor of $q = |\mathcal{I}(\ell)|$ if the residuals at the beginning of the iteration were employed instead. Therefore he investigated this idea by trying some numerical experiments. He found in most cases that the number of iterations to achieve any prescribed accuracy was about the same as before. Then consideration of a theoretical explanation provided the relation between the procedure and Newton's interpolation method that has been described in this section.

Recent work has included attempts to construct positions of the data points $\underline{x}_i$, $i = 1, 2, \dots, n$, so that one or both of the versions of the iterative procedure cause the residuals $f_i - s(\underline{x}_i)$, $i = 1, 2, \dots, n$, to diverge. The year old version seems to be very robust, but some examples of failure have been found for the newer version that usually requires much less computation. One of the bad cases is when $n = 1000$ and the data points are equally spaced on two concentric quarter circles of radii 1 and $1+10^{-5}$. Therefore research has started on trying to develop an intermediate version that combines the reliability of the year old procedure with some of the savings that can be achieved by calculating fewer residuals. In particular, we have tried with some success to calculate the first half of the coefficients (2.16) using the residuals at the beginning of the iteration. Then the first half of the sum of expression (2.12) is added to s, which gives new residuals that are employed in the calculation of the third quarter of the coefficients (2.16). Further, we alternate between updating s and calculating about half of the remainder of the coefficients. Thus the ordering of the data points $\underline{x}_i$, $i = 1, 2, \dots, n$, becomes important, so several questions require attention. Another current topic of investigation is the replacement of the thin plate spline radial function by the multiquadric $\phi(r) = (r^2 + c^2)^{1/2}$, $r \geq 0$, in the iterative procedure. Several numerical experiments have been tried, most of them being very successful.

3 A semi-norm for global optimization

The author has become very interested in an application of radial basis function interpolation to global optimization due to an inspiring paper of Jones (1996). The global optimization problem is to find the least value of a function $f(\underline{x})$, $\underline{x} \in \mathcal{X}$, where $\mathcal{X}$ is a subset of $\mathcal{R}^d$, when the value $f(\underline{x})$ can be calculated for any $\underline{x}$ in $\mathcal{X}$, but no derivatives are available. Let the values that have been generated already at the beginning of an iteration of a recursive procedure be $f(\underline{x}_i)$, $i = 1, 2, \dots, n$.

Then the main task of an iteration is to use these data to select $\underline{x}_{n+1}$, which is the next point at which the value of the objective function will be calculated. Jones described the following technique for making the selection automatically. It may demand much routine computation, which is tolerable if function values are very expensive.

A way of defining a unique function s that satisfies the interpolation equations

$$s(\underline{x}_i) = f(\underline{x}_i) = f_i, \qquad i = 1, 2, \ldots, n, \tag{3.1}$$

is required, for general n, $\underline{x}_i \in \mathcal{X}$ and f_i, $i = 1, 2, \ldots, n$. Further, s should be the "smoothest" function from a linear space $\mathcal{A}$ that is allowed by the interpolation equations, where the smoothness of $a \in \mathcal{A}$ is a nonnegative real number, $\sigma(a)$ say, that is defined for every a in $\mathcal{A}$. Therefore, if a is any function in $\mathcal{A}$ with $a(\underline{x}_i) = f_i$, $i = 1, 2, \ldots, n$, then the smoothness property of s is the inequality $\sigma(s) \le \sigma(a)$. Now $\mathcal{A}$ has to be large enough to include a function $\hat{s}$ that satisfies the conditions

$$\hat{s}(\underline{x}_{n+1}) = f_* \quad \text{and} \quad \hat{s}(\underline{x}_i) = f_i, \quad i = 1, 2, \ldots, n, \tag{3.2}$$

for any choice of $\underline{x}_{n+1}$ in $\mathcal{X}$ that is different from $\underline{x}_i$, $i = 1, 2, \ldots, n$, and any $f_* \in \mathcal{R}$. Further, we pick the $\hat{s}$ that minimizes $\sigma(\hat{s})$ subject to the constraints (3.2), and we denote the dependence of $\sigma(\hat{s})$ on $\underline{x}_{n+1} \in \mathcal{X}$ and $f_* \in \mathcal{R}$ by the notation

$$\sigma(\hat{s}) = \hat{\sigma}(\underline{x}_{n+1}, f_*). \tag{3.3}$$

The technique sets $\underline{x}_{n+1}$ to the vector of variables in $\mathcal{X}$ that minimizes $\hat{\sigma}(\underline{x}, f_*)$, $\underline{x} \in \mathcal{X}$, for some chosen value of f_*.

The technique has some nice features due to the dependence of this value of $\underline{x}_{n+1}$ on f_*, assuming the bound

$$f_* < \min\{s(\underline{x}) : \underline{x} \in \mathcal{X}\}, \tag{3.4}$$

where s is defined at the beginning of the previous paragraph. Indeed, because this bound implies $f_* < f_i$, $i = 1, 2, \ldots, n$, the smoothness of $\hat{s}$ keeps $\underline{x}_{n+1}$ away from the points $\underline{x}_i$, $i = 1, 2, \ldots, n$. Furthermore, if f_* is much less than the right hand side of expression (3.4), the value $f_* = -\infty$ being used occasionally, then there is a tendency for the position of $\underline{x}_{n+1}$ to provide a relatively large value of the nearest neighbour distance $\min\{\|\underline{x}_{n+1} - \underline{x}_i\| : i = 1, 2, \ldots, n\}$. Thus $\underline{x}_{n+1}$ can be placed automatically in a large gap between the interpolation points $\underline{x}_i$, $i = 1, 2, \ldots, n$, which is important to global optimization. On the other hand, if the global minimum of $s(\underline{x})$, $\underline{x} \in \mathcal{X}$, is close to the required global minimum, which is usual for sufficiently large n, then it may be possible to achieve some good convergence properties by letting the difference between the two sides of expression (3.4) be tiny. The idea of alternating between such choices of f_* on successive iterations is a subject of current research at Cambridge. Hence we hope to develop some useful new algorithms for global optimization.

We will find in the remainder of this section that the radial basis function interpolation method provides the smoothness properties that are required in a form that is suitable for practical computation. In particular, the work of Duchon (1977) suggests a choice of $\mathcal{A}$ and the smoothness measure $\sigma(a)$, $a \in \mathcal{A}$, for thin plate spline interpolation when $d = 2$. Specifically, letting $\mathcal{A}$ be the linear space of functions from $\mathcal{R}^2$ to $\mathcal{R}$ with square integrable second derivatives, letting $\sigma(a)$ be the integral

$$\sigma(a) = \frac{1}{8\pi} \int_{\mathcal{R}^2} \left(\frac{\partial^2 a(\underline{x})}{\partial x_1^2} \right)^2 + 2 \left(\frac{\partial^2 a(\underline{x})}{\partial x_1 \partial x_2} \right)^2 + \left(\frac{\partial^2 a(\underline{x})}{\partial x_2^2} \right)^2 d\underline{x}, \qquad a \in \mathcal{A}, \quad (3.5)$$

and letting s be the thin plate spline solution of the interpolation equations (3.1) that is studied in Section 2, it is proved by Duchon (1977) that s has the property $\sigma(s) \leq \sigma(a)$, for every $a \in \mathcal{A}$ that satisfies the interpolation conditions. Moreover, the dependence of $\sigma(\hat{s}) = \hat{\sigma}(\underline{x}_{n+1}, f_*)$ on $\underline{x}_{n+1}$ is required, where $\hat{s}$ is the thin plate spline solution of the equations (3.2), but it is not convenient to extract this dependence by estimating the integral (3.5). Integration by parts avoids this difficulty. Indeed, because the bi-Laplacian of $\phi(\|\underline{x}\|)$, $\underline{x} \in \mathcal{R}^2$, in the case (1.4) is a delta function, one can deduce the identity

$$\sigma(a) = \sum_{i=1}^{n(a)} \mu_i \, a(\underline{y}_i), \qquad (3.6)$$

when a is any element of $\mathcal{A}$ that has the form

$$a(\underline{x}) = \sum_{j=1}^{n(a)} \mu_j \, \phi(\|\underline{x} - \underline{y}_j\|) + q(\underline{x}), \qquad \underline{x} \in \mathcal{R}^2. \qquad (3.7)$$

Here the condition $a \in \mathcal{A}$ imposes the constraints $\sum_{j=1}^{n(a)} \mu_j = 0$ and $\sum_{j=1}^{n(a)} \mu_j \, \underline{y}_j = 0$, but there are no other restrictions on the parameters $\mu_j \in \mathcal{R}$ and $\underline{y}_j \in \mathcal{R}^2$, $j = 1, 2, \ldots, n(a)$, and q is any polynomial of degree at most one. For example, equations (1.1), (3.1) and (3.6) give the formula

$$\sigma(s) = \sum_{i=1}^{n} \lambda_i \, s(\underline{x}_i) = \sum_{i=1}^{n} \lambda_i \, f_i. \qquad (3.8)$$

A useful and obvious property of the smoothness measure (3.5) is that the inequality

$$\sigma(\hat{s}) = \hat{\sigma}(\underline{x}_{n+1}, f_*) \geq \sigma(s), \qquad (3.9)$$

holds, because $\hat{s}$ is an element of $\mathcal{A}$ that satisfies the equations (3.2), while $s \in \mathcal{A}$ minimizes $\sigma(s)$ subject to the constraints (3.1). It is not straightforward, however, to extend the smoothness measure (3.5) to other choices of the radial function

ϕ and the number of variables d. An extension suggests itself, however, if we let expression (3.6) be the definition of the smoothness instead, which is equivalent to the definition (3.5) in the $d=2$ thin plate spline case, provided that the linear space $\mathcal{A}$ is reduced. Specifically, the new $\mathcal{A}$ is composed of all functions of the form (3.7) whose parameters satisfy $\sum_{j=1}^{n(a)} \mu_j = 0$ and $\sum_{j=1}^{n(a)} \mu_j \underline{y}_j = 0$. Thus $\mathcal{A}$ includes all the functions that are required by the application to global optimization that has been described. We are going to gain some understanding of expression (3.6) as a measure of smoothness by using it to deduce the property (3.9).

It is sufficient to prove that, if the generic function (3.7) of the linear space $\mathcal{A}$ satisfies the interpolation conditions $a(\underline{x}_i) = f_i$, $i = 1, 2, \ldots, n$, then the inequality $\sigma(a) \geq \sigma(s)$ is achieved. There is no loss of generality in assuming that $\{\underline{x}_i : i = 1, 2, \ldots, n\}$ is a subset of $\{\underline{y}_j : j = 1, 2, \ldots, n(a)\}$, because we can enlarge the latter set if necessary, letting the corresponding coefficients μ_j be zero in expression (3.7). Further, we let Ψ be the $n(a) \times n(a)$ matrix with the elements $\Psi_{ij} = \phi(\|\underline{y}_i - \underline{y}_j\|)$, $1 \leq i, j \leq n(a)$. It follows from equation (3.7) that a takes the values

$$a(\underline{y}_i) = \sum_{j=1}^{n(a)} \Psi_{ij} \mu_j + q(\underline{y}_i), \qquad i = 1, 2, \ldots, n(a), \tag{3.10}$$

so the smoothness measure (3.6) can be expressed in the form

$$\sigma(a) = \sum_{i=1}^{n(a)} \sum_{j=1}^{n(a)} \mu_i \Psi_{ij} \mu_j = \underline{\mu}^T \Psi \underline{\mu}. \tag{3.11}$$

We see that the nonnegativity of $\sigma(a)$ can be attributed to the conditional positive definiteness of Φ (or Ψ) that is stated in the second paragraph of Section 1.

We continue the proof of $\sigma(a) \geq \sigma(s)$ by adding some zero terms to formula (1.1) for the thin plate spline solution of the equations (3.1). Indeed, we assume without loss of generality that $\underline{x}_i = \underline{y}_i$ holds for $i = 1, 2, \ldots, n$, and we set λ_j to zero for every integer j in $[n+1, n(a)]$, which gives the identity

$$s(\underline{x}) = \sum_{j=1}^{n(a)} \lambda_j \phi(\|\underline{x} - \underline{y}_j\|) + p(\underline{x}), \qquad \underline{x} \in \mathcal{R}^2. \tag{3.12}$$

Further, equation (3.11) shows that $\sigma(s)$ is the nonnegative number $\underline{\lambda}^T \Psi \underline{\lambda}$, where $\underline{\lambda} \in \mathcal{R}^{n(a)}$ has the components λ_j, $j = 1, 2, \ldots, n(a)$. Thus the conditional positive definiteness of Ψ provides the inequality

$$\sigma(a) - \sigma(s) = \underline{\mu}^T \Psi \underline{\mu} - \underline{\lambda}^T \Psi \underline{\lambda} = (\underline{\mu} - \underline{\lambda})^T \Psi (\underline{\mu} - \underline{\lambda}) + 2 \underline{\lambda}^T \Psi (\underline{\mu} - \underline{\lambda})$$

$$\geq 2 \underline{\lambda}^T \Psi (\underline{\mu} - \underline{\lambda}). \tag{3.13}$$

Now equations (3.10) and (3.12) imply that the i-th component of the product $\Psi(\underline{\mu} - \underline{\lambda})$ has the value

$$[a(\underline{y}_i) - q(\underline{y}_i)] - [s(\underline{y}_i) - p(\underline{y}_i)], \qquad i = 1, 2, \ldots, n(a). \tag{3.14}$$

Moreover, the second part of the system (1.3) gives $\sum_{i=1}^{n} \lambda_i [p(\underline{y}_i) - q(\underline{y}_i)] = 0$. Therefore, because the last $n(a) - n$ components of $\underline{\lambda} \in \mathcal{R}^{n(a)}$ are zero, inequality (3.13) can be written in the form

$$\sigma(a) - \sigma(s) \geq 2 \sum_{i=1}^{n} \lambda_i [a(\underline{y}_i) - s(\underline{y}_i)] = 2 \sum_{i=1}^{n} \lambda_i [a(\underline{x}_i) - s(\underline{x}_i)]. \qquad (3.15)$$

It follows from the interpolation conditions $s(\underline{x}_i) = f_i$ and $a(\underline{x}_i) = f_i$, $i = 1, 2, \ldots, n$, that the right hand side of expression (3.15) is zero, which completes the proof.

It is straightforward to extend the measure of smoothness (3.6) to the radial basis functions that are introduced at the beginning of Section 1. Recalling the purpose of m_0, we pick a constant integer m that satisfies $m \geq m_0$, and, as in Schaback (1993), we let a general element of $\mathcal{A}$ have the form

$$a(\underline{x}) = \sum_{j=1}^{n(a)} \mu_j \, \phi(\|\underline{x} - \underline{y}_j\|) + q(\underline{x}), \qquad \underline{x} \in \mathcal{R}^d, \qquad (3.16)$$

where the coefficients μ_j can have any values that satisfy $\sum_{j=1}^{n(a)} \mu_j \pi_k(\underline{y}_j) = 0$, $k = 1, 2, \ldots, \hat{m}$, and where q is any polynomial of degree at most m from $\mathcal{R}^d$ to $\mathcal{R}$. As before, $\{\pi_k : k = 1, 2, \ldots, \hat{m}\}$ is a basis of the space of these polynomials. These conditions imply the identity

$$\sum_{i=1}^{n(a)} \mu_i \, a(\underline{y}_i) = \underline{\mu}^T \Psi \underline{\mu}, \qquad (3.17)$$

where Ψ is still the matrix with the elements $\Psi_{ij} = \phi(\|\underline{y}_i - \underline{y}_j\|)$, $1 \leq i, j \leq n(a)$. Now it is claimed in Section 1 that, if $\underline{\mu}$ is nonzero, then $(-1)^{m_0+1} \underline{\mu}^T \Psi \underline{\mu}$ is positive. Therefore we pick the measure of smoothness

$$\sigma(a) = (-1)^{m_0+1} \sum_{i=1}^{n(a)} \mu_i \, a(\underline{y}_i), \qquad a \in \mathcal{A}. \qquad (3.18)$$

Further, we consider the arguments of the previous two paragraphs in this more general setting. It follows that, if s is the element of $\mathcal{A}$ of the form (1.1) whose coefficients are defined by the system (1.3), then $\sigma(s)$ is the least value of $\sigma(a)$, $a \in \mathcal{A}$, subject to the interpolation conditions $a(\underline{x}_i) = f_i$, $i = 1, 2, \ldots, n$. This result is Theorem 4 of Schaback (1993). It provides the highly interesting possibility that several versions of the radial basis function interpolation method may be suitable for our application to global optimization.

Furthermore, there is a scalar product (a, b), between any two elements a and b of $\mathcal{A}$, such that the measure of smoothness is just $\sigma(a) = (a, a)$, $a \in \mathcal{A}$, and such that $(a, a)^{1/2}$, $a \in \mathcal{A}$, is a semi-norm of the linear space $\mathcal{A}$. In particular, if $b \in \mathcal{A}$ is the function

$$b(\underline{x}) = \sum_{j=1}^{n(a)} \nu_j \, \phi(\|\underline{x} - \underline{y}_j\|) + r(\underline{x}), \qquad \underline{x} \in \mathcal{R}^d, \tag{3.19}$$

where r is a polynomial as usual, then equation (3.18) and the homogeneity of the quadratic function (3.17) suggest the value

$$(a, b) = (-1)^{m_0 + 1} \underline{\mu}^T \Psi \underline{\nu} = (-1)^{m_0 + 1} \sum_{i=1}^{n(a)} \mu_i \, b(\underline{y}_i) = (-1)^{m_0 + 1} \sum_{i=1}^{n(a)} \nu_i \, a(\underline{y}_i), \tag{3.20}$$

where the last two identities follow easily from equations (3.19) and (3.16). In general, however, $b \in \mathcal{A}$ is the function

$$b(\underline{x}) = \sum_{j=1}^{n(b)} \nu_j \, \phi(\|\underline{x} - \underline{z}_j\|) + r(\underline{x}), \qquad \underline{x} \in \mathcal{R}^d, \tag{3.21}$$

where the sets $\{\underline{z}_j : j = 1, 2, \ldots, n(b)\}$ and $\{\underline{y}_j : j = 1, 2, \ldots, n(a)\}$ are different. In this case we replace both sets by their union, multiplying the new radial terms of expressions (3.16) and (3.21) by zero coefficients, in order that equation (3.20) becomes valid after increasing $n(a)$. Further, all of the new zero coefficient terms can be removed from the sums of this formula. Thus we derive the bilinear form

$$(a, b) = (-1)^{m_0 + 1} \sum_{i=1}^{n(a)} \mu_i \, b(\underline{y}_i) = (-1)^{m_0 + 1} \sum_{i=1}^{n(b)} \nu_i \, a(\underline{z}_i), \tag{3.22}$$

even if the sets $\{\underline{y}_i : i = 1, 2, \ldots, n(a)\}$ and $\{\underline{z}_i : i = 1, 2, \ldots, n(b)\}$ have no points in common. It follows that (a, b) is a scalar product with the property that $(a, a)^{1/2}$, $a \in \mathcal{A}$, is a semi-norm of the linear space $\mathcal{A}$ (Schaback, 1993). Further, the sums of the definition (3.22) are convenient for the practical computation of the scalar product.

These remarks imply that expression (3.3) has some very useful properties. In order to describe them, we let $\underline{x}_{n+1}$ be any point of $\mathcal{X}$ that is different from $\underline{x}_i$, $i = 1, 2, \ldots, n$, and we let $\chi(\underline{x})$, $\underline{x} \in \mathcal{X}$, be the Lagrange function of radial basis function interpolation at the points $\underline{x}_i$, $i = 1, 2, \ldots, n+1$, that satisfies the conditions

$$\chi(\underline{x}_{n+1}) = 1 \quad \text{and} \quad \chi(\underline{x}_i) = 0, \quad i = 1, 2, \ldots, n. \tag{3.23}$$

Then equations (1.1) and (3.22) give the identity

$$(s, \chi) = (-1)^{m_0 + 1} \sum_{i=1}^{n} \lambda_i \, \chi(\underline{x}_i) = 0, \tag{3.24}$$

which shows that χ is orthogonal to s. Now the function

$$\hat{s}(\underline{x}) = s(\underline{x}) + [f_* - s(\underline{x}_{n+1})]\,\chi(\underline{x}), \qquad \underline{x} \in \mathcal{X}, \tag{3.25}$$

is the radial basis function solution of the interpolation equations (3.2). It follows from the orthogonality property (3.24) that the measure of smoothness (3.3) has the value

$$\sigma(\hat{s}) = (\hat{s}, \hat{s}) = (s, s) + [f_* - s(\underline{x}_{n+1})]^2\,(\chi, \chi). \tag{3.26}$$

Further, defining $\mu(\underline{x}_{n+1})$ to be the coefficient of $\phi(\|\underline{x} - \underline{x}_{n+1}\|)$, $\underline{x} \in \mathcal{X}$, in χ, equations (3.22) and (3.23) provide $(\chi, \chi) = (-1)^{m_0+1}\mu(\underline{x}_{n+1})$. Thus we deduce the formula

$$\hat{\sigma}(\underline{x}_{n+1}, f_*) = \sigma(s) + (-1)^{m_0+1}\,[f_* - s(\underline{x}_{n+1})]^2\,\mu(\underline{x}_{n+1}), \tag{3.27}$$

which shows the dependence of $\hat{\sigma}$ on $\underline{x}_{n+1}$ and f_*. This result is very helpful to the current research on global optimization at Cambridge, not only for seeking a point $\underline{x}_{n+1}$ that minimizes $\hat{\sigma}(\underline{x}_{n+1}, f_*)$ approximately, but also for providing information that assists the choice of f_*.

4   On thin plate spline interpolation in one dimension

Throughout this section we consider interpolation to a function $f(x)$, $0 \le x \le 1$, of only one variable at the equally spaced points $x_i = ih$, $i = 0, 1, \ldots, n$, where n is a positive integer and $h = 1/n$. The interpolant has the form

$$s(x) = \sum_{j=0}^{n} \lambda_j\,\phi(|x - x_j|) + p(x), \qquad 0 \le x \le 1, \tag{4.1}$$

where ϕ and p are still the thin plate spline radial function (1.4) and a polynomial of degree at most one, respectively. Corresponding to the system (1.3), the coefficients of the usual interpolant are defined by the equations

$$s(x_i) = f(x_i), \qquad i = 0, 1, \ldots, n, \tag{4.2}$$

and by the constraints

$$\sum_{j=0}^{n} \lambda_j = 0 \quad \text{and} \quad \sum_{j=0}^{n} \lambda_j x_j = 0, \tag{4.3}$$

but later we will study the replacement of the constraints (4.3) by two other conditions. We give particular attention to the error function

$$e(x) = f(x) - s(x), \qquad 0 \le x \le 1, \tag{4.4}$$

n	$f(x)=x^2$	$f(x)=x^3$
80	0.00013546	0.00009982
160	0.00004780	0.00003554
320	0.00001689	0.00001261
640	0.00000597	0.00000447

Table 1: The usual order of accuracy

n	$f(x)=x^3-\frac{3}{4}x^2$	$f(x)=(4.8)$
80	0.0000017728	0.00005749
160	0.0000003149	0.00002069
320	0.0000000558	0.00000738
640	0.0000000099	0.00000262

Table 2: Attempts at higher order convergence

when f has bounded fourth derivatives. In this case the number $\max\{|e(x)| : \varepsilon \le x \le 1-\varepsilon\}$, is bounded above by a multiple of h^3, where ε is any positive constant and the multiplying factor depends on f and ε but not on h (Bejancu, 1997). Unfortunately, the factor grows so rapidly as $\varepsilon \to 0$ that the maximum error

$$\|e\|_\infty = \max\{|e(x)| : 0 \le x \le 1\} \tag{4.5}$$

is hardly ever of magnitude h^3.

Numerical results show that typically the greatest value of $|e(x)|$, $0 \le x \le 1$, occurs when x is in the interval $[0, h]$ or $[1-h, 1]$. Further, if $|e(\frac{1}{2}h)|$ tends to zero more slowly than a multiple of h^3 as $h \to \infty$, which happens in all the numerical experiments of this section, then $|e(\frac{1}{2}h)|$ indicates the magnitude of $\max\{|e(x)| : 0 \le x \le \frac{1}{2}\}$. Therefore the main parts of the displays of the given tables are just values of $|e(\frac{1}{2}h)|$ for a range of choices of n and f. These results provide some interesting conjectures. Work has started at Cambridge on trying to establish them analytically.

The most usual behaviour is shown in Table 1 for two functions f, where s is defined by the equations (4.2) and (4.3). We see that, when h is halved, then $|e(\frac{1}{2}h)|$ is reduced by a factor of about $2\sqrt{2}$, which suggests the order of accuracy

$$\|e\|_\infty = \mathcal{O}(h^{3/2}). \tag{4.6}$$

This conjecture has been corroborated by many other calculations.

The entries in the third column of Table 1 are about 3/4 of the entries in the second column. Furthermore, if f is any function such that $|e(\frac{1}{2}h)| \sim h^{3/2}$,

n	$f(x)=x^2$	$f(x)=x^3$
80	0.00000668	0.00001120
160	0.00000121	0.00000202
320	0.00000022	0.00000036
640	0.00000004	0.00000006

Table 3: The first derivative conditions (4.9)

then it is usual for $e(x)$, $0 \le x \le h$, to be independent of f approximately, apart from a scaling factor. Therefore the $\mathcal{O}(h^{3/2})$ term of e in Table 1 may become zero if we pick $f(x) = x^3 - \frac{3}{4}x^2$, $0 \le x \le 1$. This possibility was investigated numerically, the results being shown in the second column of Table 2. We find that the condition $|e(\frac{1}{2}h)| = \mathcal{O}(h^{5/2})$ is achieved, which raises the question of characterising the smooth functions f that enjoy an order of convergence that is higher than usual. It is answered in Theorem 4.2 of Beatson and Powell (1992) for multiquadric interpolation, and there it is sufficient if f satisfies the equations

$$f(0) = f(1) = f'(0) = f'(1) = 0. \tag{4.7}$$

Therefore we tested the possibility of a similar situation in the thin plate spline case by making the choice

$$f(x) = \begin{cases} 0, & 0 \le x \le \frac{1}{4}, \\ 10^5 \left(x - \frac{1}{4}\right)^4 \left(x - \frac{3}{4}\right)^4, & \frac{1}{4} \le x \le \frac{3}{4}, \\ 0, & \frac{3}{4} \le x \le 1. \end{cases} \tag{4.8}$$

The resultant values of $|e(\frac{1}{2}h)|$ are shown in the last column of Table 2. We see that the $\mathcal{O}(h^{3/2})$ magnitude of Table 1 occurs again. Therefore we are still seeking a characterisation of the smooth functions f that provide $\|f - s\|_\infty = \mathcal{O}(h^{5/2})$, when s is the thin plate spline radial function that is given by the conditions (4.2) and (4.3).

The form of $e(x)$, $0 \le x \le h$, in the case $|e(\frac{1}{2}h)| \sim h^{3/2}$, mentioned at the beginning of the previous paragraph, has to be due mainly to the thin plate spline interpolation method. Indeed, it seems that $e(x)$, $0 \le x \le h$, is dominated by a function of the form (4.1) that is zero at all the interpolation points. Now different multiples of such functions occur in s if we retain the interpolation conditions (4.2) but alter the constraints (4.3). Further, the magnitude of the error near the ends of the interval $[0, 1]$ is likely to be smaller if the two constraints are replaced by the equations

$$s'(0) = f'(0) \quad \text{and} \quad s'(1) = f'(1). \tag{4.9}$$

Therefore this replacement was investigated numerically for the functions of Table 1, the new values of $|e(\frac{1}{2}h)|$ being shown in Table 3. It is clear that the accuracy is much better than before, and there is strong evidence that $\|e\|_\infty = \mathcal{O}(h^{5/2})$ is achieved instead of the order (4.6).

These results suggest that usually the constraints (4.3) cause unnecessary loss of accuracy, but the derivatives $f'(0)$ and $f'(1)$ may not be available. Therefore another technique from Beatson and Powell (1992) was also investigated. It takes up the two degrees of freedom that remain in the coefficients λ_j, $j = 0, 1, \ldots, n$, of the function (4.1), after the interpolation conditions (4.2) are satisfied, by minimizing the sum of squares $\sum_{j=0}^n \lambda_j^2$. The resultant values of $|e(\frac{1}{2}h)|$ were about four times the corresponding entries in Table 3, so again the order of accuracy seems to be $\mathcal{O}(h^{5/2})$, which was corroborated by other choices of f.

There is a fundamental objection to the methods of the last two paragraphs, however, due to the log term of $\phi(r) = r^2 \log r$, $r \geq 0$. We explain it by scaling the variable x by a positive constant α say, to give the new variable $\hat{x} = \alpha x$. Then the function $f(x)$, $0 \leq x \leq 1$, becomes $f(\alpha^{-1}\hat{x})$, $0 \leq \hat{x} \leq \alpha$, and we seek an interpolant of the form

$$\hat{s}(\hat{x}) = \sum_{j=0}^n \hat{\lambda}_j \, \phi(|\hat{x} - \alpha x_j|) + \hat{p}(\hat{x}), \qquad 0 \leq \hat{x} \leq \alpha, \tag{4.10}$$

that satisfies the conditions

$$\hat{s}(\alpha x_i) = f(x_i), \qquad i = 0, 1, \ldots, n, \tag{4.11}$$

where the notation $x_i = ih$, $i = 0, 1, \ldots, n$, has not changed. The fundamental objection is that, although $\hat{s}(\alpha x)$, $0 \leq x \leq 1$, is a thin plate spline function that interpolates the original data, one may not be able to express it in the form (4.1). Indeed, equation (4.10) implies the identity

$$\hat{s}(\alpha x) = \sum_{j=0}^n \hat{\lambda}_j \, (\alpha x - \alpha x_j)^2 \log |\alpha x - \alpha x_j| + \hat{p}(\alpha x)$$

$$= \sum_{j=0}^n \alpha^2 \hat{\lambda}_j \, \phi(|x - x_j|) + \alpha^2 \log \alpha \sum_{j=0}^n \hat{\lambda}_j \, (x - x_j)^2 + \hat{p}(\alpha x), \tag{4.12}$$

which differs from the form (1.1) by the quadratic polynomial $(\alpha^2 \log \alpha \sum_{j=0}^n \hat{\lambda}_j) x^2$, $0 \leq x \leq 1$. Therefore the abandonment of the first of the constraints (4.3) has caused the interpolation method to lose an important invariance property under simple scaling of the variable $x \in \mathcal{R}$.

Equation (4.12) shows that the objection does not occur if we allow p and $\hat{p}$ in expressions (4.1) and (4.10) to be quadratic instead of linear polynomials. Then three degrees of freedom remain in the coefficients of s after the interpolation equations (4.2) are satisfied, and they can also be fixed by minimizing $\sum_{j=0}^n \lambda_j^2$. This technique was tested for many smooth functions f, and again it was found that the order of accuracy is $\|e\|_\infty = \mathcal{O}(h^{5/2})$. Further, it is usual for the values of

$|e(x)|$, $0 \leq x \leq 1$, to be smaller than those that occur in all the other versions of thin plate spline interpolation that have been addressed in this section. We do not recommend, however, that the new method should be used in practice instead of cubic spline interpolation.

On the other hand, thin plate splines provide some strong advantages over piecewise polynomials when $d=2$. Therefore the main purpose of the work of this section is to assist understanding of the errors that occur due to edge effects. It is hoped that these investigations will lead to some new versions of thin plate spline interpolation in two dimensions that will give better accuracy than usual near the edges of the range of data points.

Acknowledgements: George Goodsell carried out most of the calculations that guided the development of the method of Section 2, and he pioneered the highly useful idea of not updating the residuals during an iteration. The topics that are mentioned in the last paragraph of Section 2 and the global optimization technique of Section 3 are under investigation by Anita Faul and Hans-Martin Gutmann, respectively. The numerical results of Section 4 were found by Aurelian Bejancu, and he proposed the inclusion of a quadratic polynomial term in thin plate spline interpolation. It is a great pleasure to express my thanks to all of these people for many very helpful discussions and for their contributions to the research that is described.

References

R.K. Beatson and G.N. Newsam (1992), "Fast evaluation of radial basis functions: I", *Comput. Math. Applic.*, Vol. 24, pp. 7–19.

R.K. Beatson and M.J.D. Powell (1992), "Univariate interpolation on a regular finite grid by a multiquadric plus a linear polynomial", *IMA J. Numer. Anal.*, Vol. 12, pp. 107–133.

R.K. Beatson and M.J.D. Powell (1994), "An iterative method for thin plate spline interpolation that employs approximations to Lagrange functions", in *Numerical Analysis 1993*, eds. D.F. Griffiths and G.A. Watson, Longman Scientific & Technical (Burnt Mill), pp. 17–39.

A. Bejancu (1997), "Local accuracy for radial basis function interpolation on finite uniform grids", Report No. DAMTP 1997/NA19, University of Cambridge, accepted for publication in the *Journal of Approximation Theory*.

M.D. Buhmann (1990), "Multivariate cardinal interpolation with radial-basis functions", *Constr. Approx.*, Vol. 6, pp. 225–255.

J. Duchon (1977), "Splines minimizing rotation-invariant seminorms in Sobolev spaces", in *Constructive Theory of Functions of Several Variables, Lecture*

Notes in Mathematics 571, eds. W. Schempp and K. Zeller, Springer-Verlag (Berlin), pp. 85–100.

R. Franke (1982), "Scattered data interpolation: tests of some methods", *Math. Comp.*, Vol. 38, pp. 181–200.

R.L. Hardy (1990), "Theory and applications of the multiquadric-biharmonic method", *Comput. Math. Applic.*, Vol. 19, pp. 163–208.

D. Jones (1996), "Global optimization with response surfaces", presented at the Fifth SIAM Conference on Optimization, Victoria, Canada.

C.A. Micchelli (1986), "Interpolation of scattered data: distance matrices and conditionally positive definite functions", *Constr. Approx.*, Vol. 2, pp. 11–22.

M.J.D. Powell (1997), "A new iterative method for thin plate spline interpolation in two dimensions", *Annals of Numerical Mathematics*, Vol. 4, pp. 519–527.

R. Schaback (1993), "Comparison of radial basis function interpolants", in *Multivariate Approximations: From CAGD to Wavelets*, eds. K. Jetter and F. Utreras, World Scientific (Singapore), pp. 293–305.

Department of Applied Mathematics and Theoretical Physics,
University of Cambridge,
Silver Street,
Cambridge CB3 9EW,
England.
Email address: M.J.D.Powell@damtp.cam.ac.uk

International Series of Numerical Mathematics
Vol. 132, © 1999 Birkhäuser Verlag Basel/Switzerland

Representation of quasi-interpolants as differential operators and applications

Paul Sablonnière

Abstract

Most of the best known positive linear operators are isomorphisms of the maximal subspace of polynomials that they preserve. We give here the differential forms of these isomorphisms and of their inverses for Bernstein and Szàsz-Mirakyan operators, and their Durrmeyer and Kantorovitch extensions. They allow to define families of intermediate left and right quasi-interpolants of which we study some properties like Voronowskaya type asymptotic error estimates and uniform boundedness of norms. In the Durrmeyer case, the polynomial coefficients of the associated linear differential operators are nicely connected with Jacobi or Laguerre orthogonal polynomials.

1   Introduction

Most of the classical quasi-interpolants of order n such as Bernstein or Szàsz-Mirakyan operators (see e.g. [14], chapter 10 or [20]) and their Durrmeyer-type extensions (see e.g. [2], [7], [8], [9], [10], [15], [23], [31], [33], [34], [39], [41], [42]) leave invariant the space Π_n of polynomials of degree at most n. Let $\mathcal{B}_n$ denote such an operator : $\mathcal{B}_n$ and $\mathcal{A}_n = \mathcal{B}_n^{-1}$ are linear isomorphisms of Π_n and can be expressed as linear differential operators with polynomial coefficients in the following form : $\mathcal{B}_n = \sum_{k=0}^n \beta_n^k D^k$ and $\mathcal{A}_n = \sum_{k=0}^n \alpha_k^n D^k$ with $D = d/dx$ and $D^0 = id$.

In general, the polynomial sequences $\{\beta_k^n\}$ and $\{\alpha_k^n\}$ can be computed through recurrence relationships involving at most three consecutive polynomials and their first derivatives. We give them for some classes of operators in section 2 (see also Kageyama [24] and [25] for some recent generalizations). In section 3, we show that for Durrmeyer type operators, the coefficients are expressed explicitly in terms of classical orthogonal polynomials like Jacobi or Laguerre polynomials. In section 4, we define new families of intermediate operators obtained by composition with $\mathcal{B}_n$ of its truncated inverses $\mathcal{A}_n^{(r)} = \sum_{k=0}^r \alpha_k^n D^k$, $0 \leq r \leq n$.

So, we obtain two families of operators :

(i) a family of *left quasi-interpolants* (LQI) defined by :

$$\mathcal{B}_n^{(r)} = \mathcal{A}_n^{(r)} \circ \mathcal{B}_n, \ 0 \leq r \leq n$$

(ii) a family of *right quasi-interpolants* (RQI) defined by :

$$\mathcal{B}_n^{[r]} = \mathcal{B}_n \circ \mathcal{A}_n^{(r)}, \ 0 \le r \le n.$$

Of course, $\mathcal{B}_n^{(0)} = \mathcal{B}_n^{[0]} = \mathcal{B}_n$, and $\mathcal{B}_n^{(n)} = \mathcal{B}_n^{[n]} = id$ on Π_n. Moreover, for $0 \le r \le n$, $\mathcal{B}_n^{(r)}$ and $\mathcal{B}_n^{[r]}$ are exact on Π_r, i.e $\mathcal{B}_n^{(r)}p = \mathcal{B}_n^{[r]}p = p$ for all $p \in \Pi_r$. These operators have extensions to spaces of smooth functions and they provide better approximation properties than $\mathcal{B}_n$. More specific results are obtained via Voronowskaya-type relations such as :

$$\lim_{n \to +\infty} n^{[r/2+1]}(\mathcal{B}_n^{(r)} f - f) = D^{(r)}f.$$

In section 5, we prove a theorem about the uniform boundedness of left Bernstein-Durrmeyer quasi-interpolants. In section 6, we give an application to numerical integration. Finally, section 7 is devoted to the proof of a part of theorem 2 .

Throughout the paper, we use the following notations : $\{\ell_i^n, 0 \le i \le n\}$ is the *Lagrange basis* of Π_n associated with the uniform partition of $[0,1]$, $\{\nu_i^n, 0 \le i \le n\}$ is the *Newton basis* of Π_n associated with the same partition, $J_k^{(\alpha,\beta)}(x) = P^{(\alpha,\beta)}(2x-1)$ and $L_k^{(\alpha)}(x)$ are respectively the *shifted Jacobi polynomial* and the *Laguerre polynomial* of degree k, $\{e_k(x) = x^k, \ k \ge 0\}$ is the monomial basis (see e.g. [43], chapters 4 and 5 for orthogonal polynomials).

2 Recurrence relationships for the computation of the differential forms of direct and inverse operators

Due to space limitation, we only consider the following examples of operators $\mathcal{B}_n$, but our method can be extended to other families of operators (e.g. [22], [23], [26], [29], [30], [31], [42]).

(B1) *Classical Bernstein operators*

$$B_np = \sum_{i=0}^{n} p(i/n)b_i^n, \text{ where } b_i^n(x) = \binom{n}{k} x^i (1-x)^{n-i}.$$

(B2) *Bernstein-Durrmeyer operators with Jacobi weight* :

$$M_np = \sum_{i=0}^{n} \langle p, \tilde{b}_i^n \rangle b_i^n,$$

where $\langle f, g \rangle = \int_0^1 w(t) f(t) g(t)dt$ and $\tilde{b}_i^n = b_i^n / \langle e_0, b_i^n \rangle$. Here $w(t) = t^\beta(1-t)^\alpha$, with $\alpha, \beta > -1$, is the Jacobi weight shifted to the interval $[0,1]$.

(B3) *Bernstein-Kantorovitch operators* :

$$K_n p = \sum_{i=0}^{n} \mu_i^n(p) b_i^n, \text{ where } \mu_i^n(p) = (n+1) \int_{i/(n+1)}^{(i+1)(/(n+1)} p(t) dt.$$

(B4) *Classical Szàsz-Mirakyan operators* :

$$S_n p = \sum_{i \geq 0} p(i/n) s_i^n, \text{ where } s_i^n(x) = e^{-nx} (nx)^i / i!.$$

(B5) *Szàsz-Mirakyan-Durrmeyer operators with Laguerre weight* :

$$T_n p = \sum_{i \geq 0} \langle p, \tilde{s}_i^n \rangle \, s_i^n, \text{ where } \langle f, g \rangle = \int_0^{+\infty} w(t) \, f(t) \, g(t) dt,$$

$w(t) = t^\alpha e^{-t}$ (Laguerre weight, $\alpha > -1$), and $\tilde{s}_i^n = s_i^n / \langle e_0, s_i^n \rangle$.

(B6) *The Szàsz-Mirakyan-Kantorovitch operators* :

$$U_n p = \sum_{i \geq 0} \omega_i^n(p) s_i^n, \text{ where } \omega_i^n(p) = \int_{i/n}^{(i+1)/n} p(t) dt.$$

Theorem 2.1 *The six above operators are linear isomorphisms of the space* Π_n.

Proof. For B_n, this results from $B_n \ell_i^n = b_i^n$ for $0 \leq i \leq n$. For M_n, it is well known (e.g. [2], [7], [8], [33]) that $M_n J_k^{(\alpha, \beta)} = \lambda_{n,k}^{(\alpha, \beta)} J_k^{(\alpha, \beta)}$ for $0 \leq k \leq n$, where :

$$\lambda_{k,n}^{(\alpha, \beta)} = \frac{\Gamma(\alpha + \beta + n + 2) \, n!}{\Gamma(\alpha + \beta + k + n + 2) \, (n-k)!}.$$

Since $S_n \nu_k^n = e_k$ for $0 \leq k \leq n$ and $T_n L_k^{(\alpha)} = \left(\frac{n}{n+1}\right)^k L_k^{(\alpha)}$ (see [17] and [33]) we see that the result is also true for these operators. Finally, similar results hold for the Kantorovitch versions of these operators (see e.g. [14] and [20]). ∎

Theorem 2.2 *For the classical or Durrmeyer forms of the above operators, the polynomial coefficients* $\gamma_k^n = \beta_k^n$ *or* α_k^n *of the corresponding differential operators and of their inverses satisfy the following recurrence relationships, where* $\pi(x) = X = x(1-x)$ *(resp.* $\pi(x) = x$*) for Bernstein (resp. Szàsz-Mirakyan) operators* :

$$(k+1) a_{k+1}^n \gamma_{k+1}^n(x) = \pi(x) \left(\lambda \, D\gamma_k^n(x) + \mu \, \gamma_{k-1}^n(x) \right) + \sigma_k(x) \gamma_k^n(x).$$

The constants a_k^n, λ, μ *and polynomials* $\sigma_k(x) \in \Pi_1$ *are given in table 1. Moreover,* $\alpha_0^n = \beta_0^n = 1$ *for all operators.*

Proof. (i) For all polynomials β_k^n, the relationship is obtained by differentiating the expression $\beta_k^n(x) = B_n\left[(.-x)^k/k!\right]$. For (B1), the proof is given in DeVore-Lorentz ([14], chapter 10, see also [5], [6], [27]) and for (B2) in [8] (in the Legendre case $\alpha = \beta = 0$). For (B4), see [20] and for (B5), see e.g. [15].

   (ii) For polynomials α_k^n, the proof is much less obvious since in general $\mathcal{A}_n$ has no simple expression as a discrete or integral operator. For (B1), a proof is given in Kageyama [24] for a more general class of operators : a direct proof is given in section 7. For (B2) (B4) and (B5), see also the same section. ∎

Corollary 2.3 *For the Kantorovitch families of operators, we have the following identities :*

(B3) Let $K_n = \sum\limits_{k=0}^{n} \hat{\beta}_k^n D^k$ *and* $K_n^{-1} = \sum\limits_{k=0}^{n} \hat{\alpha}_k^n D^k$,

then $\hat{\beta}_k^n(x) = \beta_k^{n+1}(x) + D\beta_{k+1}^{n+1}(x)$ *and* $\hat{\alpha}_k^n(x) = \alpha_k^{n+1}(x) + D\alpha_{k+1}^{n+1}(x)$,
where $\{\beta_k\}$ *and* $\{\alpha_k\}$ *are the coefficients of* B_n *and* B_n^{-1} *respectively.*

(B6) Let $U_n = \sum\limits_{k=0}^{n} \tilde{\beta}_k^n D^k$ *and* $U_n^{-1} = \sum\limits_{k=0}^{n} \tilde{\alpha}_k^n D^k$, *then* $\tilde{\beta}_k^n(x) = \beta_k^n(x) + D\beta_{k+1}^n(x)$
and $\tilde{\alpha}_k^n(x) = \alpha_k^n(x) + D\alpha_{k+1}^n(x)$ *where* $\{\beta_k^n\}$ *and* $\{\alpha_k^n\}$ *are the coefficients of* S_n
and S_n^{-1} *respectively.*

Proof. It is a direct consequence of the fact than $K_n p = D(B_{n+1}P)$ and $U_n p = D(S_n P)$ for $p \in \Pi_n$ and $P(x) = \int_0^x p(t)dt$. ∎

3   Durrmeyer-type operators and their inverses

For Durrmeyer-type operators, the coefficients β_k^n and α_k^n can be expressed in terms of Jacobi or Laguerre polynomials. We use the notation $(n)_0 = 1$ and $(n)_r = n(n-1)....(n-r+1)$ for $r \geq 1$.

Theorem 3.1 *(i) For Bernstein-Durrmeyer operators (B2), the polynomials* β_k^n *and* α_k^n *have the following representations :*

$$\beta_k^n(x) = (-1)^k \sum_{s=0}^{[k/2]} X^s\, J_{k-2s}^{(\alpha+s,\beta+s)}(x)\, /s!(n+\alpha+\beta+k-s+1)_{k-s}$$

$$\alpha_k^n(x) = \sum_{s=0}^{[k/2]} (-1)^s\, X^s\, J_{k-2s}^{(\alpha+s,\beta+s)}(x)\, /s!(n)_{k-s}\ .$$

Operator	γ_k^n	γ_1^n	a_{k+1}^n	λ	μ	$\sigma_k(x)$
Classical Bernstein	β_k^n	0	n	1	1	0
	α_k^n	0	$n-k$	0	-1	$-k(1-2x)$
Bernstein Durrmeyer	β_k^n	$J_1^{(\alpha,\beta)}(x)$	$n+k+\alpha+\beta+2$	1	2	$J_1^{(\alpha+k,\beta+k)}(x)$
	α_k^n	$-J_1^{(\alpha,\beta)}(x)$	$n-k$			$-J_1^{(\alpha+k,\beta+k)}(x)$
Classical Szàsz-Mirakyan	β_k^n	0	n	1	1	0
	α_k^n	0	n	0	-1	$-k$
Szàsz-Mirakyan Durrmeyer	β_k^n	$\frac{\alpha+1-x}{n+1}$	$n+1$	1	2	$L_1^{(\alpha+k)}(x)$
	α_k^n	$\frac{x-\alpha-1}{n+1}$	n	-1	-2	$-L_1^{(\alpha+k)}(x)$

Table 1 : $J_1^{(\alpha,\beta)}(x) = (\alpha+\beta+2)x - (\beta+1)$, $L_1^{(\alpha)}(x) = 1+\alpha - x$

(ii) For Szàsz-Mirakyan-Durrmeyer operators (B5), the polynomials β_k^n and α_k^n have the following representations :

$$\beta_k^n(x) = \sum_{s=0}^{[k/2]} X^s \, L_{k-2s}^{(\alpha+s)}(x) \, / s! (n+1)^{k-s}$$

$$\alpha_k^n(x) = (-1)^k \sum_{s=0}^{[k/2]} (-1)^s \, x^s \, L_{k-2s}^{(\alpha+s)}(x) \, / s! \, n^{k-s}.$$

Proof. Due to space limitation, we only give the proof for polynomials $\alpha_k = \alpha_k^n$ in the Szàsz-Mirakyan-Durrmeyer case (the proof is similar for other polynomials). They satisfy the following relationship (theorem 2.2, table 1) :

$$n(k+1)\alpha_{k+1}(x) + x(D\alpha_k(x) + 2\alpha_{k-1}(x)) + (k+\alpha+1-x)\alpha_k(x) = 0$$

with $\alpha_0 = 1$ and $\alpha_1(x) = -\frac{1}{n}L_1^\alpha(x) = -\frac{1}{n}(1+\alpha-x)$.

Let us assume that $k = 2r$ (the proof is similar for $k = 2r+1$). Thus we have :

$$\alpha_{2r}(x) = \sum_{i=0}^{r} \frac{(-1)^i}{n^{2r-i}} \frac{x^i}{i!} L_{2r-2i}^{(\alpha+i)}(x)$$

$$\alpha_{2r+1}(x) = \sum_{i=0}^{r} \frac{(-1)^{i+1}}{n^{2r+1-i}} \frac{x^i}{i!} L_{2r+1-2i}^{(\alpha+i)}(x)$$

$$D\alpha_{2r}(x) = \sum_{i=0}^{r} \frac{(-1)^i}{n^{2r-i}} \frac{ix^{i-1}}{i!} L_{2r-2i}^{(\alpha+i)}(x) + \sum_{i=0}^{r} \frac{(-1)^{i+1}}{n^{2r-i}} \frac{x^i}{i!} L_{2r-2i-1}^{(\alpha+i+1)}(x)$$

since $DL_k^{(\alpha)}(x) = -L_{k-1}^{(\alpha+1)}(x)$ (see Szegö [43], chap. 5) with the convention $L_{-1}^{(\alpha)}(x) = 0$. Substituting in the recurrence relationship, we get :

$$\sum_{i=0}^{r} \frac{(-1)^{i+1}}{n^{2r-i}} \frac{x^i}{i!} \left\{ (2r+1) \, L_{2r+1-2i}^{(\alpha+i)}(x) - iL_{2r-2i}^{(\alpha+i)}(x) \right.$$

$$\left. + xL_{2r-2i-1}^{(\alpha+i-1)}(x) - 2iL_{2r+1-2i}^{(\alpha+i-1)}(x) - (\alpha+2r+1-x)L_{2r-2i}^{(\alpha+i)}(x) \right\}$$

Since Laguerre polynomials satisfy the two following identities (see Szegö [43], p.102)

$$kL_k^{(\alpha)}(x) = (\alpha+k+x)L_{k-1}^{(\alpha)}(x) - xL_{k-2}^{(\alpha+1)}(x)$$

$$\text{and} \quad L_k^{(\alpha+1)}(x) = L_k^{(\alpha)}(x) + L_{k-1}^{(\alpha+1)}(x),$$

we can write :

$$(2r+1)L^{(\alpha+i)}_{2r+1-2i}(x) - (\alpha+2r+1-x)L^{(\alpha+i)}_{2r-2i}(x) =$$

$$-xL^{(\alpha+i+1)}_{2r-2i-1}(x) + 2iL^{(\alpha+i)}_{2r+1-2i}(x) - iL^{(\alpha+i)}_{2r-2i}(x).$$

and the quantity between braces becomes :

$$2i\left(L^{(\alpha+i)}_{2r+1-2i}(x) - L^{(\alpha+i-1)}_{2r+1-2i}(x) - L^{(\alpha+i)}_{2r-2i}(x)\right)$$

which is equal to zero because of the second identity above, q.e.d. ∎

4 Dual bases and Voronowskaya-type results

The definitions and first properties of left and right quasi-interpolants have been given in the introduction. We first give here an application to the construction of dual forms of the bases $\{b^n_i\}$ and $\{s^n_i\}$.

Theorem 4.1 *(i) Let* $p = \sum\limits_{i=0}^{n} c(i)b^n_i$ *be a polynomial expressed in the Bernstein basis of* Π_n. *Then the B-coefficients* $c(i)$ *of* p *can be expressed in either of the following forms :*

(a) $c(i) = B^{-1}_n p\left(\frac{i}{n}\right) = \sum\limits_{k=0}^{n} \alpha^n_k\left(\frac{i}{n}\right)D^k p\left(\frac{i}{n}\right)$, *where the* $\{\alpha^n_k\}$ *are defined in theorem 2 and table 1 (second row).*

(b) $c(i) = \left\langle M^{-1}_n p, \tilde{b}^n_i \right\rangle = \sum\limits_{k=0}^{n} \int_0^1 w(t)\,\alpha^n_k(t)\,\tilde{b}^n_i(t)\,D^k p(t)dt$, *where*

$$w(t) = t^\beta(1-t)^\alpha, \; \tilde{b}^n_i(t) = b^n_i(t)\,/\int_0^1 w\,b^n_i = \frac{\Gamma(\alpha+\beta+n+2)\,t^i(1-t)^{n-i}}{\Gamma(\alpha+n-i+1)\,\Gamma(\beta+i+1)} \; \text{and the}$$

$\{\alpha^n_k\}$ *are defined in table 1 (fourth row).*

(ii) Similarly, let $p = \sum\limits_{i\geq 0} \gamma(i)s^n_i \in \Pi_n$, *then its coefficients can be expressed in either of the following forms :*

(c) $\gamma(i) = S^{-1}_n p\left(\frac{i}{n}\right) = \sum\limits_{k=0}^{n} \alpha^n_k\,D^k p\left(\frac{i}{n}\right)$, *the coefficients being defined in table 1 (row 6)*

(d) $\gamma(i) = \left\langle T^{-1}_n p, \tilde{s}^n_i \right\rangle = \sum\limits_{k=0}^{n} \int_0^{+\infty} w(t)\,\alpha^n_k(t)\,\tilde{s}^n_i(t)\,D^k p(t)dt$ *where*

$$w(t) = t^\alpha e^{-t}, \; \tilde{s}^n_i = s^n_i(t)\,/\int_0^{+\infty} w(t)\,s^n_i(t)dt = (n+1)^{\alpha+i+1}\,t^i e^{-nt}/\Gamma(\alpha+i+1)$$

Proof. It is a direct consequence of the fact that $p = \mathcal{B}_n(\mathcal{A}_n p)$ for all $p \in \Pi_n$, where $\mathcal{B}_n$ is any of the four operators B_n, M_n, S_n or T_n. ∎

Now, we compute some limits which are useful for establishing the Voronowskaya-type results given below (theorem 4.3).

Theorem 4.2 *(i) For all operators, the following limits exist, for $\gamma_k^n = \beta_k^n$ or α_k^n, when n tends to infinity :*

$$\lim n^s \, \gamma_{2s}^n(x) = \bar{\gamma}_{2s}(x) \in \Pi_{2s}, \quad \lim n^{s+1} \gamma_{2s+1}^n(x) = \bar{\gamma}_{2s+1}(x) \in \Pi_{2s+1}.$$

(ii) Moreover, the polynomials $\bar{\gamma}^k$ satisfy the following recurrence relationships :

$$2s \, \bar{\gamma}_{2s}(x) = \mu.\pi(x) \, \bar{\gamma}_{2s-2}(x)$$
$$(2s + 1) \, \bar{\gamma}_{2s+1}(x) = \pi(x) \, (\lambda \, D\bar{\gamma}_{2s}(x) + \mu \, \bar{\gamma}_{2s-1}(x)) + \sigma_{2s}(x) \, \bar{\gamma}_{2s}(x)$$

with $\bar{\gamma}_0 = 0$ and $\bar{\gamma}_1 = \sigma_0$.

Proof. It is a straightforward consequence of theorem 2.2 and of the fact that $\lim\limits_{n \to +\infty} n^{-1} a_{k+1}^n = 1$ (see table 1)   ■

Table 2 gives the polynomials $\bar{\alpha}_k$ and $\bar{\beta}_k$ for the six operators.

Now, using theorem 4.2 we can prove the following Voronowskaya-type theorem for polynomials.

Theorem 4.3 *For any polynomial p, the following limits hold when n tends to infinity :*

(i) $\lim n^{s+1} \left(p - \mathcal{B}_n^{(2s)}p\right) = \bar{\alpha}_{2s+1} D^{2s+1}p + \bar{\alpha}_{2s+2} D^{2s+2}p$

(ii) $\lim n^{s+1} \left(p - \mathcal{B}_n^{(2s+1)}p\right) = \bar{\alpha}_{2s+2} D^{2s+2}p.$

The same result is valid for right quasi-interpolants.

Proof. For $n \geq \deg p$, there holds :

$$p - \mathcal{B}_n^{(r)}p = \left(\mathcal{A}_n - \mathcal{A}_n^{(r)}\right) \mathcal{B}_n p = \sum_{k \geq r+1} \alpha_k^n D^k \mathcal{B}_n p,$$

this sum being finite since $\mathcal{B}_n p$ is a polynomial. Therefore, we obtain :

$$n^{[(r+2)/2]} \left(p - \mathcal{B}_n^{(r)}p\right) = \sum_{k \geq r+1} n^{[(r+2)/2]} \alpha_k^n \, D^k \mathcal{B}_n p$$

and the result follows from theorem 4.2 and from the fact that :

$$\lim_{n \to +\infty} D^k \mathcal{B}_n p = D^k p, \; \forall p \in \Pi$$

for all the operators that we consider. (see e.g. [27], [14]), chapter 10) for $\mathcal{B}_n$, [7] for M_n, [20] for S_n and U_n. For right quasi-interpolants, the result follows from the identity

$$n^{[(r+2)/2]} \left(p - \mathcal{B}_n^{[r]}p\right) = \mathcal{B}_n \left(\sum_{k \geq r+1} n^{[(r+2)/2]} \alpha_k^n \, D^k p\right),$$

from theorem 4.2 and from the fact that $\lim\limits_{n \to +\infty} \mathcal{B}_n D^k p = D^k p$ for all types of operators $\mathcal{B}_n$.   ■

Operator	$\bar{\beta}_{2s-1}$	$\bar{\beta}_{2s}$	$\bar{\alpha}_{2s-1}$	$\bar{\alpha}_{2s}$
Classical Bernstein	$\dfrac{(1-2x)^{s-1}}{6.2^{s-2}(s-2)!}$	$\dfrac{X^s}{s!}$	$\dfrac{(-1)^s(1-2x)X^{s-1}}{3.2^{s-2}(s-2)!}$	$\dfrac{(-1)^s X^s}{2^s s!}$
Bernstein Jacobi	$\dfrac{-X^{s-1}J_1^{(\alpha+s-1,\beta+s-1)}(x)}{(s-1)!}$	$\dfrac{X^s}{s!}$	$\dfrac{(-X)^{s-1}J_1^{(\alpha+s-1,\beta+s-1)}(x)}{(s-1)!}$	$\dfrac{(-1)^s X^s}{2^s s!}$
Bernstein-Kantorovitch	$\dfrac{(2s+1)(1-2x)X^{s-1}}{6.2^{s-1}(s-1)!}$	$\dfrac{X^s}{2^s s!}$	$\dfrac{(-1)^s(4s-1)(1-2x)X^{s-1}}{3.2^s(s-1)!}$	$\dfrac{(-1)^s X^s}{2^s s!}$
Classical Szàsz Mirakyan	$\dfrac{x^{s-1}}{6.2^{s-2}(s-2)!}$	$\dfrac{x^s}{2^s s!}$	$\dfrac{(-1)^s x^{s-1}}{3.2^{s-2}(s-2)!}$	$\dfrac{(-1)^s x^s}{2^s s!}$
Szàsz-Mirakyan Laguerre	$\dfrac{x^{s-1}L_1^{(\alpha+s-1)}(x)}{(s-1)!}$	$\dfrac{x^s}{s!}$	$\dfrac{(-1)^s x^{s-1}L_1^{(\alpha+s-1)}(x)}{(s-1)!}$	$\dfrac{(-1)^s x^s}{s!}$
Szàsz-Mirakyan Kantorovitch	$\dfrac{(2s+1)x^{s-1}}{6.2^{s-1}(s-1)!}$	$\dfrac{x^s}{2^s s!}$	$\dfrac{(-1)^s(4s-1)x^{s-1}}{3.2^s(s-1)!}$	$\dfrac{(-1)^s x^s}{2^s s!}$

Table 2

Remark 4.4 *With some care, the proof can be extended to smooth functions (see e.g. [16] and [38] for quasi-interpolants associated with B_n, [15] and [17] for quasi-interpolants associated with S_n). For this purpose, we need the uniform boundedness of left quasi-interpolants.*

5 Uniform boundedness of left quasi-interpolants

Theoretical and practical considerations lead to the following conjecture, which is already proved for some classes of operators :

For each $r \geq 0$, there exists a constant $K(r) > 0$ such that $\left\| \mathcal{B}_n^{(r)} \right\| \leq K(r)$ **for all $n \geq r$.**

Of course, the operator norm has to be specified in each case and the constant $K(r)$ depends on the family of operators. For *classical Bernstein operators* and the infinite norm, $|f|_\infty = \max\limits_{x \in [0,1]} |f(x)|$, it is proved in [35] that $\left\| \mathcal{B}_n^{(2)} \right\| \leq 3$ and the general result is proved in [45]. For *Szàsz-Mirakyan operators,* the general result is proved in [15], where S_n is considered as linear operator in a weighted space of functions on R^+.(see also [17] for further results)

For *Bernstein-Durrmeyer-Jacobi quasi-interpolants,* we give here a short proof based on the following lemma (see e.g. [14], chapter 10 and also [4], [19], [28]).

Lemma 5.1 *Let $p = \sum_{i=0}^{n} c(i) b_i^n \in \Pi_n$ whose coefficients satisfy $|c(i)| \leq M$ for some constant $M > 0$. Then for all $r \geq 0$, there exists a constant $C_r > 0$, independent of p, such that the following Bernstein type inequality is satisfied :*

$$\left| X^r D^{2r} p \right|_\infty \leq C_r \, M . n^r$$

Theorem 5.2 *For all $r \geq 0$, there exists a constant $K(r) > 0$ such that $\left\| M_n^{(r)} \right\|_\infty \leq K(r)$ for all $n \geq r$, M_n being considered as a bounded linear operator on the space $C [0,1]$ equipped with the uniform norm.*

Proof. According to theorem 3.1, we can write (the proof is similar for $M_n^{(2r+1)}$) :

$$M_n^{(2r)} = \sum_{s=0}^{2r} \alpha_s^n \, D^s M_n = \sum_{k=0}^{r} \alpha_{2k}^n \, D^{2k} M_n + \sum_{k=0}^{r-1} \alpha_{2k+1}^n D^{2k+1} M_n$$

where $\alpha_{2k}^n(x) = \sum_{j=0}^{k} \lambda_{2k,j}^n(x) \dfrac{X^{k-j}}{n^{k+j}}$, $\alpha_{2k+1}^n(x) = \sum_{j=0}^{k} \lambda_{2k+1,,j}^n(x) \dfrac{X^{k-j}}{n^{k+j+1}}$,

and $\begin{cases} \lambda_{2k+1}^n = (-1)^{k-j} \, n^{k+j} \, J_{2j}^{(\alpha+k-j, \beta+k-j)}(x) \, / \, (n)_{k+j} \, (k-j)! \\[2mm] \lambda_{2k+1,j}^n(x) = (-1)^{k-j} \, n^{k+j+1} \, J_{2j+1}^{(\alpha+k-j, \beta+k-j)}(x) \, / \, (n)_{k+j+1} \, (k-j)! \end{cases}$

Setting $p = M_n f$, for $f \in C[0, 1]$, we obtain :

$$M_n^{(2r)} f = \sum_{k=0}^{r} \sum_{j=0}^{k} \lambda_{2k,j}^n(x) \, n^{-(k+j)} \, X^{k-j} \, D^{2k-2j}(D^{2j}p)$$

$$+ \sum_{k=0}^{r-1} \sum_{j=0}^{k} \lambda_{2k+1,j}^n(x) \, n^{(-k+j+1)} \, X^{k-j}(D^{2j+1}p).$$

Writing $p = \sum_{i=0}^{n} c(i) \, b_i^n$ and $|f|_\infty = M$, we see that $|c(i)| = \left| \left\langle f, \tilde{b}_i^n \right\rangle \right| \le M$, since $\int_0^1 w \, \tilde{b}_i^n = 1$. Therefore $D^m p = \dfrac{n!}{(n-m)!} \sum_{i=0}^{n-m} \Delta^m c(i) \, b_i^{n-m}$, and the B-coefficients of this polynomial are bounded by $n^m |\Delta^m c(i)| \le 2^m \, n^m M$, for $0 \le m \le n$. From lemma 5.1, there exists constants $C_{k-j} > 0$ such that :

$$\left| X^{k-j} \, D^{2k-2j}(D^{2j}p) \right|_\infty \le C_{k-j} \, 4^j \, M \, n^{k+j}$$
$$\left| X^{k-j} \, D^{2k-2j}(D^{2j+1}p) \right|_\infty \le C_{k-j} \, 2 \, 4^j \, M \, n^{k+j+1}.$$

Since Jacobi polynomials are bounded on $[0, 1]$ and $(n)_{k+j} \sim n^{k+j}$ when n tends to infinity, all polynomials $\lambda_{s,j}^n$ are uniformly bounded in n and there exists constants $\kappa_{s,j} > 0$ such that

$$\left| \lambda_{s,j}^n(x) \right| \le \kappa_{s,j} \quad \text{for} \quad x \in [0, 1] \quad \text{and} \quad n \ge r.$$

Therefore, we obtain :

$$\left| M_n^{(2r)} f \right|_\infty \le M \left\{ \sum_{k=0}^{r} \sum_{j=0}^{k} 4^j C_{k-j} \, \kappa_{2k,j} + 2 \sum_{k=0}^{r-1} \sum_{j=0}^{k} 4^j C_{k-j} \, \kappa_{2k+1,j} \right\} = M.K(2r),$$

whence $\left\| M_n^{(2r)} \right\|_\infty \le K(2r)$ for all $n \ge 2r$. ∎

Remarks

1. *With some care, it is possible to extend this theorem to L^p norms for $1 \le p < +\infty$.*

2. *These results are also extendable to the Durrmeyer form of Szàsz-Mirakyan operators by using theorem 3.1 and convenient Bernstein-type inequalities.*

3. *Finally, corollary 2.3 and Wu-Zhenchang's result [45] (resp. Diallo's result [15]) show that theorem 5.2 is also valid for Bernstein-Kantorovitch quasi-interpolants (resp. for Szàsz-Mirakyan-Kantorovitch operators).*

4. *We refer to [16], and [17] for the study of the convergence of Bernstein and Szàsz-Mirakyan quasi-interpolants*

6 Quadrature formulae associated with classical left Bernstein quasi-interpolants

Theorem 6.1 Let $f \in C^{2s+2}[0,1]$, $I(f) = \int_0^1 f(x)dx$ and
$I_n^{(r)}(f) = \int_0^1 B_n^{(r)}f(x)dx$ for $r = 2s$ or $2s+1$. Then there exists $\xi \in [0,1]$ such that :

$$\lim_{n \to +\infty} n^{s+1} \left[I(f) - I_n^{(2s)}(f) \right] = (-1)^s C_s^* D^{2s+2}f(\xi)$$

$$\lim_{n \to +\infty} n^{s+1} \left[I(f) - I_n^{(2s+1)}(f) \right] = (-1)^{s+1} C_s D^{2s+2}f(\xi)$$

where $C_s = \dfrac{(s+1)!}{2^{s+1}(2s+3)!}$ and $C_s^* = \dfrac{1}{3}(4s-3)C_s$.

Therefore, for n large enough, the errors have opposite signs and the two quadrature formulae give upper and lower approximations to the exact value $I(f)$.

Proof. For sufficiently smooth functions f, we know that

$$\lim n^{s+1} \left\{ f(x) - B_n^{(2s)}f(x) \right\} = \bar{\alpha}_{2s+1} D^{2s+1}f(x) + \bar{\alpha}_{2s+2} D^{2s+2}f(x),$$

$$\lim n^{s+1} \left\{ f(x) - B_n^{(2s+1)}f(x) \right\} = \bar{\alpha}_{2s+2} D^{2s+2}f(x)$$

where the limits are uniform on $[0,1]$. Hence by integrating on $[0,1]$, we obtain successively :

$$\int_0^1 \bar{\alpha}_{2s+2}(x) D^{2s+2}f(x)dx = \frac{(-1)^{s+1}}{2^{s+1}(s+1)!} D^{2s+2}f(\xi) \int_0^1 X^{s+1}dx$$

$$= (-1)^{s+1} C_s D^{2s+2}f(\xi) \text{ for some } \xi \in [0,1].$$

Moreover, since $D(X^{s+1}) = (s+1)(1-2x)X^s$, an integration by parts gives :

$$\int_0^1 \bar{\alpha}_{2s+1}(x) D^{2s+1}f(x)dx = \frac{(-1)^{s+1}4s}{3.2^{s+1}(s+1)!} \int_0^1 D^{2s+1}f(x)\,DX^{s+1}dx,$$

$$= \frac{(-1)^s 4s}{3.2^{s+1}(s+1)!} \int_0^1 X^{s+1}D^{2s+2}f(x)dx = (-1)^s \frac{4s}{3} C_s D^{2s+2}f(\xi),$$

with the same ξ as above. Therefore, we obtain finally, with $C_s^* = \dfrac{1}{3}(4s-3)C_s$:

$\int_0^1 \left\{ \bar{\alpha}_{2s+1}(x) D^{2s+1}f(x) + \bar{\alpha}_{2s+2}(x) D^{2s+2}f(x) \right\} dx = (-1)^s C_s^* D^{2s+2}f(\xi)$. ∎

The computation of $I_n^{(r)}(f)$ needs the evaluation of $\int_0^1 \alpha_k^n(x) D^k B_n f(x)dx$ for $0 \leq k \leq r$ which we cannot detail here and will be given elsewhere together with some numerical experiments.

7 Proof of theorem 2 (inverse operators)

7.1 Coefficients of A_n in the classical Bernstein case

Taylor's formula gives $p = \sum\limits_{k=0}^{n} \dfrac{1}{k!}(. - x)^k D^k p(x)$, hence :

$$A_n p(x) = \sum_{k=0}^{n} \frac{1}{k!} A_n(. - x)^k D^k p(x),$$

from which we deduce (for sake of clarity, we omit the upper index n, which is kept fixed in the proof) the expansion :

$$k!\,\alpha_k(x) = A_n(. - x)^k = \sum_{i=0}^{k}(-1)^i \binom{k}{i} x^i\, p_{k-i}(x) \tag{1}$$

where $p_0(x) = 1$ and $p_k(x) = \dfrac{nx(nx - 1)...(nx - k + 1)}{n(n - 1)...(n - k + 1)}$ for $k \geq 1$. This is due to the fact that $B_n\, p_k = e_k$ or equivalently $A_n e_k = p_k$ for $0 \leq k \leq n$ (see e.g. Davis [6], DeVore-Lorentz [14]). Assume that the recurrence on α_k is true for $1 \leq k \leq r - 1$, i.e. by multiplying by $k!$:

$$E_k(x) = (n - k)(k + 1)!\,\alpha_{k+1}(x) + k.k!\,(1 - 2x)\alpha_k(x) + k!\,X\,\alpha_{k-1}(x) = 0. \tag{2}$$

We have to prove that $E_r(x) = 0$; in fact, it is equivalent to prove that the following polynomial vanishes :

$$F_r(x) = \sum_{k=1}^{r} \binom{r}{k} x^{r-k} E_k(x) = 0. \tag{3}$$

By substituting (1) and (2) in (3), we get :

$$F_r(x) = \sum_{k=1}^{r} \binom{r}{k} x^{r-k} \left\{ (n - k) \sum_{i=0}^{k+1}(-1)^i \binom{k+1}{i} x^i p_{k+1-i}(x) \right.$$
$$\left. + k(1 - 2x) \sum_{i=0}^{k}(-1)^i \binom{k}{i} x^i p_{k-i}(x) + k X \sum_{i=0}^{k-1}(-1)^i \binom{k-1}{i} x^i p_{k-1-i}(x) \right\}.$$

The coefficients of p_0 and p_1 are respectively equal to :

$$x^r \sum_{k=1}^{r}(-1)^{k+1} \binom{r}{k} \left\{ (n - k)x - k\,(1 - 2x) + k(1 - x) \right\} =$$
$$n\,x^{r+1} \sum_{k=1}^{r}(-1)^{k+1} \binom{r}{k} = n x^{r+1}$$

and to

$$x^{r-1} \sum_{k=1}^{r} (-1)^k \begin{pmatrix} r \\ k \end{pmatrix} \{(n-k)(k+1)x - k^2(1-2x) + k(k-1)(1-x)\}$$

$$= nx^r \sum_{k=1}^{r} (-1)^r (k+1) \begin{pmatrix} r \\ k \end{pmatrix} - x^{r-1} \sum_{k=1}^{r} (-1)^r k \begin{pmatrix} r \\ k \end{pmatrix}$$

$$= -nx^r, \text{ by using classical binomial identities which we recall below.}$$

Therefore, the first two terms of $F_r(x)$ vanish for $p_0(x) = 1$, $p_1(x) = x$, hence $nx^{r+1}p_0(x) - nx^r p_1(x) = 0$. Now, for $2 \le s \le r+1$, the coefficient of p_s in (3) is equal to

$$(-1)^{s-1}x^{r-s} \sum_{k=1}^{r}(-1)^k \begin{pmatrix} r \\ k \end{pmatrix} \left\{ (n-k) \begin{pmatrix} k+1 \\ s \end{pmatrix} x - k \begin{pmatrix} k \\ s \end{pmatrix} (1-2x) \right.$$

$$\left. +k \begin{pmatrix} k-1 \\ s \end{pmatrix} (1-x) \right\}$$

$$= (-1)^{s-1}nx^{r-s+1}a(r,s) + (-1)^s(s-1)x^{r-s+1}b(r,s) + (-1)^s x^{r-s}c(r,s), \text{ where}$$

$$a(r,s) = \sum_{k=s-1}^{r} (-1)^k \begin{pmatrix} r \\ k \end{pmatrix} \begin{pmatrix} k+1 \\ s \end{pmatrix},$$

$$b(r,s) = \sum_{k=s-1}^{r} (-1)^k \begin{pmatrix} r \\ k \end{pmatrix} \begin{pmatrix} k \\ s-1 \end{pmatrix}$$

$$c(r,s) = \sum_{k=s}^{r} (-1)^k \begin{pmatrix} r \\ k \end{pmatrix} \begin{pmatrix} k \\ s \end{pmatrix}.$$

Using binomial identities 45 and 47 given in [32], p. 619, we obtain :

$$a(r,s) = b(r,s) = c(r,s) = 0 \text{ for } s \ne r, r+1.$$

Then, $a(r,r) = a(r,r+1) = (-1)^r$, $b(r,r) = 0$, $b(r,r+1) = c(r,r) = (-1)^r$. Finally, the remaining terms of $F_r(x)$ are simply equal to :

$$F_r(x) = (r-nx)p_r(x) + (n-r)p_{r+1}(x) = 0, \text{ q.e.d.} \qquad \blacksquare$$

7.2 Coefficients of M_n^{-1} in the Bernstein-Durrmeyer Jacobi case

In this section, we prove the recurrence relationship satisfied by the coefficients α_k^n of the differential form of the inverse of the Bernstein-Durrmeyer-Jacobi operator :

$$(n-m)(m+1)\alpha_{m+1}(x) = X(D\alpha_m(x) + 2\alpha_{m-1}(x))$$

$$+ ((\beta+m+1)(1-x) - (\alpha+m+1)x)\alpha_m(x),$$

where the upper indices n have been omitted, for sake of clarity. We shall use the following expansions of M_n and M_n^{-1} :

$$M_n f = \sum_{k=0}^{n} \lambda_{n,k} c_k \langle f, J_k \rangle J_k , \qquad M_n^{-1} f = \sum_{k=0}^{n} \lambda_{n,k}^{-1} c_k \langle f, J_k \rangle J_k$$

where $J_k(x) = J_k^{(\alpha,\beta)}(x)$, $w(x) = x^\beta (1-x)^\alpha$, and from Szegö [43], chapter 4,

$$c_k^{-1} = |J_k|_{L_w^2[0,1]}^2 = \int_0^1 w.J_k^2 = \frac{1}{k!} \frac{\Gamma(\alpha+k+1)\Gamma(\beta+k+1)}{(\alpha+\beta+2k+1)\Gamma(\alpha+\beta+k+1)}$$

$$\lambda_{n,k}^{-1} = \frac{(n-k)!}{n!} \frac{\Gamma(\alpha+\beta+k+n+2)}{\Gamma(\alpha+\beta+n+2)}$$

We also recall the following identities (Szegö, p.72) satisfied by shifted Jacobi polynomials :

$$X D J_k(x) = A_k J_{k-1}(x) + B_k J_k(x) + c_k J_{k+1}(x), \text{where :}$$

$$\begin{cases} A_k = (\alpha+k)(\beta+k)(\alpha+\beta+k+1) / (\alpha+\beta+2k)(\alpha+\beta+2k+1) \\ B_k = k(\alpha-\beta)(\alpha+\beta+k+1) / (\alpha+\beta+2k)(\alpha+\beta+2k+2) \\ C_k = -k(k+1)(\alpha+\beta+k+1) / (\alpha+\beta+2k+1)(\alpha+\beta+2k+2) \end{cases}$$

$$((2x-1) + E_k) J_k(x) = D_k J_{k-1}(x) + F_k J_{k+1}(x)$$

where $\begin{cases} D_k = 2(\alpha+k)(\beta+k) / (\alpha+\beta+2k)(\alpha+\beta+2k+1) \\ E_k = (\alpha^2-\beta^2) / (\alpha+\beta+2k)(\alpha+\beta+2k+2) \\ F_k = 2(k+1)(\alpha+\beta+k+1) / (\alpha+\beta+2k+1)(\alpha+\beta+2k+2). \end{cases}$

Finally, we need the following result :

Lemma 7.1 *Let* $G_n(x,t) = \sum_{k=0}^{n} \lambda_{n,k}^{-1} c_k X D J_k(x) J_k(t)$ *and*
$H_n(x,t) = \sum_{k=0}^{n} \lambda_{n,k}^{-1} c_k (1-2x) J_k(x) J_k(t)$. *Then the following identities hold :*

(i) $K_n(x,t) = G_n(x,t) - G_n(t,x)$

$$= -(\alpha+\beta+n+2) \sum_{k=0}^{n} \lambda_{n,k}^{-1} c_k (t-x) J_k(t) J_k(x)$$

(ii) $L_n(x,t) = H_n(t,x) - H_n(x,t) = 2 \sum_{k=0}^{n} \lambda_{n,k}^{-1} c_k (t-x) J_k(t) J_k(x)$.

Proof. (i) The coefficient of $J_s(t)$ in $K_n(x,.)$ is equal to

$c_s \{ \langle G_n(x,.), J_s \rangle - \langle G_n(.,x), J_s \rangle \}$.

We obtain successively :

$$c_s \langle G_n(x,.), J_s \rangle = c_s \lambda_{n,s}^{-1} X D J_s(x) = c_s \lambda_{n,s}^{-1}(A_s J_{s-1}(x) + B_s J_s(x) + C_s J_{s+1}(x))$$

$$c_s \langle G_n(x,.), J_s \rangle = c_s \sum_{k=0}^{n} \lambda_{n,k}^{-1} c_k J_k(x) \{A_k \langle J_{k-1}, J_s \rangle + B_k \langle J_k, J_s \rangle + C_k \langle J_{k+1}, J_s \rangle\}$$

$$= c_{s+1} \lambda_{n,s+1}^{-1} A_{s+1} J_{s+1}(x) + c_s \lambda_{n,s}^{-1} B_s J_s(x) + c_{s-1} \lambda_{n,s-1}^{-1} C_{s-1} J_{s-1}(x).$$

Therefore, we obtain, after some calculations :

$$K_n(x,t) = \sum_{k=0}^{n} \left\{ J_k(t) J_{k-1}(x) \left[c_k \lambda_{n,k}^{-1} A_k - c_{k-1} \lambda_{n,k-1}^{-1} C_{k-1} \right] \right.$$

$$\left. + J_{k+1}(x) \left[c_k \lambda_{n,k}^{-1} C_k - c_{k-1} \lambda_{n,k+1}^{-1} A_{k+1} \right] \right\}$$

$$= (\alpha + \beta + n + 2) \sum_{k=0}^{n} \{\omega_{n,k} J_{k-1}(x) J_k(t) - \omega_{n,k+1} J_k(t) J_{k+1}(x)\}$$

where $\omega_{n,k} = \dfrac{(n-k)k!}{n!} \dfrac{\Gamma(\alpha+\beta+k+n+1)\Gamma(\alpha+\beta+k+1)}{\Gamma(\alpha+\beta+n+2)\Gamma(\alpha+k)\Gamma(\beta+k)}$.

(ii) In a similar way, we obtain the following expansion of L_n :

$$L_n(x,t) = 2 \sum_{k=0}^{n} \lambda_{n,k}^{-1} c_k (t-x) J_k(x) J_k(t) = H_n(t,x) - H_n(x,t)$$

$$= \sum_{k=0}^{n} c_k \lambda_{n,k}^{-1} \{J_k(x) [D_k J_{k-1}(t) + F_k J_{k+1}(t)]$$

$$- J_k(t) [D_k J_{k-1}(x) + F_k J_{k+1}(x)]\}.$$

The coefficient of $J_s(t)$ in this sum is equal to :

$$c_s \{\langle H_n(\cdot,x), \rangle - \langle H_n(x,\cdot), J_s \}$$

$$= c_s \sum_{k=0}^{n} c_k \lambda_{n,k}^{-1} \{[D_k \langle J_{k-1}, J_s \rangle + F_k \langle J_{k+1}, J_s \rangle] J_k(x)$$

$$- [D_k J_{k-1}(x) + F_k J_{k+1}(x)] \langle J_k, J_s \rangle\}$$

$$= c_{s+1} \lambda_{n,s+1}^{-1} D_{s+1} J_{s+1}(x) + c_{s-1} \lambda_{n,s-1}^{-1} F_{s-1} J_{s-1}(x)$$

$$- c_s \lambda_{n,s}^{-1} D_s J_{s-1}(x) - c_s \lambda_{n,s}^{-1} F_s J_{s+1}(x)$$

$$= J_{s-1}(x) \left(c_{s-1} \lambda_{n,s-1}^{-1} F_{s-1} - c_s \lambda_{n,s}^{-1} D_s \right)$$

$$+ J_{s+1}(x) \left(c_{s+1} \lambda_{n,s+1}^{-1} D_{s+1} - c_s \lambda_{n,s}^{-1} F_s \right)$$

$$= -2\omega_{n,s} J_{s-1}(x) + 2\omega_{n,s+1} J_{s+1}(x).$$

By comparison of the final expressions of K_n and L_n, we conclude that (i) and (ii) are satisfied. ∎

Let us come back to the main proof. Since

$$\alpha_m(x) = \frac{1}{m!} M_n^{-1} [(. - x)] (x) = \frac{1}{m!} \sum_{k=0}^{n} c_k \lambda_{n,k}^{-1} \left(\int_0^1 w(t) (t-x)^m J_k(t) dt \right) J_k(x),$$

we obtain, by differentiation with respect to x :

$$D\alpha_m(x) = -\alpha_{m-1}(x) + \frac{1}{m!} \sum_{k=0}^{n} c_k \lambda_{n,k}^{-1} \left(\int_0^1 w(t)(t-x)^m J_k(t) dt \right) D J_k(x)$$

therefore :

$$X\left(D\alpha_m(x) + \alpha_{m-1}(x)\right) = \frac{1}{m!}\sum_{k=0}^{n} c_k \lambda_{n,k}^{-1} \langle (.-x)^m, J_k \rangle \, XDJ_k(x)$$

$$= \frac{1}{m!}\int_0^1 w(t)(t-x)^m G_n(x,t)dt.$$

Now, by lemma 7.1,

$$G_n(x,t) = K_n(x,t) + G_n(t,x) = G_n(t,x) - \frac{1}{2}(\alpha+\beta+n+2)L_n(x,t).$$

Moreover, we have, with $T = t(1-t)$:

$$\int_0^1 w(t)(t-x)^m G_n(t,x)dt = \sum_{k=0}^{n} c_k \lambda_{n,k}^{-1}\left(\int_0^1 w(t)(t-x)^m T J_k(t)\right)J_k(x)$$

and, by integration by parts, the integral in the right member is equal to :

$\int_0^1 t^{\beta+1}(1-t)^{\alpha+1}(t-x)^m DJ_k(t)dt$
$= -\int_0^1 w(t)J_k(t)\left\{\left[(\beta+1)(1-t) - (\alpha+1)t\right](t-x)^m + mt(1-t)(t-x)^{m-1}\right\}dt$

The quantity between { } is also equal to

$$-(\alpha+\beta+2)(t-x) + (\beta+1)(1-x) - (\alpha+1)x$$

and we can write $t(1-t) = x(1-x) + (1-2x(t-x) - (t-x)^2$, therefore the above integral is also equal to :

$(\alpha+\beta+m+2)\int_0^1 w(t)(t-x)^{m+1}J_k(t)dt - \left[(\beta+m+1)(1-x) - (\alpha+m+1)x\right]$
$\int_0^1 w(t)(t-x)^m J_k(t)dt - mX\int_0^1 w(t)(t-x)^{m-1}J_k(t)dt.$

On the other hand, in the integral $\int_0^1 w(t)(t-x)^m L_n(x,t)dt$, lemma 7.1 implies that we have to evaluate $\int_0^1 w(t)(t-x)^{m+1}J_k(t)dt$. Therefore, we obtain :
$X\left(D\alpha_m(x) + \alpha_{m-1}(x)\right)$
$= (m+1)(\alpha+\beta+m+2)\alpha_{m+1}(x) - \left[(\beta+m+1)(1-x) - (\alpha+m+1)x\right]$
$\alpha_m(x) - X\,\alpha_{m-1}(x) - (m+1)(\alpha+\beta+n+2)\alpha_{m+1}(x),$

which is the recurrence relationship we had to prove, q.e.d.

8 Final remarks

Most of the results presented in this paper have extensions to the multivariate setting, either in tensor product form or on simplices. Partial results in this direction are given in [10], [36], [37], [41]. See also [11] for orthogonal polynomials on a triangle. Other extensions are possible to discrete operators (see e.g. [34], [39]) and to some generalizations of classical operators (see [29], [30], [40]).

References

[1] H. Berens, G.G. Lorentz G.G., *Inverse theorems for Bernstein polynomials,* Indiana Univ. Math. J. **21**,(1972), 693-708.

[2] H. Berens, Y. Xu, *On Bernstein-Durrmeyer polynomials with Jacobi weights,* in Approximation Theory and Functional analysis, (1991),C.K. Chui (ed), Academic Press, Boston, 25-46.

[3] S. Bernstein, *Complément à l'article de E. Voronowskaja,* C.R. Acad. Sci. U.R.S.S., (1932), 86-92.

[4] P. Borwein, T. Erdélyi, *Polynomial and polynomial inequalities,* Springer Verlag, New-York (1995).

[5] F. Costabile, M.I. Gualtieri, S. Serra, *Asymptotic expansion and extrapolation for Bernstein polynomials with applications,* BIT **36:4**, (1986), 676-687.

[6] P.J. Davis, *Interpolation and Approximation,* Dover, New York (1975).

[7] M.M. Derriennic, *Sur l'approximation des fonctions d'une ou plusieurs variables par des polynômes de Bernstein modifiés et application au problème des moments,* Thèse, Université de Rennes (1981),.

[8] M.M. Derriennic, *Sur l'approximation de fonctions intégrables sur* [0, 1] *par des polynômes de Bernstein modifiés,* J. of Approximation Theory **31**, (1981), 325-343.

[9] M.M. Derriennic, *Polynômes de Bernstein modifiés sur un simplexe T de R^ℓ. Problème des moments,* in Polynômes orthogonaux et applications, C. Brezinski et al. (eds), LNM n°**1171**, Springer-Verlag, (1984), 296-301.

[10] M.M. Derriennic, *On multivariate approximation by Bernstein-type polynomials,* J. of Approximation Theory **45**, (1985), 155-166.

[11] M.M. Derriennic, *Polynômes orthogonaux de type Jacobi sur un triangle,* C.R. Acad. Sci. Paris, t. 300, Série I, n°**14**, (1985), 471-474.

[12] M.M. Derriennic, *Linear combinations of derivatives of Bernstein-type polynomials on a simplex,* in Approximation Theory (Kecskemét), Colloquia Mathematica Societatis Janos Bolyai, vol. **58**, (1990), 197-220.

[13] M.M. Derriennic, *De La Vallée-Poussin and Bernstein-type operators,* in Approximation Theory (IDoMAT 95), M.W. Müller, M. Felten, D.H. Mache (eds), Mathematical Research, vol. **86**, Akademie Verlag, Berlin, (1995), 71-84.

[14] R. DeVore, G.G. Lorentz, *Constructive Approximation,* Springer Verlag, New-York (1993).

[15] A.T. Diallo, *Szàsz-Mirakyan quasi-interpolants*, in Curves and Surfaces, P.J. Laurent, A. Le Méhauté, L.L. Schumaker (eds), Academic Press, New York, (1991), 149-156.

[16] A.T. Diallo, *Rate of convergence of Bernstein quasi-interpolants*, Report IC/**95**/**295**, International Centre for Theoretical Physics, Miramare-Trieste (1995).

[17] A.T. Diallo, *Rate of convergence of Szàsz-Mirakyan quasi-interpolants*, Report IC/**97**/**138**, International Center for theoretical Physics, Miramare-Trieste (1997).

[18] Z. Ditzian, *A global inverse theorem for combinations of Bernstein polynomials*, J. of Approximation Theory **26**, (1979), 277-292.

[19] Z. Ditzian, K. Ivanov, *Bernstein-type operators and their derivatives*, J. of Approximation Theory **56**, (1989), 72-90.

[20] Z. Ditzian, V. Totik, *Moduli of smoothness*, Springer-Verlag, New-York (1987).

[21] J.L. Durrmeyer, *Une formule d'inversion de la transformée de Laplace : application au problème des moments*, Thèse, Université de Paris (1967).

[22] I. Gavrea, D.H. Mache, *Generalization of Bernstein-type approximation methods*, in Approximation Theory (IDoMAT 95), M.W. Müller, M. Felten, D.H. Mache (eds), Mathematical Research, vol. **86**, Akademie Verlag, Berlin, (1995), 115-126.

[23] M. Heilmann, *Direct and converse results for operators of Baskakov-Durrmeyer type*, Approx. Theory & its Appl. **5:1**, (1989), 105-127.

[24] Y. Kageyama, *Generalization of the left Bernstein quasi-interpolants*, J. of Approximation Theory **94**, (1998), 306-329.

[25] Y. Kageyama, *A new class of modified Bernstein operators*, (to appear).

[26] H.S. Kasana, P.N. Agrawal and V. Gupta, *Inverse and saturation theorems for linear combinations of modified Baskakov operators*, Approximation Theory & its Appl. **7:2**, (1991), 65-82.

[27] G.G. Lorentz, *Bernstein polynomials*, University of Toronto Press, (1953).

[28] G.G. Lorentz, *The degree of approximation by polynomials with positive coefficients*, Math. Annalen **151**, (1963), 239-250.

[29] A. Lupaş, *The approximation by mean of some linear positive operator*, in Approximation Theory (IDoMAT 95), M.W. Müller, M. Felten, D.H. Mache (eds), Mathematical Research, vol. **86**, Akademie Verlag, Berlin, (1995), 201-229.

[30] A. Lupaş, *Approximation operators of binomial type*, 175–198 (in this book).

[31] D.H. Mache, *A link between Bernstein polynomials and Durrmeyer polynomials with Jacobi weights,* in Approximation Theory VIII, vol. I, C.K. Chui & L.L. Schumaker (eds), World Scientific Publ., (1995), 403-410.

[32] A.P. Prudnikov, Yu.A Brychkov and O.I. Marichev, *Integrals and series,* vol. 1, Gordon and Breach, New York (1986).

[33] P. Sablonnière, *Opérateurs de Bernstein-Jacobi,* Publ. ANO **37**, *Opérateurs de Bernstein-Laguerre,* Publ. ANO *38*, Université de Lille (1981).

[34] P. Sablonnière, *Hahn polynomials as eigenvectors of positive operators,* Second International Symposium on Orthogonal Polynomials & their Applications (Segovia 1986), published in Monografias Acad. Ci. Exact. Fis-Quim. Nat. Zaragoza, 1988, (1986), 139-146.

[35] P. Sablonnière, *Bernstein quasi-interpolants on* [0, 1], in Multivariate Approximation Theory IV, C.K. Chui, W. Schempp, K. Zeller (eds), ISNM, vol. 90, Birkhäuser-Verlag, Basel, (1989),287-294.

[36] P. Sablonnière, *Bernstein quasi-interpolants on a simplex,* Konstructive Approximationstheorie (Oberwolfach meeting, july 30-august 5, 1989), Publ. LANS 21, INSA de Rennes, (1989).

[37] P. Sablonnière, *Bernstein-type quasi-interpolants,* in Curves and Surfaces, P.J. Laurent, A. Le Méhauté, L.L. Schumaker (eds), Academic Press, New York, (1991), 421-426.

[38] P. Sablonnière, *A family of Bernstein quasi-interpolants on* [0, 1], Approx. Theory & its Appl. 8:3, 62-76.

[39] P. Sablonnière(1993), Discrete Bernstein bases and Hahn polynomials, J. of Comp. & Appl. Math. 49, (1992), 233-241.

[40] P. Sablonnière, *Positive Bernstein-Sheffer operators,* J. of Approximation Theory 83, (1995), 330-341.

[41] T. Sauer, *The genuine Bernstein-Durrmeyer operator on a simplex,* Results in Mathematics, vol. **26**, Birkhäuser-Verlag, Basel,(1994), 99-130.

[42] Song Li, *Local smoothness of functions and Baskakov-Durrmeyer operators,* J. of Approximation Theory **88**, (1997), 139-153.

[43] G. Szegö, *Orthogonal polynomials,* AMS Colloquium Publications, vol. **23** (4th. edition) (1939).

[44] E. Voronowskaja, *Détermination de la forme asymptotique d'approximation des fonctions par les polynômes de M. Bernstein,* C.R. Acad. Sci. U.R.S.S., (1932), 79-85.

[45] Wu-Zhengchang, *Norm of the Bernstein left-quasi-interpolant operator,* J. of Approximation Theory **66**, (1991), 36-43.

Acknowledgement. *The author thanks very much Marie-Madeleine Derriennic for the proof of the recurrence relationship between polynomial coefficients of the inverse Bernstein-Durrmeyer operator associated with the ultraspherical weight and orthogonal polynomials.*

INSA de Rennes
20, avenue des Buttes de Coësmes
CS 14315, 35043 Rennes Cédex, France
*Email address:*sablonni@perceval.univ-rennes1.fr

International Series of Numerical Mathematics
Vol. 132, © 1999 Birkhäuser Verlag Basel/Switzerland

Native Hilbert Spaces for Radial Basis Functions I

Robert Schaback

Abstract

This contribution gives a partial survey over the native spaces associated to (not necessarily radial) basis functions. Starting from reproducing kernel Hilbert spaces and invariance properties, the general construction of native spaces is carried out for both the unconditionally and the conditionally positive definite case. The definitions of the latter are based on finitely supported functionals only. Fourier or other transforms are not required. The dependence of native spaces on the domain is studied, and criteria for functions and functionals to be in the native space are given. Basic facts on optimal recovery, power functions, and error bounds are included.

1   Introduction

For the numerical treatment of functions of many variables, *radial basis functions* are useful tools. They have the form $\phi(\|x - y\|_2)$ for vectors $x, y \in \mathbb{R}^d$ with a *univariate* function ϕ defined on $[0, \infty)$ and the Euclidean norm $\|\cdot\|_2$ on $\mathbb{R}^d$. This allows to work efficiently for large dimensions d, because the function boils the multivariate setting down to a univariate setting. Usually, the multivariate context comes back into play by picking a large number M of points $x_1, \ldots, x_M$ in $\mathbb{R}^d$ and working with linear combinations

$$s(x) := \sum_{j=1}^{M} \lambda_j \phi(\|x_j - x\|_2).$$

In certain cases, low–degree polynomials have to be added, but we give details later. Typical examples for radial functions $\phi(r)$ on $r = \|x - y\|_2$, $x, y \in \mathbb{R}^d$ are

$$
\begin{array}{rl}
\text{thin–plate splines:} & r^\beta \log r,\ \beta > 0,\ \beta \in 2\,\mathbb{N}\ \ [1] \\
& r^\beta,\ \beta > 0,\ \beta \notin 2\,\mathbb{N}\ \ [1] \\
\text{multiquadrics:} & (r^2 + c^2)^{\beta/2},\ \beta > 0,\ \beta \notin 2\,\mathbb{N}\ \ [6] \\
\text{inverse multiquadrics:} & (r^2 + c^2)^{\beta/2},\ \beta < 0,\ \ [6] \\
\text{Gaussians:} & \exp(-\beta r^2),\ \beta > 0, \\
\text{Sobolev splines:} & r^{k-d/2} K_{k-d/2}(r),\ \ k > d/2 \\
\text{Wendland function:} & (1 - r)_+^4 (1 + 4r),\, d \le 3
\end{array}
$$

Another important case are *zonal* functions on the $(d-1)$–dimensional sphere $S^{d-1} \subset \mathbb{R}^d$. These have the form $\phi(x^T y) = \phi(\cos(\alpha(x,y)))$ for points x, y on the sphere spanning an angle of $\alpha(x,y) \in [0,\pi]$ at the origin. Here, the symbol ^{T} denotes vector transposition, and the function ϕ should be defined on $[-1, 1]$. Periodic multivariate functions can also be treated, e.g. by reducing them to products of univariate periodic functions.

All of these cases of *basis functions* share a common theoretical foundation which forms the main topic of this paper. The functions all have a unique associated "native" Hilbert space of functions in which they act as a generalized reproducing kernel. The different special cases (radiality, zonality) are naturally related to geometric invariants of the native spaces. The paper will thus start in section 2 with reproducing kernel Hilbert spaces and look at geometric invariants later in section 3.

But most basis functions are constructed directly and do not easily provide information on their underlying native space. Their main properties are symmetry and (strict) positive definiteness (SPD) or conditionally positive definiteness (CPD). These notions are defined without any relation to a Hilbert space, and we then have to show how to construct the native space, prove its uniqueness, and find its basic features. We do this for SPD functions in section 4 and for CPD functions in section 5. The results mostly date back to classical work on reproducing kernel Hilbert spaces and positive definite functions (see e.g. [12], [17]). We compile the necessary material here to provide easy access for researchers and students. Some new results are included, and open problems are pointed out. In particular, we show how to modify the given basis function in order to go over from the conditionally positive definite case to the (strictly) positive definite case. There are different ways to define native spaces (see [10] for comparisons), but here we want to provide a technique that is general enough to unify different constructions (e.g. on the sphere [3] or on Riemannian manifolds [2],[13]). We finish with a short account of optimal recovery of functions in native spaces from given data, and provide the corresponding error bounds based on power functions.

The notation will strictly distinguish between functions $f, g, \ldots$ and functionals $\lambda, \mu, \ldots$ as real–valued linear maps defined on functions. Spaces of functions will be denoted by uppercase letters like $F, G, \ldots$, and calligraphic letters $\mathcal{F}, \mathcal{G}, \ldots$ occur as soon as the spaces are complete. Spaces with an asterisk are dual spaces, while an asterisk at lowercase symbols indicates optimized quantities.

2 Reproducing Kernel Hilbert Spaces

Let $\Omega \subseteq \mathbb{R}^d$ be a quite general set on which we consider real–valued functions forming a real Hilbert space $\mathcal{H}$ with inner product $(.,.)_\mathcal{H}$. Assume further that for all $x \in \Omega$ the point evaluation functional $\delta_x : f \to f(x)$ is continuous in $\mathcal{H}$, i.e.

$$\delta_x \in \mathcal{H}^* \text{ for all } x \in \Omega \tag{2.1}$$

with the dual of $\mathcal{H}$ denoted by $\mathcal{H}^*$. This is a reasonable assumption if we want to apply numerical methods using function values. Note, however, that techniques like the Rayleigh–Ritz method for finite elements work in Hilbert spaces where point evaluaton functionals are not continuous. We shall deal with this more general situation later.

If (2.1) is satisfied, the Riesz representation theorem implies

Theorem 2.1 *If a Hilbert space of functions on Ω allows continuous point evaluation functionals, it has a symmetric reproducing kernel $\Phi : \Omega \times \Omega \to \mathbb{R}$ with the properties*

$$\Phi(x, \cdot) \quad \in \mathcal{H}$$
$$f(x) \quad = (f, \Phi(x, \cdot))_{\mathcal{H}}$$
$$\Phi(x, y) \quad = (\Phi(x, \cdot), \Phi(y, \cdot))_{\mathcal{H}} = \Phi(y, x) \tag{2.2}$$
$$\Phi(x, y) \quad = (\delta_x, \delta_y)_{\mathcal{H}^*}$$

for all $x, y \in \Omega$, $f \in \mathcal{H}$.

The theory of reproducing kernel Hilbert spaces is well covered in [12], for instance. In the terminology following below, a reproducing kernel Hilbert space is the *native space* with respect to its reproducing kernel. This is trivial as long as we start with a Hilbert space, but it is not trivial if we start with a function $\Phi : \Omega \times \Omega \to \mathbb{R}$.

3   Invariance Properties

In many cases, the domain Ω of functions allows a group $\mathbb{T}$ of geometric transformations, and the Hilbert space $\mathcal{H}$ of functions on Ω is invariant under this group. This means

$$f \circ T \quad \in \mathcal{H}$$
$$(f \circ T, g \circ T)_{\mathcal{H}} \quad = (f, g)_{\mathcal{H}} \tag{3.1}$$

for all $f, g \in \mathcal{H}$, $T \in \mathbb{T}$. The following simple result has important implications for the basis functions on various domains:

Theorem 3.1 *If a Hilbert space $\mathcal{H}$ of functions on a domain Ω is invariant under a group $\mathbb{T}$ of transformations on Ω in the sense of (3.1), and if $\mathcal{H}$ has a reproducing kernel Φ, then Φ is invariant under $\mathbb{T}$ in the sense*

$$\Phi(x, y) = \Phi(Tx, Ty) \quad \text{for all } x, y \in \Omega, \, T \in \mathbb{T}.$$

Proof. The assertion easily follows from

$$f(x) \quad = (f, \Phi(x, \cdot))_{\mathcal{H}}$$

$$= (f \circ T^{-1})(Tx) \quad = (f \circ T^{-1}, \Phi(Tx, \cdot))_{\mathcal{H}}$$

$$= (f \circ T^{-1} \circ T, \Phi(Tx, T \cdot))_{\mathcal{H}}$$

$$= (f, \Phi(Tx, T \cdot))_{\mathcal{H}}$$

for all $x \in \Omega$, $T \in \mathbb{T}$, $f \in \mathcal{H}.$ ∎

By some easy additional arguments one can read off the following invariance properties inherited by reproducing kernels Φ from their Hilbert spaces $\mathcal{H}$ on Ω:

- Invariance on $\Omega = \mathbb{R}^d$ under translations from $\mathbb{R}^d$ leads to *translation–invariant* functions $\Phi(x, y) = \phi(x - y)$ with $\phi(x) = \phi(-x)$: $\mathbb{R}^d \to \mathbb{R}$.

- In case of additional invariance under all orthogonal transformations we get *radial* functions $\Phi(x, y) = \phi(\|x - y\|_2)$ with ϕ : $[0, \infty) \to \mathbb{R}$. Thus radial basis functions arise naturally in all Hilbert spaces on $\mathbb{R}^d$ which are invariant under Euclidean rigid–body motions.

- Invariance on the sphere S^{d-1} under all orthogonal transformations leads to *zonal* functions $\Phi(x, y) = \phi(x^T y)$ for ϕ : $[-1, 1] \to \mathbb{R}$.

- Spaces of periodic functions induce periodic reproducing kernels.

See [5] for basis functions on topological groups, and see [3] for a review of results on the sphere. The paper [13] introduces the theory of basis functions on general manifolds, and corresponding error bounds are in [2].

4   Native Spaces of Positive Definite Functions

Instead of a single point $x \in \Omega$ with a single evaluation functional $\delta_x \in \mathcal{H}^*$ we now consider a set $\{x_1, \ldots, x_M\}$ of M distinct points in Ω and look at the point evaluation functionals $\delta_{x_1}, \ldots, \delta_{x_M}$.

Theorem 4.1 *In a real vector space $\mathcal{H}$ of functions on some domain Ω the following properties concerning a set $X = \{x_1, \ldots, x_M\}$ of M distinct points are equivalent:*

1. *There are functions $f \in \mathcal{H}$ which attain arbitrary values at the points $x_j \in \{x_1, \ldots, x_M\}$.*

2. *The points in X can be **separated**, i.e.: for all $x_j \in X$ there is a function $f_j \in \mathcal{H}$ vanishing on X except for x_j.*

3. *The point evaluation functionals $\delta_{x_1}, \ldots, \delta_{x_M} \in \mathcal{H}^*$ are linearly independent.*

Now let $\mathcal{H}$ be a reproducing kernel Hilbert space of real–valued functions on Ω. Furthermore, let one of the properties in Theorem 4.1 be satisfied for all finite sets $X = \{x_1, \ldots, x_M\} \subseteq \Omega$. Then the matrix

$$A_{\Phi,X} = (\Phi(x_k, x_j))_{1 \leq j, k \leq M} = ((\delta_{x_j}, \delta_{x_k})_{\mathcal{H}^*})_{1 \leq j, k \leq M} \qquad (4.1)$$

is a Gramian matrix formed of linearly independent elements. Thus it is symmetric and positive definite.

But the above property can be reformulated independent of the Hilbert space setting:

Definition 4.2 *A function* $\Phi \; : \; \Omega \times \Omega \to \mathbb{R}$ *is* symmetric and (strictly) positive definite *(SPD), if for arbitrary finite sets* $X = \{x_1, \ldots, x_M\} \subseteq \Omega$ *of distinct points the matrix* $A_{\Phi,X} = (\Phi(x_k, x_j))_{1 \leq j, k \leq M}$ *is symmetric and positive definite.*

Definition 4.3 *If a symmetric (strictly) positive definite function* $\Phi \; : \; \Omega \times \Omega \to \mathbb{R}$ *is the reproducing kernel of a real Hilbert space* $\mathcal{H}$ *of real–valued functions on* Ω, *then* $\mathcal{H}$ *is the* native space *for* Φ.

We can collect the above arguments into

Theorem 4.4 *If a real Hilbert space* $\mathcal{H}$ *of real–valued functions on some domain* Ω *allows continuous point evaluation functionals which are linearly independent when based on distinct points, the space has a symmetric and (strictly) positive definite reproducing kernel* Φ *and is the native space for* Φ.

Except for the unicity stated above, this is easy since we started from a given Hilbert space. Things are more difficult when we start with an SPD function Φ and proceed to construct its native space. The basic idea for this will first appear in the uniqueness proof for the native space:

Theorem 4.5 *The native space for a given SPD function* Φ *is unique if it exists, and it then coincides with the closure of the space of finite linear combinations of functions* $\Phi(x, \cdot)$, $x \in \Omega$ *under the inner product defined via*

$$(\Phi(x, \cdot), \Phi(y, \cdot))_{\mathcal{H}} = \Phi(x, y) \quad \text{for all } x, y \in \Omega. \qquad (4.2)$$

Proof. Let $\mathcal{H}$ be a Hilbert space of functions on Ω which has Φ as a symmetric positive definite kernel. Clearly all finite linear combinations of functions $\Phi(x, \cdot)$ are in $\mathcal{H}$, and the inner product on these functions depends on Φ alone because of (2.2). We can thus use (4.2) as a redefinition for the inner product on the subspace of linear combinations of functions $\Phi(x, \cdot)$. If $\mathcal{H}$ were larger than the closure of the span of these functions, there would be a nonzero $f \in \mathcal{H}$ which is orthogonal to all $\Phi(x, \cdot)$. But then $f(x) = (f, \Phi(x, \cdot))_{\mathcal{H}} = 0$ for all $x \in \Omega$.∎

Theorem 4.5 shows that the native space is the closure of the functions we work with in applications, i.e.: the functions $\Phi(x, \cdot)$ for $x \in \Omega$ fixed. Everything that can be approximated by functions $\Phi(x, \cdot)$ is in the native space. But the above result leaves us to show existence of the native space for any SPD function Φ. To do this, we mimic (4.2) to *define* an inner product

$$(\Phi(x, \cdot), \Phi(y, \cdot))_\Phi = \Phi(x, y) \quad \text{for all } x, y \in \Omega. \tag{4.3}$$

on functions $\Phi(x, \cdot)$. This inner product depends on Φ alone, and we thus use a slightly different notation. It clearly extends to an inner product on the space

$$F_\Phi(\Omega) := \left\{ \sum_{j=1}^{M} \lambda_j \Phi(x_j, \cdot) \,\Big|\, \lambda_j \in \mathbb{R}, \ M \in \mathbb{N}, \ x_j \in \Omega \right\} \tag{4.4}$$

of all finite linear combinations of such functions, because Φ is an SPD function. We keep Φ and Ω in the notation, because we want to study later how native spaces depend on Φ and Ω. The abstract Hilbert space completion $\mathcal{F}_\Phi(\Omega)$ of this space then is a Hilbert space with an inner product that we denote by $(\cdot, \cdot)_\Phi$ again, but we still have to interpret the abstract elements of $\mathcal{F}_\Phi(\Omega)$ as functions on Ω. But this is no problem since the point evaluation functionals δ_x extend continuously to the completion, and the equation

$$\delta_x(f) = (f, \Phi(x, \cdot))_\Phi \quad \text{for all } x \in \Omega, \ f \in \mathcal{F}_\Phi(\Omega) \tag{4.5}$$

makes sense there. We just *define* $f(x)$ to be the right–hand side of (4.5). Altogether we have

Theorem 4.6 *Any SPD function Φ on some domain Ω has a unique native space. It is the closure of the space $F_\Phi(\Omega)$ of (4.4) under the inner product (4.3). The elements of the native space can be interpreted as functions via (4.5).*

There are various other techniques to define the native space. See [10] for a comparison and embedding theorems.

It is one of the most challenging research topics to deduce properties of the native space from properties of Φ. For instance, Corollary 8.3 below will show that continuity of Φ on $\Omega \times \Omega$ implies that all functions in the native space are continuous on Ω. Other interesting topics are embedding theorems and density results for native spaces. See [10] for a starting point.

5 Native Spaces of Conditionally Positive Definite Functions

Several interesting functions of the form $\Phi(x, y)$ given in section 1, e.g.: Duchon's thin–plate splines or Hardy's multiquadrics are well–defined and symmetric on $\Omega = \mathbb{R}^d$ but not positive definite there. The quadratic form defined by the matrix in (4.1) is only positive definite on a certain subspace of $\mathbb{R}^M$. For later use, we make the corresponding precise definition somewhat more technical than necessary at first sight.

Let $\mathcal{P}$ be a finite–dimensional subspace of real–valued functions on Ω. In applications on $\Omega \subseteq \mathbb{R}^d$ we shall usually consider $\mathcal{P} = \mathbb{P}_m^d$, the space of polynomials of order at most m, while in periodic cases we use trigonometric polynomials, or spherical harmonics on the sphere. Then $L_\mathcal{P}(\Omega)$ denotes the space of all linear functionals with finite support in Ω that vanish on $\mathcal{P}$. For convenience, we describe such functionals by the notation

$$\lambda_{X,M} : f \to \sum_{j=1}^{M} \lambda_j f(x_j), \quad \lambda_{X,M}(\mathcal{P}) = \{0\} \tag{5.1}$$

for finite sets $X = \{x_1, \ldots, x_M\} \subseteq \Omega$ and coefficients $\lambda \in \mathbb{R}^M$. Note that these functionals form a vector space over $\mathbb{R}$ under the usual operations.

Definition 5.1 *A function* $\Phi : \Omega \times \Omega \to \mathbb{R}$ *is* **symmetric and conditionally positive definite** *(CPD) with respect to* $\mathcal{P}$, *if for all* $\lambda_{X,M} \in L_\mathcal{P}(\Omega) \setminus \{0\}$ *the value of the quadratic form*

$$\sum_{j,k=1}^{M} \lambda_j \lambda_k \Phi(x_j, x_k) = \lambda_{X,M}^x \lambda_{X,M}^y \Phi(x, y)$$

is positive.

Here, the superscript x denotes application of the functional with respect to the variable x. The classical definition of conditional positive definiteness of some order m is related to the special case $\mathcal{P} = \mathbb{P}_m^d$ on $\Omega \subseteq \mathbb{R}^d$.

From now on we assume $\Phi : \Omega \times \Omega \to \mathbb{R}$ to be CPD with respect to a finite–dimensional space $\mathcal{P}$ of functions on Ω. Because Φ is conditionally positive definite, we can define an inner product

$$(\lambda_{X,M}, \mu_{Y,N})_\Phi := \sum_{j=1}^{M} \sum_{k=1}^{N} \lambda_j \mu_k \Phi(x_j, y_k) = \lambda_{X,M}^x \mu_{Y,N}^y \Phi(x, y) \tag{5.2}$$

on the space $L_\mathcal{P}(\Omega)$. Furthermore, we can complete $L_\mathcal{P}(\Omega)$ to a Hilbert space $\mathcal{L}_{\Phi,\mathcal{P}}(\Omega)$, and we denote the extended inner product on $\mathcal{L}_{\Phi,\mathcal{P}}(\Omega)$ by $(\cdot, \cdot)_\Phi$ again.

Note that $L_{\mathcal{P}}(\Omega)$ as a vector space does not depend on Φ, but the completion $\mathcal{L}_{\Phi,\mathcal{P}}(\Omega)$ does, because Φ enters into the inner product. Now we can form inner products $(\lambda, \mu)_\Phi$ for all abstract elements λ, μ of the space $\mathcal{L}_{\Phi,\mathcal{P}}(\Omega)$, but we still have no functions on Ω, because we cannot evaluate $(\lambda, \delta_x)_\Phi$ since δ_x is in general not in $\mathcal{L}_{\Phi,\mathcal{P}}(\Omega)$.

A simple and direct, but not ultimately general workaround for this problem uses brute force to construct for all $x \in \Omega$ a substitute $\delta_{(x)} \in L_{\mathcal{P}}(\Omega)$ for a point evaluation functional. We start with the assumption of existence of a fixed set $\Xi = \{\xi_1, \ldots, \xi_q\} \subseteq \Omega$ of $q = \dim \mathcal{P}$ points of Ω which is unisolvent for $\mathcal{P}$. This means that any function $p \in \mathcal{P}$ can be uniquely reconstructed from its values on Ξ. This is no serious restriction to any application. In case of classical multiquadrics or thin–plate splines in two dimensions it suffices to fix three points in Ω which are not on a line. By picking a Lagrange–type basis $p_1, \ldots, p_q$ of $\mathcal{P}$ one can write the reconstruction as

$$p(x) = \sum_{j=1}^{q} p_j(x)p(\xi_j), \quad \text{for all } p \in \mathcal{P}, \ x \in \Omega. \tag{5.3}$$

This defines for all $x \in \Omega$ a very useful variation

$$\delta_{(x)}(f) := f(x) - \sum_{j=1}^{q} p_j(x)f(\xi_j) = \left(\delta_x - \sum_{j=1}^{q} p_j(x)\delta_{\xi_j} \right)(f) \tag{5.4}$$

of the standard point evaluation functional at x defined on all functions f on Ω. This functional annihilates functions from $\mathcal{P}$ and lies in $L_{\mathcal{P}}(\Omega)$ because it is finitely supported. Furthermore, we have

$$\delta_{(\xi_j)} = 0 \text{ for the points } \xi_j \in \Xi. \tag{5.5}$$

We now can go on with our previous argument, because we can look at the function

$$R_{\Phi,\Omega}(\lambda)(x) := (\lambda, \delta_{(x)})_\Phi, \ x \in \Omega \tag{5.6}$$

which is well–defined for all abstract elements $\lambda \in \mathcal{L}_{\Phi,\mathcal{P}}(\Omega)$. It defines a map $R_{\Phi,\Omega}$ from the abstract space $\mathcal{L}_{\Phi,\mathcal{P}}(\Omega)$ into some space of functions on Ω. We have chosen the notation $R_{\Phi,\Omega}$ because the mapping will later be continuously extended to the Riesz map on the dual of the native space. Let us look at the special situation for $\lambda = \lambda_{X,M} \in L_{\mathcal{P}}(\Omega)$. Then

$$\begin{aligned}
R_{\Phi,\Omega}(\lambda_{X,M})(x) &= (\lambda_{X,M}, \delta_{(x)})_\Phi \\
&= \sum_{j=1}^{M} \lambda_j \left(\Phi(x, x_j) - \sum_{k=1}^{q} p_k(x)\Phi(\xi_k, x_j) \right) \\
&= \lambda_{X,M}^y \Phi(x, y) - \sum_{k=1}^{q} p_k(x)\lambda_{X,M}^y \Phi(\xi_k, y)
\end{aligned} \tag{5.7}$$

shows the fundamental relation

$$\mu_{Y,N} R_{\Phi,\Omega}(\lambda_{X,M}) = (\lambda_{X,M}, \mu_{Y,N})_\Phi \tag{5.8}$$

for all $\lambda_{X,M}, \mu_{Y,N} \in L_\mathcal{P}(\Omega)$. It shows immediately that $R_{\Phi,\Omega}$ is injective on $L_\mathcal{P}(\Omega)$. We want to generalize this identity to hold on the completion $\mathcal{L}_{\Phi,\mathcal{P}}(\Omega)$, but for this we have to define a norm or inner product on the range of $R_{\Phi,\Omega}$.

This is easy, since $R_{\Phi,\Omega}$ is injective on $L_\mathcal{P}(\Omega)$. We can define an inner product on the range by

$$(R_{\Phi,\Omega}(\lambda), R_{\Phi,\Omega}(\mu))_\Phi := (\lambda, \mu)_\Phi \text{ for all } \lambda, \mu \in \mathcal{L}_{\Phi,\mathcal{P}}(\Omega),$$

where we extended the definition already to the completion $\mathcal{L}_{\Phi,\mathcal{P}}(\Omega)$ and used the same notation again, because we will never mix up functions with functionals here.

The space $\mathcal{F}_{\Phi,\mathcal{P}}(\Omega) := \overline{R_{\Phi,\Omega}(L_\mathcal{P}(\Omega))}$ for a CPD function Φ with respect to a space $\mathcal{P}$ of functions on Ω will be the major part of the native space to be constructed. It is a Hilbert space by definition, and we have

$$R_{\Phi,\Omega} : \mathcal{L}_{\Phi,\mathcal{P}}(\Omega) := \overline{L_\mathcal{P}(\Omega)} \to \mathcal{F}_{\Phi,\mathcal{P}}(\Omega) \tag{5.9}$$

as the Riesz mapping and can generalize (5.8) to

$$\mu(R_{\Phi,\Omega}(\lambda)) = (R_{\Phi,\Omega}(\mu), R_{\Phi,\Omega}(\lambda))_\Phi = (\mu, \lambda)_\Phi \text{ for all } \lambda, \mu \in \mathcal{L}_{\Phi,\mathcal{P}}(\Omega) \tag{5.10}$$

by going continuously to the completions. We thus have an interpretation of the Hilbert space $\mathcal{F}_{\Phi,\mathcal{P}}(\Omega)$ as a space of functions and the Hilbert space $\mathcal{L}_{\Phi,\mathcal{P}}(\Omega)$ as a space of functionals on $\mathcal{F}_{\Phi,\mathcal{P}}(\Omega)$. Furthermore, $R_{\Phi,\Omega}$ is the Riesz map and the spaces form a dual pair.

But we still do not have a reproduction property like (2.2), and the space $\mathcal{F}_{\Phi,\mathcal{P}}(\Omega)$ has the additional and quite superficial property that all its functions vanish on Ξ due to (5.5) and (5.6). But the latter property shows that $\mathcal{P}$ and $\mathcal{F}_{\Phi,\mathcal{P}}(\Omega)$ form a direct sum of spaces.

Definition 5.2 *The **native space** $\mathcal{N}_{\Phi,\mathcal{P}}(\Omega)$ for a conditionally positive definite function Φ on some domain Ω with respect to a finite–dimensional function space $\mathcal{P}$ consists of the sum of $\mathcal{P}$ with the Hilbert space $\mathcal{F}_{\Phi,\mathcal{P}}(\Omega)$ from (5.9).*

Note that this coincides with Definition 4.3 for $\mathcal{P} = \{0\}$, if we take Theorem 4.6 into account. To derive further properties of the native space, including a generalized notion of the reproduction equation (4.5), we use (5.3) to define a projector

$$\Pi_\mathcal{P}(f)(x) := \sum_{j=1}^{q} p_j(x) f(\xi_j) \text{ for all } f : \Omega \to \mathbb{R}, \ x \in \Omega$$

onto $\mathcal{P}$ with the property that $f - \Pi_\mathcal{P}(f)$ always vanishes on Ξ. Thus $Id - \Pi_\mathcal{P}$ projects functions in the native space onto the range of $R_{\Phi,\Omega}$, i.e. on $\mathcal{F}_{\Phi,\mathcal{P}}(\Omega)$. For all functions $f \in \mathcal{N}_{\Phi,\mathcal{P}}(\Omega)$ there is some $\lambda_f \in \mathcal{L}_{\Phi,\mathcal{P}}(\Omega)$ such that we have $f - \Pi_\mathcal{P} f = R_{\Phi,\Omega}(\lambda_f)$, and the value of this function at some point $x \in \Omega$ is $(\lambda_f, \delta_{(x)})_\Phi = (R_{\Phi,\Omega}(\lambda_f), R_{\Phi,\Omega}(\delta_{(x)}))_\Phi$ by definition. This implies

Theorem 5.3 *Every function f in the native space of a conditionally positive definite function Φ on some domain Ω with respect to a finite–dimensional function space $\mathcal{P}$ has the representation*

$$f(x) = (\Pi_{\mathcal{P}} f)(x) + (f - \Pi_{\mathcal{P}} f, R_{\Phi,\Omega}(\delta_{(x)}))_{\Phi} \text{ for all } x \in \Omega \qquad (5.11)$$

which is a generalized Taylor–type reproduction formula.

Definition 5.4 *The* **dual** $\mathcal{N}^*_{\Phi,\mathcal{P}}(\Omega)$ *of the native space consists of all linear functionals λ defined on the native space $\mathcal{N}_{\Phi,\mathcal{P}}(\Omega)$ such that the functional $\lambda - \lambda \circ \Pi_{\mathcal{P}}$ is continuous on the Hilbert subspace $\mathcal{F}_{\Phi,\mathcal{P}}(\Omega)$.*

Note that the functionals $\lambda \in \mathcal{N}^*_{\Phi,\mathcal{P}}(\Omega)$ are just linear forms on $\mathcal{N}_{\Phi,\mathcal{P}}(\Omega)$ in the sense of linear algebra. They are not necessarily continuous on $\mathcal{F}_{\Phi,\mathcal{P}}(\Omega)$ with respect to the norm $\| \cdot \|_{\Phi}$, because in that case they would have to vanish on $\mathcal{P}$. This would rule out point–evaluation functionals, for instance. Our special continuity requirement avoids this pitfall and makes point–evaluation functionals δ_x well–defined in the dual, but in general not continuous. However, the functional $\delta_{(x)} = \delta_x - \delta_x \circ \Pi_{\mathcal{P}}$ is continuous instead.

For all $\lambda \in \mathcal{N}^*_{\Phi,\mathcal{P}}(\Omega)$ we can consider the function $f_\lambda := R_{\Phi,\Omega}(\lambda - \lambda \circ \Pi_{\mathcal{P}})$ in $\mathcal{F}_{\Phi,\mathcal{P}}(\Omega)$. By (5.6) and (5.2) it can be explicitly calculated via

$$
\begin{aligned}
f_\lambda(x) \quad &= R_{\Phi,\Omega}(\lambda - \lambda \circ \Pi_{\mathcal{P}})(x) \\
&= (\lambda - \lambda \circ \Pi_{\mathcal{P}}, \delta_{(x)})_{\Phi} \qquad (5.12) \\
&= (\lambda - \lambda \circ \Pi_{\mathcal{P}})^y \delta^z_{(x)} \Phi(y, z)
\end{aligned}
$$

By (5.8) for $f = R_{\Phi,\Omega}(\mu)$, it represents the functional λ in the sense

$$(\lambda - \lambda \circ \Pi_{\mathcal{P}})(f) = (f, f_\lambda)_{\Phi} \text{ for all } f \in \mathcal{F}_{\Phi,\mathcal{P}}(\Omega). \qquad (5.13)$$

Altogether, the action of functionals can be described by

Theorem 5.5 *Each functional λ in the dual $\mathcal{N}^*_{\Phi,\mathcal{P}}(\Omega)$ of the native space $\mathcal{N}_{\Phi,\mathcal{P}}(\Omega)$ of a conditionally positive definite function Φ on some domain Ω with respect to a finite–dimensional function space $\mathcal{P}$ acts via*

$$
\begin{aligned}
\lambda(f) \quad &= (\lambda \circ \Pi_{\mathcal{P}})f + (f - \Pi_{\mathcal{P}} f, R_{\Phi,\Omega}(\lambda - \lambda \circ \Pi_{\mathcal{P}}))_{\Phi} \\
&= (\lambda \circ \Pi_{\mathcal{P}})f + (f - \Pi_{\mathcal{P}} f, \lambda^x R_{\Phi,\Omega}(\delta_{(x)}))_{\Phi}
\end{aligned}
$$

on all functions $f \in \mathcal{N}_{\Phi,\mathcal{P}}(\Omega)$.

Proof. The first formula follows easily from (5.12) and (5.13). For the second, we only have to use (5.6), (5.5), and (5.8) to prove

$$
\begin{aligned}
\lambda^x R_{\Phi,\Omega}(\delta_{(x)})(t) &= \lambda^x R_{\Phi,\Omega}(\delta_{(t)})(x) \\
&= (\lambda - \lambda \circ \Pi_{\mathcal{P}})^x R_{\Phi,\Omega}(\delta_{(t)})(x) \\
&= (\delta_{(t)}, R_{\Phi,\Omega}(\lambda - \lambda \circ \Pi_{\mathcal{P}}))_\Phi \\
&= R_{\Phi,\Omega}(\lambda - \lambda \circ \Pi_{\mathcal{P}})(t) = f_\lambda(t)
\end{aligned}
\tag{5.14}
$$

for all $t \in \Omega$. ∎

Note how Theorem 5.5 generalizes Theorem 5.3 to arbitrary functionals from the dual of the native space. The second form is somewhat easier to apply, because the representer f_λ of λ can be calculated via (5.14).

6   Modified kernels

The kernel function occurring in Theorem 5.3 is by definition

$$
R_{\Phi,\Omega}(\delta_{(x)})(y) = (\delta_{(x)}, \delta_{(y)})_\Phi =: \Psi(x,y),
\tag{6.1}
$$

and we call Ψ the **reduction** of Φ. It has the explicit representation

$$
\begin{aligned}
\Psi(x,y) &= \Phi(x,y) - \sum_{j=1}^{q} p_j(x)\Phi(\xi_j, y) - \sum_{k=1}^{q} p_k(y)\Phi(x, \xi_k) \\
&\quad + \sum_{j=1}^{q}\sum_{k=1}^{q} p_j(x)p_k(y)\Phi(\xi_j, \xi_k) \\
&= (Id - \Pi_{\mathcal{P}})^x (Id - \Pi_{\mathcal{P}})^y \Phi(x,y)
\end{aligned}
\tag{6.2}
$$

for all $x,y \in \Omega$ because we can do the evaluation of (6.1) via (5.2) and (5.4). Furthermore, equation (6.1) implies

$$
\Psi(x,y) = (R_{\Phi,\Omega}\delta_{(x)}, R_{\Phi,\Omega}\delta_{(y)})_\Phi = (\Psi(x,\cdot), \Psi(y,\cdot))_\Phi
$$

as we would expect from (2.2). A consequence of (5.5) is

$$
\Psi(\xi_j, \cdot) = \Psi(\cdot, \xi_j) = 0.
\tag{6.3}
$$

The bilinear form

$$
(f,g)_\Psi := (f - \Pi_{\mathcal{P}} f, g - \Pi_{\mathcal{P}} g)_\Phi \text{ for all } f, g \in \mathcal{N}_{\Phi,\mathcal{P}}(\Omega)
\tag{6.4}
$$

is positive semidefinite on the native space $\mathcal{N}_{\Phi,\mathcal{P}}(\Omega)$ for Φ. Its nullspace is $\mathcal{P}$, and the reproduction property (5.11) takes the simplified form

$$f(x) = (\Pi_{\mathcal{P}} f)(x) + (f, \Psi(x, \cdot))_{\Psi} \text{ for all } x \in \Omega, \tag{6.5}$$

where we used (6.3). The representer f_{λ} of a functional $\lambda \in \mathcal{N}^*_{\Phi,\mathcal{P}}(\Omega)$ in the sense of (5.13) takes the simplified form

$$f_{\lambda}(x) = \lambda^y \Psi(x, y), \; x, y \in \Omega$$

and Theorem 5.5 goes over into

Theorem 6.1 *Each functional λ in the dual $\mathcal{N}^*_{\Phi,\mathcal{P}}(\Omega)$ of the native space $\mathcal{N}_{\Phi,\mathcal{P}}(\Omega)$ of a conditionally positive definite function Φ on some domain Ω with respect to a finite–dimensional function space $\mathcal{P}$ acts via*

$$\lambda(f) = (\lambda \circ \Pi_{\mathcal{P}})f + (f, \lambda^x \Psi(x, \cdot))_{\Psi}$$

on all functions $f \in \mathcal{N}_{\Phi,\mathcal{P}}(\Omega)$.

Note that the reduction is easy to calculate. It coincides with the original function Φ if the latter is unconditionally positive definite. There also is a connection to the preconditioning technique of Jetter and Stöckler [7].

Theorem 6.2 *The reduction Ψ of Φ with respect to Ξ is (strictly) positive definite on $\Omega \setminus \Xi$.*

Proof. Let $\lambda_{X,M}$ be a functional with support $X = \{x_1, \ldots, x_M\} \subseteq \Omega \setminus \Xi$. Then the functional $\lambda_{X,M}(Id - \Pi_{\mathcal{P}})$ is finitely supported on Ω and vanishes on $\mathcal{P}$. Thus we can use the conditional positive definiteness of Φ and get from (6.2) that

$$\lambda^x_{X,M} \lambda^y_{X,M} \Psi(x, y) = (\lambda_{X,M}(Id - \Pi_{\mathcal{P}}))^x (\lambda_{X,M}(Id - \Pi_{\mathcal{P}}))^y \Phi(x, y)$$

is nonnegative and vanishes only if $\lambda_{X,M}(Id - \Pi_{\mathcal{P}})$ is the zero functional in $L_{\mathcal{P}}(\Omega)$. Its representation is

$$\begin{aligned}
\lambda_{X,M}(Id - \Pi_{\mathcal{P}})(f) &= \sum_{j=1}^{M} \lambda_j \left(f(x_j) - \sum_{k=1}^{q} p_k(x_j) f(\xi_k) \right) \\
&= \sum_{j=1}^{M} \lambda_j f(x_j) - \sum_{k=1}^{q} f(\xi_k) \sum_{j=1}^{M} \lambda_j p_k(x_j),
\end{aligned}$$

and since the sets $X = \{x_1, \ldots, x_M\}$ and Ξ are disjoint, the coefficients must vanish. ∎

There is an easy possibility used in [9] to go over from here to a fully positive definite case. Using an early idea from Golomb and Weinberger [4] we form a new kernel function $K : \Omega \times \Omega \to \mathbb{R}$ by

$$K(x, y) := \Psi(x, y) + \sum_{j=1}^{q} p_j(x) p_j(y) \tag{6.6}$$

and a new inner product

$$(f, g)_\Phi := \sum_{j=1}^{q} f(\xi_j) g(\xi_j) + (f - \Pi_\mathcal{P} f, g - \Pi_\mathcal{P} g)_\Phi \tag{6.7}$$

on the whole native space $\mathcal{N}_{\Phi,\mathcal{P}}(\Omega)$.

Theorem 6.3 *Under the new inner product (6.7) the native space $\mathcal{N}_{\Phi,\mathcal{P}}(\Omega)$ for a CPD function Φ on Ω is a Hilbert space with reproducing kernel defined by (6.6). In other words $\mathcal{N}_{\Phi,\mathcal{P}}(\Omega) = \mathcal{N}_K(\Omega)$ as vector spaces but with different though very similar topologies.*

Proof. It suffices to prove the reproduction property for some $f \in \mathcal{N}_{\Phi,\mathcal{P}}(\Omega)$ at some $x \in \Omega$ via

$$
\begin{aligned}
(f, K(x, \cdot))_\Phi \; &= \sum_{j=1}^{q} f(\xi_j) p_j(x) + (f - \Pi_\mathcal{P} f, \Psi(x, \cdot))_\Phi \\
&= (\Pi_\mathcal{P} f)(x) + (f - \Pi_\mathcal{P} f)(x) \\
&= f(x).
\end{aligned}
$$

Theorem 6.3 is the reason why we do not consider the CPD case to be more complicated than the SPD case. We call K the **regularized** kernel with respect to the original CPD function Φ. ∎

7 Numerical treatment of modified kernels

Here, we want to describe the numerical implications induced by reduction or regularization of a kernel Φ. Consider first the standard setting of interpolation of point–evaluation data $s_{|X} \in \mathbb{R}^M$ in some set $X = \{x_1, \ldots, x_M\} \subset \Omega$ by a CPD function Φ, where we additionally assume that point–evaluation of Φ is possible. A generalization to Hermite–Birkhoff data will be given in section 10 below. The interpolant takes the form

$$s = p_X + R_{\Phi,\Omega}(\lambda_{X,M}), \; p_X \in \mathcal{P}, \; \lambda_{X,M} \in L_\mathcal{P}(\Omega) \tag{7.1}$$

and this gives the linear system

$$\begin{pmatrix} A_{\Phi,X} & P_X \\ P_X^T & 0 \end{pmatrix} \begin{pmatrix} \lambda \\ \rho \end{pmatrix} = \begin{pmatrix} s|_X \\ 0 \end{pmatrix},$$ (7.2)

where P_X contains the values $p_j(x_k)$, $1 \leq j \leq q$, $1 \leq k \leq M$ for some arbitrary basis $p_1, \ldots, p_q$ of $\mathcal{P}$. The set $X = \{x_1, \ldots, x_M\}$ must be $\mathcal{P}$–unisolvent to make the system nonsingular, and thus we can assume $\Xi \subset X$ and number the points of X such that $x_j = \xi_j$, $1 \leq j \leq q = \dim \mathcal{P}$. This induces a splitting

$$\begin{pmatrix} A_{11} & A_{12} & P_1 \\ A_{12}^T & A_{22} & P_2 \\ P_1^T & P_2^T & 0 \end{pmatrix} \begin{pmatrix} \lambda_1 \\ \lambda_2 \\ \rho \end{pmatrix} = \begin{pmatrix} s|_\Xi \\ s|_{X \setminus \Xi} \\ 0 \end{pmatrix}$$

with $q \times q$ matrices A_{11} and P_1 and an $(M - q) \times (M - q)$ matrix A_{22}. Passing to a Lagrange basis on Ξ then means setting $\sigma = P_1 \rho$ and

$$\begin{pmatrix} A_{11} & A_{12} & I \\ A_{12}^T & A_{22} & P_2 P_1^{-1} \\ I & (P_1^{-1})^T P_2^T & 0 \end{pmatrix} \begin{pmatrix} \lambda_1 \lambda_2 \\ \sigma \end{pmatrix} = \begin{pmatrix} s|_\Xi \\ s|_{X \setminus \Xi} \\ 0 \end{pmatrix}.$$ (7.3)

Now we take a closer look at the formula (6.2) and relate it to the above matrices. Denoting the identity matrix by I, we get the result

$$A_{\Psi,X} = \begin{pmatrix} A_{11} & A_{12} \\ A_{12}^T & A_{22} \end{pmatrix} - \begin{pmatrix} I \\ P_2 P_1^{-1} \end{pmatrix} (A_{11}, A_{12})$$

$$- \begin{pmatrix} A_{11} \\ A_{12}^T \end{pmatrix} (I, (P_1^{-1})^T P_2^T) + \begin{pmatrix} I \\ P_2 P_1^{-1} \end{pmatrix} A_{11} (I, (P_1^{-1})^T P_2^T)$$

and see that everything except the lower right $(M - q) \times (M - q)$ block vanishes. Setting $Y := X \setminus \Xi$ we thus have proven

$$A_{\Psi,Y} = A_{22} - P_2 P_1^{-1} A_{12} - A_{12}^T (P_1^{-1})^T P_2^T + P_2 P_1^{-1} A_{11} (P_1^{-1})^T P_2^T$$

and this matrix is symmetric and positive definite due to Theorem 6.2. If we eliminate λ_1 and σ in the system (7.3) by

$$\lambda_1 = -(P_1^{-1})^T P_2^T \lambda_2$$

$$\sigma = s|_\Xi - A_{11} \lambda_1 - A_{12} \lambda_2$$ (7.4)

$$= s|_\Xi + (A_{11}(P_1^{-1})^T P_2^T - A_{12}) \lambda_2$$

we arrive at the system

$$A_{\Psi,Y} \lambda_2 = s|_Y - P_2 P_1^{-1} s|_\Xi = s|_Y - p|_Y$$ (7.5)

if p is calculated beforehand from the values of s on Ξ. The algebraic reduction to the above system and the transition to the case of Theorem 6.3 take at most $\mathcal{O}(qM^2)$ operations and thus are worth while when compared to the complexity of a direct solution. Note that the transformations may spoil sparsity properties of the original matrix $A_{\Phi,X}$, but they are unnecessary anyway, if compactly supported functions on $\mathbb{R}^d$ are used, because these are SPD, not CPD.

Finally, let us look at the numerical effect of a regularized kernel as in (6.6). Since we used a Lagrange basis of $\mathcal{P}$ there, and since we have a special numbering, we get

$$A_{K,X} = \begin{pmatrix} I & P_Y^T \\ P_Y & A_{K,Y} \end{pmatrix}$$

with $P_Y = P_2 P_1^{-1}$ and $A_{K,Y} = A_{\Psi,Y} + P_Y P_Y^T$, $A_{K,\Xi} = A_{\Psi,\Xi} + I = I$.

Note that the representation (7.1) is different from (5.11). In particular, if two interpolants s_Y, s_X based on different sets $Y \supseteq X \supseteq \Xi$ coincide on X, then necessarily $\Pi s_Y = \Pi s_Y$, but not $p_X = p_Y$.

8 Properties of Native Spaces

Due to the pioneering work of Madych and Nelson [11], the native space can be written in another form. It is the largest space on which all functionals from $L_\mathcal{P}(\Omega)$ (and its closure $\mathcal{L}_{\Phi,\mathcal{P}}(\Omega)$) act continuously:

Theorem 8.1 *The space*

$$\mathcal{M}_{\Phi,\mathcal{P}}(\Omega) := \{f : \Omega \to \mathbb{R} \; : \; |\lambda(f)| \leq C_f \|\lambda\|_\Phi \text{ for all } \lambda \in L_\mathcal{P}(\Omega)\}$$

coincides with the native space $\mathcal{N}_{\Phi,\mathcal{P}}(\Omega)$. *It has a seminorm*

$$|f|_\mathcal{M} := \sup \{|\lambda(f)| \; : \; \lambda \in L_\mathcal{P}(\Omega), \; \|\lambda\|_\Phi \leq 1\} \tag{8.1}$$

which coincides with $|f|_\Psi$ *defined via (6.4).*

Proof. For all functions $f = p_f + R_{\Phi,\Omega}(\lambda_f) \in \mathcal{N}_{\Phi,\mathcal{P}}(\Omega)$ with $\lambda_f \in \mathcal{L}_{\Phi,\mathcal{P}}(\Omega)$ and $p_f \in \mathcal{P}$ we can use (5.10) to prove that f lies in the space $\mathcal{M}_{\Phi,\mathcal{P}}(\Omega)$ with $C_f \leq \|\lambda_f\|_\Phi$. To prove the converse, consider an arbitrary function f in $\mathcal{M}_{\Phi,\mathcal{P}}(\Omega)$ and define a functional $F_f \in L_\mathcal{P}(\Omega)^*$ by $F_f(\lambda) := \lambda(f)$. The definition of $\mathcal{M}_{\Phi,\mathcal{P}}(\Omega)$ makes sure that F_f is continuous on $L_\mathcal{P}(\Omega)$, and thus F_f can be continuously extended to $\mathcal{L}_{\Phi,\mathcal{P}}(\Omega)$. By the Riesz representation theorem, there is some $\lambda_f \in \mathcal{L}_{\Phi,\mathcal{P}}(\Omega)$ such that

$$F_f(\mu) = (\mu, \lambda_f)_\Phi \text{ for all } \mu \in \mathcal{L}_{\Phi,\mathcal{P}}(\Omega).$$

Then we can define the function $f - R_{\Phi,\Omega}(\lambda_f)$ and apply arbitrary functionals $\mu \in L_\mathcal{P}(\Omega)$ to get

$$\mu(f - R_{\Phi,\Omega}(\lambda_f)) = \mu(f) - \mu(R_{\Phi,\Omega}(\lambda_f)) = \mu(f) - (\mu, \lambda_f)_\Phi = \mu(f) - F_f(\mu) = 0.$$

Specializing to $\mu = \delta_{(x)}$ for all $x \in \Omega$ we see that $f - R_{\Phi,\Omega}(\lambda_f)$ coincides on Ω with a function p_f from $\mathcal{P}$. Thus $f = p_f + R_{\Phi,\Omega}(\lambda_f)$ is a function in the native space $\mathcal{N}_{\Phi,\mathcal{P}}(\Omega)$. This proves that the two spaces coincide as spaces of real–valued functions on Ω.

To prove the equivalence of norms, we use the above notation and first obtain $|f|_{\mathcal{M}} \leq \|\lambda_f\|_\Phi$ from (8.1). But since we can replace $L_\mathcal{P}(\Omega)$ by its closure $\mathcal{L}_{\Phi,\mathcal{P}}(\Omega)$ in (8.1) and then use λ_f as a test functional, we also get

$$|f|_{\mathcal{M}} \geq |\lambda_f(f)|/\|\lambda_f\|_\Phi = \|\lambda_f\|_\Phi.$$

Due to $|f|_\Psi = \|\lambda_f\|_\Phi$ this proves the assertion. ∎

We now want to give a sufficient criterion due to Mark Klein [8] for differentiability of functions in the native space $\mathcal{N}_{\Phi,\mathcal{P}}(\Omega)$ of a CPD function Φ with respect to some space $\mathcal{P}$ on $\Omega \subseteq \mathbb{R}^d$. Since this is a disguised statement about functionals in the dual space $\mathcal{L}_{\Phi,\mathcal{P}}(\Omega)$, we first look at functionals:

Theorem 8.2 *Let an arbitrary functional λ be defined for functions on Ω and have the properties*

1. *The real number $\lambda^x \lambda^y \Phi(x,y)$ is well–defined and*

2. *obtainable as the "double" limit of values $\lambda_n^x \lambda_m^y \Phi(x,y)$ for a sequence $\{\lambda_k\}_k$ of finitely supported linear functionals λ_k from $L_\mathcal{P}(\Omega)$.*

3. *For any finitely supported linear functional $\rho \in L_\mathcal{P}(\Omega)$ the value $\rho^x \lambda^y \Phi(x,y)$ exists and is the limit of the values $\rho^x \lambda_n^y \Phi(x,y)$ for $n \to \infty$.*

Then the functional λ has an extension μ in the space $\mathcal{L}_{\Phi,\mathcal{P}}(\Omega)$ such that all appearances of λ in the above properties can be replaced by μ. All functionals in $\mathcal{L}_{\Phi,\mathcal{P}}(\Omega)$ can be obtained this way.

Proof. The second property means that for any $\epsilon > 0$ there is some $N \in \mathbb{N}$ such that

$$|\lambda^x \lambda^y \Phi(x,y) - \lambda_n^x \lambda_m^y \Phi(x,y)| < \epsilon$$

for $n, m \geq N$. Then for $c := \lambda^x \lambda^y \Phi(x,y)$ we get

$$\begin{aligned}
\|\lambda_n - \lambda_m\|_\Phi^2 &= \|\lambda_n\|_\Phi^2 + \|\lambda_m\|_\Phi^2 - 2(\lambda_n, \lambda_m)_\Phi \\
&\leq \left|\|\lambda_n\|_\Phi^2 - c\right| + \left|\|\lambda_n\|_\Phi^2 - c\right| + 2\left|(\lambda_n, \lambda_m)_\Phi - c\right| \\
&< 4\epsilon,
\end{aligned}$$

proving that $\{\lambda_k\}_k$ is a Cauchy sequence. It has a limit $\mu \in \mathcal{L}_{\Phi,\mathcal{P}}(\Omega)$, and by continuity we have $c = (\mu, \mu)_\Phi$. For any finitely supported functional $\rho \in L_\mathcal{P}(\Omega)$ we get

$$\rho^x \lambda^y \Phi(x,y) = \lim \rho^x \lambda_n^y \Phi(x,y) = \lim(\rho, \lambda_n)_\Phi = (\rho, \mu)_\Phi.$$

Thus the action of λ on a function $R_{\Phi,\Omega}(\rho)$ coincides with the action of μ. The final statement concerning necessity of the conditions is a simple consequence of the construction of $\mathcal{L}_{\Phi,\mathcal{P}}(\Omega)$. ■

The advantage of this result is that it does only involve limits of real numbers and values of finitely supported functionals (except for λ itself). A typical application is

Corollary 8.3 *Let $\Omega_1 \subseteq \Omega \subseteq \mathbb{R}^d$ be an open domain, and let derivatives of the form $(D^\alpha)^x\,(D^\alpha)^y\Phi(x,y)$ exist and be continuous on $\Omega_1 \times \Omega_1$ for a fixed multiindex $\alpha \in \mathbb{N}^d$. Furthermore, assume $\mathcal{P} = \mathbb{P}_m^d$ for $m < |\alpha|$ such that $D^\alpha(\mathcal{P}) = \{0\}$. Then all functions f in the native space $\mathcal{F}_{\Phi,\mathcal{P}}(\Omega)$ have a continuous derivative $D^\alpha f$ on Ω_1.*

Proof. Any pointwise multivariate derivative of order α at an interior point x can be approximated by finitely and locally supported functionals which vanish on $\mathcal{P} = \mathbb{P}_m^d$ if $m < |\alpha|$. Thus there is a functional $\delta_x^\alpha \in \mathcal{L}_{\Phi,\mathcal{P}}(\Omega)$ which acts like this derivative on the functions in the space $R_{\Phi,\Omega}(L_{\mathcal{P}}(\Omega))$. Its action on general functions $R_{\Phi,\Omega}(\rho) \in \mathcal{F}_{\Phi,\mathcal{P}}(\Omega)$ with $\rho \in \mathcal{L}_{\Phi,\mathcal{P}}(\Omega)$ is also obtainable as the limit of the action of these functionals. This proves that the pointwise derivative exists for all functions in the native space.

To prove continuity of the derivative, we first evaluate

$$\|\delta_x^\alpha - \delta_y^\alpha\|_\Phi^2 = \|\delta_x^\alpha\|_\Phi^2 + \|\delta_y^\alpha\|_\Phi^2 - 2(\delta_x^\alpha, \delta_y^\alpha)_\Phi$$

$$= (D^\alpha)_{|_x}^u\,(D^\alpha)_{|_x}^v\,\Phi(u,v) + (D^\alpha)_{|_y}^u\,(D^\alpha)_{|_y}^v\,\Phi(u,v)$$

$$-2(D^\alpha)_{|_x}^u\,(D^\alpha)_{|_y}^v\,\Phi(u,v)$$

and see that the right–hand side is a continuous function. Then

$$|(D^\alpha f)(x) - (D^\alpha f)(y)|^2 = (\delta_x^\alpha - \delta_y^\alpha, R_{\Phi,\Omega}^{-1} f)_\Phi^2 \leq \|\delta_x^\alpha - \delta_y^\alpha\|_\Phi^2 \|f\|_\Phi^2$$

proves continuity of the derivative of functions f in the native space. ■

Lower order derivatives and more general functionals λ can be treated similarly, applying Theorem 8.2 to approximations of $\lambda - \lambda \circ \Pi_{\mathcal{P}}$.

9   Extension and Restriction

We now study the dependence of the native space $\mathcal{N}_{\Phi,\mathcal{P}}(\Omega)$ on the domain. To this end, we keep the function Φ and its corresponding domain Ω fixed while looking at subdomains $\Omega_1 \subseteq \Omega$. In short,

Theorem 9.1 *Each function from a native space for a smaller domain Ω_1 contained in Ω has a canonical extension to Ω with the same seminorm, and the restriction of each function in $\mathcal{N}_{\Phi,\mathcal{P}}(\Omega)$ to Ω_1 lies in $\mathcal{N}_{\Phi,\mathcal{P}}(\Omega_1)$ and has a possibly smaller norm there.*

To prove the above theorem, we have to be somewhat more precise. Consider a subset Ω_1 with

$$\Xi \subseteq \Omega_1 \subseteq \Omega \subseteq \mathbb{R}^d$$

and first extend the functionals with finite support in Ω_1 trivially to functionals on Ω. This defines a linear map

$$\epsilon_{\Omega_1} : L_{\mathcal{P}}(\Omega_1) \to L_{\mathcal{P}}(\Omega)$$

which is an isometry because the inner products are based on (5.2) in both spaces. The map extends continuously to the respective closures, and we can use the Riesz maps to define an isometric extension map

$$e_{\Omega_1} : \mathcal{F}_{\Phi,\mathcal{P}}(\Omega_1) \to \mathcal{F}_{\Phi,\mathcal{P}}(\Omega), \; e_{\Omega_1} := R_{\Phi,\Omega} \circ \epsilon_{\Omega_1} \circ R_{\Phi,\Omega_1}{}^{-1}$$

between the nontrivial parts of the native spaces $\mathcal{N}_{\Phi,\mathcal{P}}(\Omega_1)$ and $\mathcal{N}_{\Phi,\mathcal{P}}(\Omega)$. The main reason behind this very general construction of canonical extensions of functions from "local" native spaces is that the variable x in (5.7) can vary in all of Ω while the support $X = \{x_1, \ldots, x_M\}$ of the functional $\lambda_{X,M}$ is contained in Ω_1. Of course, we define e_{Ω_1} on $\mathcal{P}$ by straightforward extension of functions in $\mathcal{P}$, and thus have e_{Ω_1} well–defined on all of $\mathcal{N}_{\Phi,\mathcal{P}}(\Omega)$.

At this point we want to mark a significant difference to the standard technique of defining local Sobolev spaces. On $\Omega = \mathbb{R}^d$ one can prove that the global Sobolev space $W_2^k(\mathbb{R}^d)$ for $k > d/2$ is the native space for the radial positive definite function

$$\Phi(x, y) = \phi(\|x - y\|_2), \; \phi(r) = r^{k-d/2} K_{r-d/2}(r)$$

with a Bessel or Macdonald function. If we go over to localized versions of the native space, we can do this for very general (even finite) subsets Ω_1 of $\Omega = \mathbb{R}^d$, and there are no boundary effects. Furthermore, any locally defined function has a canonical extension. This is in sharp contrast to the classical construction of local Sobolev spaces, introducing functions with singularities if the boundary has incoming edges. These functions have no extension to a Sobolev space on a larger domain. Functions from local Sobolev spaces only have extensions if the domain satisfies certain boundary conditions. Our definition always starts with the global function Φ and then does a local construction. Is our construction really "local"? To gain some more insight into this question, we have to look at the restrictions of functions from global native spaces.

Curiously enough, the restriction of functions from $\mathcal{N}_{\Phi,\mathcal{P}}(\Omega)$ to Ω_1 is slightly more difficult to handle than the extension. If we define $r_{\Omega_1}(f) := f_{|\Omega_1}$ for $f \in \mathcal{N}_{\Phi,\mathcal{P}}(\Omega)$, we have to show that the result is a function in $\mathcal{N}_{\Phi,\mathcal{P}}(\Omega_1)$.

Lemma 9.2 *The restriction map*

$$r_{\Omega_1} \; : \; \mathcal{N}_{\Phi,\mathcal{P}}(\Omega) \to \mathcal{N}_{\Phi,\mathcal{P}}(\Omega_1)$$

is well–defined and coincides with the formal adjoint of e_{Ω_1}. For any function f in $\mathcal{N}_{\Phi,\mathcal{P}}(\Omega)$ the "localized" seminorm $|r_{\Omega_1} f|_{\Psi,\Omega_1}$ depends only on the values of f on Ω_1. It is a monotonic function of Ω_1.

Proof. From the extension property of e_{Ω_1} and the restriction property of r_{Ω_1} we can conclude

$$(e_{\Omega_1} \lambda_{\Omega_1})(f) = \lambda_{\Omega_1}(r_{\Omega_1} f) \text{ for all } \lambda_{\Omega_1} \in \mathcal{L}_{\Phi,\mathcal{P}}(\Omega_1), \; f \in \mathcal{N}_{\Phi,\mathcal{P}}(\Omega). \qquad (9.1)$$

This is obvious for finitely supported functionals from $L_{\mathcal{P}}(\Omega_1)$ and holds in general by continuous extension.

We use this to prove that $r_{\Omega_1}(f) := f_{|\Omega_1}$ lies in $\mathcal{M}_{\Phi,\mathcal{P}}(\Omega_1)$ for each f in $\mathcal{M}_{\Phi,\mathcal{P}}(\Omega)$. This follows from

$$|\lambda_{\Omega_1}(r_{\Omega_1} f)| = |(e_{\Omega_1} \lambda_{\Omega_1})f| \le |f|_{\Psi,\Omega} \|e_{\Omega_1} \lambda_{\Omega_1}\|_{\Phi,\Omega} = |f|_{\Psi,\Omega} \|\lambda_{\Omega_1}\|_{\Phi,\Omega_1}$$

for all $\lambda_{\Omega_1} \in \mathcal{L}_{\Phi,\mathcal{P}}(\Omega_1)$ and all $f \in \mathcal{F}_{\Phi,\mathcal{P}}(\Omega)$. This also proves the inequality $|r_{\Omega_1} f|_{\Psi,\Omega_1} \le |f|_{\Psi,\Omega}$. Both inequalities are trivially satisfied for $f \in \mathcal{P}$. The monotonicity statement can be obtained by the same argument as above. Thus the norm of the restriction map is not exceeding one. Our detour via $\mathcal{M}_{\Phi,\mathcal{P}}(\Omega_1)$ implies that $|r_{\Omega_1} f|_{\Psi,\Omega_1}$ only depends on the values of f on Ω_1. For all $f \in \mathcal{F}_{\Phi,\mathcal{P}}(\Omega)$ and $f_{\Omega_1} \in \mathcal{F}_{\Phi,\mathcal{P}}(\Omega_1)$ equation (9.1) yields

$$
\begin{aligned}
(e_{\Omega_1} f_{\Omega_1}, f)_{\Phi,\Omega} &= (R_{\Phi,\Omega} e_{\Omega_1} R_{\Phi,\Omega_1}{}^{-1} f_{\Omega_1}, f)_{\Phi,\Omega} \\
&= (e_{\Omega_1} R_{\Phi,\Omega_1}{}^{-1} f_{\Omega_1})(f) \\
&= (R_{\Phi,\Omega_1}{}^{-1} f_{\Omega_1})(r_{\Omega_1} f) \\
&= (f_{\Omega_1}, r_{\Omega_1} f)_{\Phi,\Omega_1}.
\end{aligned}
\qquad (9.2)
$$

This is the nontrivial part of the proof that that r_{Ω_1} is the formal adjoint of e_{Ω_1}. ∎

Lemma 9.3 *The above extension and restriction maps satisfy*

$$r_{\Omega_1} \circ e_{\Omega_1} = Id_{\mathcal{N}_{\Phi,\mathcal{P}}(\Omega_1)}.$$

Proof. The assertion is true on $\mathcal{P}$. On $\mathcal{F}_{\Phi,\mathcal{P}}(\Omega_1)$ we use (9.2) and the fact that e_{Ω_1} is an isometry. Then for all $f_{\Omega_1}, g_{\Omega_1} \in \mathcal{F}_{\Phi,\mathcal{P}}(\Omega_1)$ we have

$$
\begin{aligned}
(f_{\Omega_1}, g_{\Omega_1})_{\Phi,\Omega_1} &= (e_{\Omega_1} f_{\Omega_1}, e_{\Omega_1} g_{\Omega_1})_{\Phi,\Omega} \\
&= (f_{\Omega_1}, r_{\Omega_1} e_{\Omega_1} g_{\Omega_1})_{\Phi,\Omega_1}
\end{aligned}
$$

Thus $g_{\Omega_1} - r_{\Omega_1} e_{\Omega_1} g_{\Omega_1}$ must be zero, because it is in $\mathcal{F}_{\Phi,\mathcal{P}}(\Omega_1)$. ∎

An interesting case of Lemma 9.2 occurs when $\Omega_1 = X = \{x_1, \ldots, x_M\}$ is finite and contains Ξ. Then the functions in $\mathcal{N}_{\Phi,\mathcal{P}}(\Omega_1)$ are of the form

$$s = p + R_{\Phi,\Omega_1}(\lambda_{X,M}), \ p \in \mathcal{P}, \ \lambda_{X,M} \in L_{\mathcal{P}}(\Omega_1)$$

and their seminorm is explicitly given by

$$|s|_{\Psi,\Omega_1} = \|\lambda_{X,M}\|_\Phi.$$

Due to Lemma 9.2 this value depends only on the values of s on $\Omega_1 = X = \{x_1, \ldots, x_M\}$, and we can read off from (7.2) how this works. The value of the norm is numerically accessible. See [14] for the special case of thin–plate splines.

If we look at the extension $e_{\Omega_1} s$ of s to all of Ω we see that it has the same data on X. From

$$e_{\Omega_1}(s - \Pi_{\mathcal{P}} s) = R_{\Phi,\Omega} \epsilon_{\Omega_1} R_{\Phi,\Omega_1}{}^{-1}(s - \Pi_{\mathcal{P}} s)$$

$$R_{\Phi,\Omega}{}^{-1} e_{\Omega_1}(s - \Pi_{\mathcal{P}} s) = \epsilon_{\Omega_1} R_{\Phi,\Omega_1}{}^{-1}(s - \Pi_{\mathcal{P}} s)$$

one can read off that this functional has support in $X = \Omega_1$ and thus is the global form of the interpolant.

This holds also for general transfinite interpolation processes. Consider an arbitrary subset Ω_1 of Ω with $\Xi \subseteq \Omega_1$. On such a set, data are admissible if they are obtained from some function $f \in \mathcal{N}_{\Phi,\mathcal{P}}(\Omega)$. Then $e_{\Omega_1} r_{\Omega_1} f$ has the same data on Ω_1. Furthermore, we assert that it is the global function with least seminorm with these data. By standard arguments this boils down to proving the variational equation

$$(e_{\Omega_1} r_{\Omega_1} f, v)_{\Psi,\Omega} = 0 \ \text{ for all } v \in \mathcal{N}_{\Phi,\mathcal{P}}(\Omega) \text{ with } r_{\Omega_1} v = 0.$$

But this is trivial for $f \in \mathcal{P}$ and follows from

$$
\begin{aligned}
(e_{\Omega_1} r_{\Omega_1} f, v)_{\Psi,\Omega} &= (R_{\Phi,\Omega} \epsilon_{\Omega_1} R_{\Phi,\Omega_1}{}^{-1} r_{\Omega_1} f, v)_{\Psi,\Omega} \\
&= (\epsilon_{\Omega_1} R_{\Phi,\Omega_1}{}^{-1} r_{\Omega_1} f)(v) \\
&= (R_{\Phi,\Omega_1}{}^{-1} r_{\Omega_1} f)(r_{\Omega_1} v)
\end{aligned}
$$

for all $f \in \mathcal{F}_{\Phi,\mathcal{P}}(\Omega)$. Now we can generalize a result that plays an important part in Duchon's [1] error analysis of polyharmonic splines:

Theorem 9.4 *For all functions $f \in \mathcal{N}_{\Phi,\mathcal{P}}(\Omega)$ and all subsets Ω_1 of Ω with $\Xi \subseteq \Omega_1 \subseteq \Omega$ the function $e_{\Omega_1} r_{\Omega_1} f \in \mathcal{N}_{\Phi,\mathcal{P}}(\Omega)$ has minimal seminorm in $\mathcal{N}_{\Phi,\mathcal{P}}(\Omega)$ under all functions coinciding with f on Ω_1.* ∎

Theorem 9.5 *The orthogonal complement of $e_{\Omega_1}(\mathcal{N}_{\Phi,\mathcal{P}}(\Omega_1))$ in $\mathcal{N}_{\Phi,\mathcal{P}}(\Omega)$ is the space of all functions that agree on Ω_1 with a function in $\mathcal{P}$.*

Proof. Use the above display again, but for $v \in \mathcal{N}_{\Phi,\mathcal{P}}(\Omega)$ being orthogonal to all functions $g = r_{\Omega_1} f$. ∎

10 Linear recovery processes

For applications of native space techniques, we want to look at general methods for reconstructing functions in the native space $\mathcal{N}_{\Phi,\mathcal{P}}(\Omega)$ from given data. The data are furnished by linear functionals $\lambda_1, \ldots, \lambda_N$ from $\mathcal{N}^*_{\Phi,\mathcal{P}}(\Omega)$, and the reconstruction uses functions $v_1, \ldots, v_N$ from $\mathcal{N}_{\Phi,\mathcal{P}}(\Omega)$. At this point, we do not assume any link between the functionals λ_j and the functions v_j. A function $f \in \mathcal{N}_{\Phi,\mathcal{P}}(\Omega)$ is to be reconstructed via a linear quasi–interpolant

$$s_f(x) := \sum_{j=1}^{N} \lambda_j(f) v_j(x), \ x \in \Omega, \ f \in \mathcal{N}_{\Phi,\mathcal{P}}(\Omega)$$

which should reproduce functions from $\mathcal{P}$ by

$$p(x) = s_p(x) = \sum_{j=1}^{N} \lambda_j(p) v_j(x), \ x \in \Omega, \ p \in \mathcal{P}. \tag{10.1}$$

Then the error functional

$$\epsilon_x : f \mapsto f(x) - s_f(x), \epsilon_x = \delta_x - \sum_{j=1}^{N} v_j(x) \lambda_j(f)$$

is in $\mathcal{L}_{\Phi,\mathcal{P}}(\Omega)$ for all $x \in \Omega$ and is continuous on $\mathcal{F}_{\Phi,\mathcal{P}}(\Omega)$. This setting covers a wide range of Hermite–Birkhoff interpolation or quasi–interpolation processes.

Theorem 10.1 *With the **power function** defined by*

$$P(x) := \|\epsilon_x\|_{\Phi}$$

the error is bounded by

$$|f(x) - s_f(x)| = |\epsilon_x(f)| \leq P(x)|f|_{\Psi} \ \text{for all } f \in \mathcal{N}_{\Phi,\mathcal{P}}(\Omega), \ x \in \Omega.$$

Proof. Just use (10.1) and evaluate

$$|f(x) - s_f(x)| = |\epsilon_x(f)| = |\epsilon_x(f - \Pi_P f)|$$

$$\leq \|\epsilon_x\|_\Phi \|f - \Pi_P f\|_\Phi = P(x)|f|_\Psi.$$

∎

Theorem 10.2 *The power function can be explicitly evaluated by*

$$P^2(x) := \|\epsilon_x\|_\Phi^2$$

$$= \Psi(x, x) - 2\sum_{j=1}^{N} v_j(x)\lambda_j^y \Psi(y, x) \tag{10.2}$$

$$+ \sum_{j=1}^{N}\sum_{k=1}^{N} v_j(x)v_k(x)\lambda_j^y \lambda_k^z \Psi(y, z).$$

Proof. Use

$$\epsilon_x(f) = \left(\delta_{(x)} - \sum_{j=1}^{N} v_j(x)\lambda_j\right)(f - \Pi_P f)$$

for all $f \in \mathcal{N}_{\Phi,P}(\Omega)$ and evaluate $\|\epsilon_x\|_\Psi^2 = \|\epsilon_x\|_\Phi^2 = P^2(x)$. ∎

Note that it is in general not feasible to evaluate $\lambda^x \Phi(x, \cdot)$ for functionals $\lambda \in \mathcal{N}_{\Phi,P}^*(\Omega)$ unless they are plain point evaluations or vanish on $\mathcal{P}$. This is why we make a detour via Ψ here and use a different representation of ϵ_x. But in many standard cases one can replace Ψ in (10.2) by Φ.

Altogether, each linear recovery process has a specific power function that describes the pointwise worst–case error for recovery of functions from the native space. The power function can be explicitly evaluated, and it would be interesting to see various examples. Now we add more information on the relation between the functionals λ_j and the functions v_j:

Theorem 10.3 *If the recovery process is* **interpolatory**, *i.e.*

$$\lambda_j(v_k) = \delta_{jk}, \ 1 \leq j, k \leq N, \tag{10.3}$$

the error is bounded by

$$|f(x) - s_f(x)| = |\epsilon_x(f)| \leq P(x)|f - s_f|_\Psi \ \text{for all } f \in \mathcal{N}_{\Phi,P}(\Omega), \ x \in \Omega.$$

Furthermore, we have

$$\lambda_j(P) := \lambda_j^x \lambda_j^y (\epsilon_x, \epsilon_y)_\Psi = 0, \ 1 \leq j \leq N. \tag{10.4}$$

Proof. The interpolation property implies $s_{f-s_f} = 0$, and then Theorem 10.1 can be applied to the difference $f - s_f$. This proves the error bound, and the second assertion follows directly from (10.3) when applied to (10.2) written out in two variables as required for (10.4). ∎

Note that the special definition (10.4) for evaluation of $\lambda_j(P)$ is necessary because P is in general not in the native space. The new kernel function $Q(x,y) := (\epsilon_x, \epsilon_y)_\Psi$ has a form similar to the reduction of Φ. It has interesting additional properties and can in particular be used for recursive construction of interpolants and orthogonal bases. See [15] for details and [14] for a special case.

11 Optimal recovery

We now want to let the functions v_j vary freely, but we keep the functionals λ_j and the evaluation point $x \in \Omega$ fixed. Since the $v_j(x)$ influence only the power function part of the error bound, we can try to minimize the quadratic form (10.2) with respect to the N real numbers $v_j(x)$, $1 \le j \le N$ under the linear constraints imposed by (10.1). We do this by application of standard techniques of optimization. Picking a basis $p_1, \ldots, p_q$ of $\mathcal{P}$ and introducing Lagrange multipliers $q_m(x)$, $1 \le m \le q$ for the q linear constraints from (10.1), we get that any critical point of the constrained quadratic form is a critical point of the unconstrained quadratic form

$$\Psi(x,x) - 2\sum_{j=1}^{N} v_j(x)\lambda_j^y \Psi(y,x)$$

$$+ \sum_{j=1}^{N}\sum_{k=1}^{N} v_j(x)v_k(x)\lambda_j^y \lambda_k^z \Psi(y,z)$$

$$- 2\sum_{m=1}^{q} q_m(x)\left(p_m(x) - \sum_{k=1}^{N} v_k(x)\lambda_k(p_m)\right)$$

to be minimized with respect to the values $v_j(x)$. The equations for a critical point of the constrained quadratic form thus are

$$\lambda_k^y \Psi(x,y) = \sum_{j=1}^{N} \lambda_j^y \lambda_k^z \Psi(y,z)v_j^*(x) + \sum_{m=1}^{q} q_m^*(x)\lambda_k(p_m)$$

$$p_m(x) = \sum_{j=1}^{N} \lambda_j(p_m)v_j^*(x), \quad 1 \le k \le N, \ 1 \le m \le q. \tag{11.1}$$

We want to prove that this system has a unique solution. Taking a solution of the homogeneous system

$$0 = \sum_{j=1}^{N} \lambda_j^y \lambda_k^z \Psi(y, z) a_j + \sum_{m=1}^{q} b_m \lambda_k(p_m)$$

$$0 = \sum_{j=1}^{N} \lambda_j(p_m) a_j, \ 1 \le k \le N, \ 1 \le m \le q,$$

we see that the functional

$$\chi := \sum_{j=1}^{N} a_j \lambda_j$$

vanishes on $\mathcal{P}$ and thus is in $\mathcal{L}_{\Phi,\mathcal{P}}(\Omega)$. Applying it to the first N equations of (11.1) yields

$$0 = \sum_{j=1}^{N} \sum_{k=1}^{N} \lambda_j^y \lambda_k^z \Psi(y, z) a_j a_k + \sum_{m=1}^{q} b_m \sum_{k=1}^{N} \lambda_k(p_m) a_k$$

$$= \sum_{j=1}^{N} \sum_{k=1}^{N} \lambda_j^y \lambda_k^z \Psi(y, z) a_j a_k + 0$$

$$= \chi^y \chi^z \Psi(y, z)$$

$$= \chi^y \chi^z \Phi(y, z).$$

Since Φ is a CPD function, we conclude that the functional χ must vanish on the native space. If we make the reasonable additional assumption that the functionals λ_j are linearly independent, we see that the coefficients a_j must vanish. This leaves us with the system

$$0 = \sum_{m=1}^{q} b_m \lambda_k(p_m), \ 1 \le k \le N,$$

and we can conclude that all the coefficients b_m are zero, if we assume that there is no nonzero function p in $\mathcal{P}$ for which all data $\lambda_k(p)$, $1 \le k \le N$ vanish.

Now we know that (11.1) has a unique solution $v_j^*(x)$, $q_m^*(x)$ for indices $1 \le j \le N$, $1 \le m \le q = \dim \mathcal{P}$. Thus the augmented unconstrained quadratic form has a unique critical point, and the same holds for the constrained quadratic form. Since the latter is nonnegative and positive definite, the critical point must be a minimum. Furthermore, we see that the optimal solution values $v_j^*(x)$, $q_m^*(x)$ are linear combinations of the right–hand values $p_m(x)$, $1 \le m \le q$ and $\lambda_k^y \Psi(x, y)$ for $1 \le k \le N$. If we apply the functional λ_j to the system, we see that the j–th column coincides with the right–hand side, and this proves that the solution must satisfy the interpolation property (10.3). Furthermore, we get $\lambda_j(q_m) = 0$ for all indices $1 \le j \le N$, $1 \le m \le q$.

But note that our solution is not just any interpolatory set of functions matching the data functionals. It is uniquely determined by the λ_j, and it is composed of functions $\lambda_j^y \Psi(y, \cdot)$ acting as the Riesz representers of the λ_j in the sense of Theorem (6.1) plus functions from $\mathcal{P}$. We summarize:

Theorem 11.1 *Assume that there is no nonzero function from $\mathcal{P}$ on which all functionals λ_j vanish, and that the λ_j are linearly independent. Then the power function at any point $x \in \Omega$ can be minimized among all other power functions using the same functionals λ_j, but possibly different functions v_j. The minimum is attained for a specific set v_j^* of functions that satisfy the interpolation conditions (10.3) and are linear combinations of functions from $\mathcal{P}$ and generalized represen-ters of the functionals λ_j.*

We add an intrinsic characterization of the optimal power function:

Theorem 11.2 *Under the above assumptions, the optimal power function $P^*(x)$ describes the maximal value that any function $f \in \mathcal{N}_{\Phi,\mathcal{P}}(\Omega)$ can attain at $x \in \Omega$ if it satisfies the restrictions $\lambda_j(f) = 0$, $1 \le j \le N$ and $|f|_\Psi \le 1$.*

Proof. For any such function, Theorem 10.3 implies $|f(x)| \le P^*(x)$ because of $s_f = 0$. For fixed $x \in \Omega$, the maximal value is attained for the special function

$$f_x = \Psi(\cdot, x) - \sum_{j=1}^N v_j^*(x) \lambda_j^y \Psi(y, \cdot) - \sum_{m=1}^q q_m^*(x) p_m(\cdot)$$

after rescaling. This is due to

$$f_x(x) = (P^*(x))^2 = |f_x|_\Psi^2$$

which needs some elementary calculations and the identity

$$\sum_{k=1}^N v_k^*(x) \lambda_k^y \Psi(x, y) = \sum_{j=1}^N \sum_{k=1}^N v_j^*(x) v_k^*(x) \lambda_j^y \lambda_k^z \Psi(y, z) + \sum_{m=1}^q q_m^*(x) p_m(x)$$

following from (11.1). ■

Under the assumptions of Theorem 11.1 one can try to rewrite the recovery function as

$$s_f^*(x) = \sum_{j=1}^N \lambda_j^y \Psi(y, x) a_j + \sum_{m=1}^q p_m(x) b_m, \tag{11.2}$$

where the coefficients a_j satisfy

$$\sum_{j=1}^N \lambda_j(p_m) a_j = 0, \ 1 \le m \le q = \dim \mathcal{P}$$

in order to form a functional from $L_{\mathcal{P}}(\Omega)$. The equations for a generalized inter-polation of a function f then are

$$\lambda_k(f) = \lambda_k(s_f^*) = \sum_{j=1}^{N} \lambda_j^y \lambda_k^z \Psi(y, z) a_j + \sum_{m=1}^{q} \lambda_k(p_m) b_m$$

$$0 = \sum_{j=1}^{N} \lambda_j(p_m) a_j, \ 1 \le k \le N, \ 1 \le m \le q,$$

(11.3)

and the coefficient matrix coincides with the matrix in (11.1). There is much similarity to (7.2), but we are more careful here and put Ψ instead of Φ into the definition (11.2) of the recovery function, because $\lambda_j^y \Phi(y, \cdot)$ may not make sense while $\delta_{x_j}^y \Phi(y, \cdot)$ in (7.1) always is feasible.

If we take the Ψ–inner product of s_f^* with an arbitrary function g from the native space, we get

$$(s_f^*, g)_\Psi = \sum_{j=1}^{N} a_j \left(\lambda_j^y \Psi(y, \cdot), g \right)_\Psi + 0$$

$$= \sum_{j=1}^{N} a_j \lambda_j^y \left(g(y) - (\Pi_{\mathcal{P}} g)(y) \right)$$

$$= \sum_{j=1}^{N} a_j \lambda_j(g).$$

By a standard variational argument this implies an extension of Theorem 9.4 to more general data functionals:

Theorem 11.3 *Under the hypotheses of Theorem 11.1 the function s_f^* solves the minimization problem*

$$\min \left\{ |g|_\Psi \ : \ g \in \mathcal{N}_{\Phi, \mathcal{P}}(\Omega), \ \lambda_j(f - g) = 0, \ 1 \le j \le N \right\}.$$

Furthermore, the orthogonality relation

$$(f - s_f^*, s_f)_\Psi = 0$$

holds and implies

$$|f|_\Psi^2 = |f - s_f^*|_\Psi^2 + |s_f^*|_\Psi^2.$$

From section 9 we already know that the second term on the right–hand side can be calculated explicitly. The minimization principle of Theorem 11.3 implies that this value increases when more and more data functionals are used to recover the same function f. In case of reconstruction of a function from the native space, there is the upper bound $|f|_\Psi^2$ for these values, but for general given functions there might be no upper bound. However, the following related characterization of native spaces was proven in [16]:

Theorem 11.4 *The native space for a continuous CPD function* Φ *on* Ω *can be characterized as the set of all real–valued functions* f *on* Ω *for which there is a fixed upper bound* $C_f \geq |s_f^*|_\Psi$ *for all interpolants* s_f^* *based on arbitrary point–evaluation data* $\lambda_j = \delta_{x_j}$, $x_j \in \Omega$.

Acknowledgements: Special thanks go to Holger Wendland for careful proofreading and various improvements.

References

[1] J. Duchon, *Interpolation des fonctions de deux variables suivant le principe de la flexion des plaques minces*, Rev. Française Automat. Informat. Rech. Opér., Anal. Numer. **10** (1976), 5–12

[2] N. Dyn, F. Narcowich and J. Ward, *Variational principles and Sobolev–type estimates for generalized interpolation on a Riemannian manifold*, to appear in Constr. Approx.

[3] W. Freeden, R. Franke and M. Schreiner, *A survey on spherical spline approximation*, Surveys on Mathematics for Industry **7** (1997), 29–85

[4] M. Golomb, H.F. Weinberger, *Optimal approximation and error Bounds*, in: On Numerical Approximation, The University of Wisconsin Press, Madison, R.E. Langer (1959), 117–190

[5] T. Gutzmer, *Interpolation by positive definite functions on locally compact groups with application to SO(3)*, Resultate Math. **29** (1996), 69–77

[6] R.L. Hardy, *Multiquadric equations of topography and other irregular surfaces*, J. Geophys. Res. **76** (1971), 1905–1915

[7] K. Jetter, J. Stöckler, *A Generalization of de Boor's stability result and symmetric preconditioning*, Advances in Comp. Math. **3** (1995), 353–367

[8] M. Klein, *Spezielle Probleme der Rekonstruktion multivariater Funktionen*, Diplom thesis, University of Göttingen, 1998

[9] W.L. Light, H. Wayne, *On power functions and error estimates for radial basis function interpolation*, J. Approx. Theory **92** (1998), 245–267

[10] Lin–Tian Luh, *Native spaces for radial basis functions*, Dr. rer. nat. dissertation, Göttingen 1998

[11] W.R. Madych, S.A. Nelson, *Multivariate Interpolation and conditionally positive definite functions*, Approx. Theory Appl. **4** (1988), 77–89

[12] H. Meschkowski, *Hilbertsche Räume mit Kernfunktion*, Springer, Berlin (1962)

[13] F.J. Narcowich, *Generalized Hermite interpolation by positive definite kernels on a Riemannian manifold*, J. Math. Anal. Appl. **190** (1995), 165–193

[14] M.J.D. Powell, Recent research at Cambridge on radial basis functions, 215–232 (in this book).

[15] R. Schaback, *Reconstruction of multivariate functions from scattered data*, manuscript, available via

http://www.num.math.uni-goettingen.de/schaback/research/group.html

[16] R. Schaback, H. Wendland, *Inverse and saturation theorems for radial basis function interpolation*, Preprint 1998, Univ. Göttingen

[17] J. Stewart, *Positive definite functions and generalizations, an historical survey*, Rocky Mountain J. Math. **6** (1976), 409–434

Robert Schaback
Universität Göttingen, Lotzestraße 16-18
D-37083 Göttingen, Germany
Email address: schaback@math.uni-goettingen.de

International Series of Numerical Mathematics
Vol. 132, © 1999 Birkhäuser Verlag Basel/Switzerland

Adaptive Approximation with Walsh-similar Functions

Bl. Sendov

Abstract

The classical orthonormal system of Walsh functions is generalized in a new direction, called Walsh-similar functions, different from the already well-known ones [5], [7], [9], [14], [21] and from the Generalized Walsh-like functions [10], [11] . The definition of the Walsh-similar functions involves real parameters and allows adaptation of the orthonormal system to a particular function by an appropriate choice of these parameters. A Walsh-similar function may have any given fractal dimension.

1 Introduction

A basic principle in the Approximation theory is to choose the instrument for approximation according to the properties of the objects to be approximated. This principle was theoretically developed in the Kolmogorov's theory of widths [8]. Let X be a linear functional space and $A \subset X$ be a set of functions with given properties. For every natural n A. N. Kolmogorov defined the optimal n-dimensional linear subspace $A_n \subset X$ for linear approximation of the elements of A. In other words, if we have to use a linear method for approximation of functions from A, the best instruments for this are the elements of A_n.

In the classical approximation theory, the objects for approximation are sets A of functions characterized by some properties of smoothness. A proper instrument for approximation has to satisfy the same property of smoothness and to reflect some peculiarities of the functions belonging to A. A typical example is the theory for *local approximation* and its application for *adaptive approximation* (see [4] and the references given there), developed mainly by Yu. A. Brudnyĭ.

The splines and the wavelets, which occupy the minds of the approximation theorists in the last decades, are also useful instruments for approximation, because they reflect some features of the sets of functions, which are important in applications and have to be approximated. The use of the approximation theory methods in the signal and image processing requires approximation of functions which are far for smooth and even resemble fractal functions. It is not a surprise, that the *fractal approximation theory*, developed by the group around M. F. Barnsley [1] is so helpful in image compression [2].

Is it reasonable to choose a particular instrument for approximation of a particular function ? In fact, the methods in the fractal approximation theory [2] are based on the so called *Iterated Function Systems* (IFS), which are chosen differently for different target functions for approximation. Also, some methods for signal and image compression, based on the so called *wavelet libraries* and *wavelet packets* [12], use in fact instruments for approximation chosen for a particular function.

In this paper we consider a particular approximation instrument for any particular function from a functional space. For this purpose a class of functions, similar to the Walsh functions, is used. The scheme is as follows.

Consider, for example, the functional space $L^2[0,1)$ of square integrable real functions defined on $[0,1)$. We define a subset of functions

$$S = \bigcup_{n=1}^{\infty} S_n \subset L^2[0,1), \quad S_n \subset S_{n+1},$$

where S_n depend on n parameters. Let us mention that the constant 1 belongs to S, S is not a linear set, and S is not dense in $L^2[0,1)$.

Every function $\varphi \in S$ has the property that *starting* with φ one may construct, "fast and easy" as the construction of the Walsh functions, an orthonormal basis

$$\varphi_0(x) = \varphi(x), \varphi_1(x), \varphi_2(x), \varphi_3(x), \ldots$$

in $L^2[0,1)$.

To approximate a particular function $f \in L^2[0,1)$ two steps are to be made:

1) Solve a "small" non linear approximation problem. Find the function $\varphi \in S_n$, which is the closest to f, in a given sense, i. e., to calculate n parameters defining φ.

2) Solve a "large" linear approximation problem. Find the best L^2 approximation of f in the N-dimensional linear subspace $A_n \subset L^2[0,1)$, spanned over the orthonormal system $\varphi_0, \varphi_1, \varphi_2, \ldots, \varphi_{N-1}$. That is to calculate the Fourier coefficients $f_i^\wedge = \int_0^1 f(t)\varphi_i(t)\, dt; \quad i = 0, 1, 2, \ldots, N-1$.

We say "small" and "large", because it is natural to have $N = 2^n$.

The first step is the choice of the particular approximation instrument, adapted to a particular function to be approximated. The aim of this step is to find an orthonormal system in respect to which the entropy

$$\mathcal{E}(\varphi; f)_N = -\sum_{i=1}^{N-1} (f_i^\wedge)^2 \log_2 (f_i^\wedge)^2$$

of the Fourier coefficients of f is smaller than the entropy in respect to an standard, non adapted, orthonormal system. This is important for the compression of the data.

It is true, that using this scheme, to represent the approximation, one needs $N + n$ numbers, instead of N. But if a smaller entropy of the Fourier coefficients is got, this may pay off. As a matter of fact, we shall show that this increase from N to $N + n$ may be avoided.

Until now we do not have a general proof that our scheme gives an optimal orthonormal system in a given sense. We have only convincing examples comparing the classical Walsh and Haar systems with the adapted ones.

The organization of the paper is the following.

In Section 2 the basic notations and define the so called relative module of continuity, which is useful for adaptive approximation is introduced.

In Section 3 we start with the definition of the set S of the starting functions. In 3.3 the set of Walsh-similar (WS) functions is defined and prove the theorem for the best L^2 approximation with Walsh-similar functions is proved.

In Section 4 we show a mechanism for adaptation by the best L^2 approximation with starting functions. In Section 5 some properties of the Walsh-similar functions are given. In Section 5 some properties of the Walsh-similar functions are given.

In Section 6 a generalization of the Walsh-similar functions is defined in the line of the generalized Walsh functions defined by N. Fine [7].

The Section 7 is devoted to the definition of Walsh-similar functions of rank 2 and 3. These definitions are suitable also for functions of 2 and 3 variables.

To simplify the presentation we limit ourselves to the spaces $L^2[0,1)$ and $L^2[0,\infty)$. In general cases, the results are true in $L^p[0,1)$ and $L^p[0,\infty)$; $1 \le p \le \infty$ respectively.

2 Basic notations

Let $y = y_{-k}2^k + y_{-k+1}2^{k-1} + \cdots + y_0 + y_1 2^{-1} + y_2 2^{-2} + \cdots$ be the dyadic representation of the non negative real number y, where $y_i = 0, 1$ are the dyadic digits of y. If y is dyadic rational, take the finite dyadic representation of y (with finite number of ones). By definition $y_i = 0$ for $i > \log_2(1 + y)$. Pay attention that the binary digits are defined in the same way for real numbers $y < 1$ and for natural numbers. The sum of the power of 2 and the index is always zero.

For the natural numbers, the following unique representation is also used:

$$m = 2^{m(0)} + 2^{m(1)} + 2^{m(2)} + \cdots + 2^{m(p)}, \tag{1}$$

where $0 \le m(0) < m(1) < m(2) < \cdots < m(p)$.

The addition $\dotplus$ (addition modulo 2) of two non negative real numbers x, y is defined as follows

$$z = x \dotplus y = z_{-l}2^l + z_{-l+1}2^{l-1} + \cdots + z_0 + z_1 2^{-1} + z_2 2^{-2} + \cdots, \tag{2}$$

where $l \le max\{\log_2(1 + x), \log_2(1 + y)\}$ and

$$z_i = x_i \dotplus y_i, \quad \text{for} \quad i = -l, 1 - l, 2 - l, \ldots \ .$$

If in the representation (2) for z, we have $z_m = 0$ and $z_i = 1$ for $i = m + 1, m+2, m+3, \ldots$, then we correct the representation (2) by the change $z_m = 1$ and $z_i = 0$ for $i = m+1, m+2, m+3, \ldots$.

We shall use the operation $x \dot{+} y$ together with the operation $x + y$. The first operation is defined only in $R^+ = [0, \infty)$. Both operations are commutative and $x \dot{+} x = 0$, or $(x \dot{+} h) \dot{+} h = x$. As usual, set $y_i \dot{+} 1 = \overline{y_i}$.

Define the operation $x \circ y$ for $x, y \geq 0$ as follows

$$x \circ y = \sum_{i=-\infty}^{\infty} x_i y_{1-i}. \tag{3}$$

Obviously the sum (3) is in fact final and $x \circ y$ is a non negative integer and $x \circ y = y \circ x$. If $[x]$ is the integer part of x, then

$$x \circ y = [x] \circ y + x \circ [y]. \tag{4}$$

If m is a non negative integer

$$m = m_0 2^0 + m_{-1} 2^1 + m_{-2} 2^2 + \cdots + m_{-p} 2^p,$$

then

$$\check{m} = m_0 2^{-1} + m_{-1} 2^{-2} + m_{-2} 2^{-3} + \cdots + m_{-p} 2^{-p-1} \in [0, 1) \tag{5}$$

is a dyadic rational.

From (5) it follows that

$$\check{m}_i = m_{1-i}; \quad i = 1, 2, 3, \ldots \quad .$$

The Rademacher functions may be defined as the transformations

$$(r_s^* f)(x) = (-1)^{x_s} f(x) = (-1)^{x_s} \quad \text{for} \quad x \in [0, 1) \tag{6}$$

of the function $f(x) = 1$.

The Walsh functions in Paley ordering [13] may be defined also as the transformations

$$(w_m^* f)(x) = \left(\prod_{i=1}^{p} r_{m(i)}^* \right) f(x) = (-1)^{x \circ m} f(x) = (-1)^{x \circ m} \tag{7}$$

of the function $f(x) = 1$. The Generalized Walsh functions [7],[14] are defined as follows:

$$w_y(x) = (-1)^{x \circ y} \quad \text{for} \quad x, y \geq 0. \tag{8}$$

The basic step in defining the Walsh-similar functions is that to add a dyadic shift to the transformations r_s^*.

Definition 2.1 *Define the* **Rademacher** *and* **Walsh transformations** *in* $L^2[0,1)$ *as follows*

$$(r_s f)(x) = (-1)^{x_s} f(x \dotplus 2^{-s}); \quad s = 0, 1, 2, \ldots \quad , \tag{9}$$

$$(w_m f)(x) = (-1)^{x \circ m} f(x \dotplus \breve{m}); \quad m = 0, 1, 2, \ldots \quad . \tag{10}$$

It is obvious that the transformations r_s and w_m produce the Walsh functions from the starting function $w_0(x) = 1$ in the same way as the transformations r_s^* and w_m^* do. The advantage of the transformations (9) and (10), compared with the transformations (6) and (7), is that the first ones produce orthonormal systems, as it will be seen, from a wider variety of functions.

It is possible to prove that if $f \in L^2[0,1)$, $\|f\|_2 = 1$ and $f_0 = f, f_1, f_2, \ldots$, where $f_m(x) = (\varphi_m f)(x)$; $m = 1, 2, 3 \ldots$, is an orthonormal system, then $f(x)^2 = 1$ for every point $x \in [0,1)$ where f is continuous, hence $f(x) = -1$ or 1 a.e.

2.1 Square integrable functions

Let, as usual, $L^2[0,1)$ and $L^2[0,\infty)$ be the set of square integrable functions defined on $[0,1)$ and on $[0,\infty)$ respectively, and

$$\langle f, g \rangle = \int_0^1 f(x)g(x) \, dx, \quad \|f\|^2 = \sqrt{\langle f, g \rangle}$$

be the scalar product and the L^2 norm.

N. Fine [6] proved that for every integrable f and every $h \in [0,1)$, the equality

$$\int_0^1 f(x \dotplus h) \, dx = \int_0^1 f(x) \, dx. \tag{11}$$

holds.

From (11) there follows that the scalar product is invariant under the transformations of Rademacher:

$$\langle r_s f, r_s g \rangle = \langle f, g \rangle, \quad \langle \varphi_m f, \varphi_m g \rangle = \langle f, g \rangle. \tag{12}$$

The orthonormality of the sequence $\{w_n\}_{n=0}^\infty$ of the Walsh functions follows easily from (12).

2.2 Module of continuity

C. Watary [20] introduced the L^p; $p \in [1, \infty]$ **dyadic module of continuity**. Use the **uniform dyadic module of continuity**

$$\hat{\omega}(f; \delta) = \sup_{h \leq \delta} \left\{ |f(x \dotplus h) - f(x)| \right\}, \tag{13}$$

for a bounded function, and the L^2 **dyadic module of continuity**

$$\hat\omega(f;\delta)_2 = \sup_{h\le\delta}\left\{\int_0^1 \left(f(x\dot+h) - f(x)\right)^2\right\}^{1/2}. \tag{14}$$

for a function $f \in L^2[0,1)$

Definition 2.2 *Call the function $f \in L^2[0,1)$ **dyadic continuous** if*

$$\lim_{\delta\to+0}\hat\omega(f;\delta) = 0.$$

It is easy to see that if the function f is dyadic continuous, then f is continuous in every non dyadic rational $x \in [0,1)$ and

$$\lim_{x\to x_0,\ x>x_0} f(x) = f(x_0)$$

for every $x_0 \in [0,1)$.

For the usual module of uniform continuity

$$\hat\omega(f;\delta) \le \omega(f;\delta) = \sup_{|h|\le\delta} \{|f(x+h) - f(x)| : x, x+h \in [0,1)\}$$

is valid.

Every Walsh function w_n is dyadic continuous as for $2^s > n$, we have

$$\hat\omega(w_n,\delta) = 0 \quad \text{for} \quad \delta \le 2^{-s}.$$

Define also the so called **relative** module of continuity.

Definition 2.3 *Let f, g be functions from $L^2[0,1)$. The L^2 **relative** dyadic module of continuity are defined as follows:*

$$\hat\omega_g(f;\delta)_2 = \hat\omega_f(g;\delta)_2 = \sup_{h\le\delta}\left\{\int_0^1 \left(g(x)f(x\dot+h) - g(x\dot+h)f(x)\right)^2\right\}^{1/2}. \tag{15}$$

Obviously, for $g(x) = 1$

$$\hat\omega_g(f;\delta)_2 = \hat\omega(f;\delta)_2 \quad \text{and} \quad \hat\omega_f(f;\delta)_2 = 0.$$

As it will be further shown, the relative module of continuity is useful for measuring the approximation when the approximation instrument is chosen for a particular function.

3   Walsh-similar functions

Starting with the function $w_0(x) = 1$ and using the Rademacher or Walsh transformations we produce the Walsh orthonormal system. A natural question is: Are there functions, which are not equal to 1 or -1 a. e., and such that the Rademacher and Walsh transformations produced from them orthonormal systems ? The answer is positive. In this section we define such **starting** functions. These functions play the role of the constant 1 in defining the so called Walsh-similar functions, which depend on a sequence of parameters and have properties similar to these of the Walsh functions.

3.1   Normal sequences

Let $\Lambda = \{\lambda_i\}_{i=1}^{\infty}$ be a sequence of real numbers, $\lambda_i^2 \leq 2$, and

$$\overline{\lambda}_i = \max\left\{\lambda_i, \sqrt{2 - \lambda_i^2}\right\}, \quad \underline{\lambda}_i = \min\left\{\lambda_i, \sqrt{2 - \lambda_i^2}\right\}, \tag{16}$$

$$\lambda_i(0) = \sqrt{2 - \lambda_i^2} = \sqrt{2}\sin\alpha_i, \quad \lambda_i(1) = \lambda_i = \sqrt{2}\cos\alpha_i. \tag{17}$$

Definition 3.1 *Call the sequence* $\Lambda = \{\lambda_k\}_{k=1}^{\infty}$ **normal** *if the products*

$$\prod_{i=1}^{\infty} \overline{\lambda}_i, \quad \text{and} \quad \prod_{i=1}^{\infty} \underline{\lambda}_i$$

are convergent.

Set the notations

$$\overline{\Lambda} = \prod_{i=1}^{\infty} \overline{\lambda}_i, \quad \overline{\Lambda}^{<s} = \prod_{i=1}^{s} \overline{\lambda}_i, \quad \overline{\Lambda}^{>s} = \prod_{i=s+1}^{\infty} \overline{\lambda}_i, \quad \underline{\Lambda}^{>s} = \prod_{i=s+1}^{\infty} \underline{\lambda}_i. \tag{18}$$

From Definition 3.1 there follows that

$$\lim_{s\to\infty} \overline{\Lambda}^{>s} = \lim_{s\to\infty} \underline{\Lambda}^{>s} = 1. \tag{19}$$

3.2   Starting functions

Definition 3.2 *Let Λ be a normal sequence, then, according to (3.1) the function*

$$\varphi(\Lambda; x) = \prod_{i=1}^{\infty} \lambda_i(x_i)$$

is defined for every $x \geq 0$ and it is periodic with period 1. Call the function $\varphi(\Lambda; x)$ **starting function** *or dyadic exponential function [15] with sequence Λ. The set of all starting functions is denoted by $\mathcal{S}$.*

From Definition 3.2, (19) and (13) there follows that every starting function φ is bounded, dyadic continuous as

$$\hat{\omega}(\varphi_m(\Lambda;\cdot);2^{-s}) \leq \overline{\Lambda}\left(\overline{\Lambda}^{>s} - \underline{\Lambda}^{>s}\right) \to 0, \quad \text{for} \quad s \to \infty,$$

and $\mathcal{S} \subset L^2[0,1)$.

The L^2 norm of every $\varphi \in \mathcal{S}$ is equal to 1 [15].

$$\|\varphi\|_2^2 = \langle \varphi, \varphi \rangle = 2 \int_0^{1/2} \prod_{i=2}^{\infty} \lambda_i(x_i)^2 \, dx = \cdots = 2^s \int_0^{2^{-s}} \prod_{i=s+1}^{\infty} \lambda_i(x_i)^2 \, dx = 1.$$

A starting function φ may have an arbitrary fractal dimension between 1 and 2. Let Λ be a normal sequence with $\lambda_i(1)/\lambda_i(0) = 2^{\alpha^i}$, where $1/2 \leq \alpha < 1$. The minimal number $N_s(\varphi(\Lambda;\cdot))$ of $2^{-s} \times 2^{-s}$ squares covering the graph of $\varphi(\Lambda;\cdot)$ is

$$N_s(\varphi) = 2^s \sum_{p=0}^{2^s-1} \sup_{x \in \Delta_p} \left\{ |\varphi(x\dot{+}h) - \varphi(x)| : h < 2^{-s} \right\} =$$

$$2^s \left(\prod_{i=s+1}^{\infty} \lambda_i(1) - \prod_{i=s+1}^{\infty} \lambda_i(0) \right) \sum_{p=0}^{2^s-1} \prod_{i=1}^{s} \lambda_i(x_i) \bigg|_{x \in \Delta_p}$$

where $\Delta_p = [p2^{-s}, (p+1)2^{-s})$, or

$$N_s(\varphi) = 2^{2s} \varphi(0) \prod_{i=1}^{s} \frac{1+2^{\alpha^i}}{2} \left(2^{\alpha^{s+1}/(\alpha-1)} - 1\right) \tag{20}$$

By definition, the fractal dimension of $\varphi(\Lambda;x)$ is

$$\kappa(\varphi) = \lim_{s \to \infty} \frac{\log_2 N_s(\varphi)}{s}. \tag{21}$$

From (20) there follows that $\kappa(\varphi) = 2 - \log_2 \alpha$, or $1 \leq \kappa(\varphi) < 2$.

3.3   Walsh-similar (WS) functions

The following theorem is proved in [15].

Theorem 3.1 *Let $\varphi \in \mathcal{S}$ and*

$$\varphi_m(x) = (\dot{w}_m\varphi)(x) = (-1)^{x \circ m} \varphi(x\dot{+}\check{m}). \tag{22}$$

Then, the sequence

$$\varphi_0(x) = \varphi(x), \varphi_1(x), \varphi_2(x), \varphi_3(x), \ldots \tag{23}$$

is an orthonormal basis in $L^2[0,1)$.

We call the functions φ_m **Walsh-similar (WS) functions**. The motivation for this name is the use of the Rademacher transformations for their definition and the representation

$$\varphi_m(x) = \varphi_m(\Lambda; x) = w_m(x) \prod_{i=1}^{\infty} \lambda_i(x_i \dot{+} m_{1-i}) = \prod_{i=1}^{\infty} (-1)^{x_i m_{1-i}} \lambda_i(x_i \dot{+} m_{1-i}),$$
(24)

where $w_m(x)$ is the classical Walsh function in Paley ordering [18], [13]. Other appropriate names as *Walsh-like functions* and *generalized Walsh functions* are already used [5], [7], [9], [10], [11] for other purposes.

From Definition 3.2, (19) and (13) it follows that every Walsh-similar function is dyadic continuous since

$$\hat{\omega}(\varphi_m(\Lambda; \cdot); 2^{-s}) \leq \overline{\Lambda} \left(\overline{\Lambda}^{>s} - \underline{\Lambda}^{>s} \right) \to 0, \quad \text{for} \quad s \to \infty.$$

Set the notations

$$\varphi_m^{<s}(\Lambda; x) = \prod_{i=1}^{s} (-1)^{x_i m_{1-i}} \lambda_i(x_i \dot{+} m_{1-i}),$$
(25)

$$\varphi_m^{>s}(\Lambda; x) = \prod_{i=s+1}^{\infty} (-1)^{x_i m_{1-i}} \lambda_i(x_i \dot{+} m_{1-i}).$$
(26)

It is easy to see that

$$\varphi_m(\Lambda; x) = \varphi_m^{<s}(\Lambda; x) \varphi_m^{>s}(\Lambda; x), \quad \|\varphi_m^{<s}(\Lambda)\|_2 = \|\varphi_m^{>s}(\Lambda)\|_2 = 1$$
(27)

and

$$\varphi_m^{>s}(\Lambda; x) = \varphi_0^{>s}(\Lambda; x) = \varphi^{>s}(x) \quad \text{for} \quad m < 2^s.$$
(28)

For the Kernel

$$K_{2^s}(\Lambda; x, t) = \sum_{m=0}^{2^s-1} \varphi_m(x) \varphi_m(t)$$

we have [15]

$$K_{2^s}(\Lambda; x, t) = \Omega_s(x, t) \varphi^{>s}(x) \varphi^{>s}(t),$$
(29)

where

$$\Omega_s(x, t) = \begin{cases} 2^s & \text{for} \quad x \dot{+} t \leq 2^{-s}, \\ 0 & \text{for} \quad x \dot{+} t > 2^{-s}. \end{cases}$$
(30)

From (30) we have

$$\int_0^1 \Omega_s(x, t) \, dt = 1.$$
(31)

Theorem 3.2 *Let*

$$B_{2^s}(f;x) = \sum_{m=0}^{2^s-1} f_m^\wedge \varphi_m(x), \quad f_m^\wedge = \int_0^1 f(t)\varphi_m(t) \, dt$$

be the 2^s-th partial Fourier sum of a function $f \in L^2[0,1)$ for the orthonormal system (23). Then

$$\|f - B_{2^s}(f)\|_2 \le \overline{\Lambda}^{>s} \hat\omega_{\varphi^{>s}}(f; 2^{-s})_2. \tag{32}$$

Proof. From (29) and (30) there follows that

$$B_{2^s}(f;x) = 2^s \varphi^{>s}(x) \int_{\Delta_s(x)} f(t)\varphi^{>s}(t) \, dt, \tag{33}$$

where $\Delta_s(x)$ is the dyadic interval of the form $[p2^s, (p+1)2^s)$ containing x. Then from (27) and (33) we have

$$\|f - B_{2^s}(f)\|_2^2 = \int_0^1 \left(f(x) - 2^s \varphi^{>s}(x) \int_{\Delta_s(x)} f(t)\varphi^{>s}(t) \, dt \right)^2 dx =$$

$$\int_0^1 2^s \left[\int_{\Delta_s(x)} \varphi^{>s}(t) \left(f(x)\varphi^{>s}(t) - f(t)\varphi^{>s}(x) \right) \, dt \right]^2 dx \le$$

$$\left(\overline{\Lambda}^{>s} \right)^2 \int_0^1 \sup_{t \in \Delta_s(x)} \left[f(x)\varphi^{>s}(t) - f(t)\varphi^{>s}(x) \right]^2 \, dx = \left(\overline{\Lambda}^{>s} \hat\omega_{\varphi^{>s}}(f; 2^{-s})_2 \right)^2,$$

which completes the proof. $\blacksquare$

Question 1. Let f be dyadic continuous and $\|f\|_2 = 1$. Let $f_0 = f, f_1, f_2, \ldots$ be an orthonormal system, where $f_m(x) = (w_m f)(x)$. Is it true that $f \in \mathcal{S}$?

4 Adaptation through best L^2 approximation

Let $\varphi(\Lambda; x) \in \mathcal{S}$ be a starting function. Consider φ as a function of $\alpha_1, \alpha_2, \alpha_3, \ldots$ defined by (17). Then, according to (17)

$$\frac{\partial}{\partial \alpha_s} \varphi(x) = (-1)^{x_s} \varphi(x \dot+ 2^{-s}) = \varphi_{2^s}(x). \tag{34}$$

From (34) we obtain that the m-th Walsh-similar function with starting function φ may be represented in the form

$$\varphi_m(x) = \frac{\partial}{\partial \alpha_{m(0)}} \frac{\partial}{\partial \alpha_{m(1)}} \cdots \frac{\partial}{\partial \alpha_{m(p)}} \varphi(x),$$

where $m = 2^{m(0)} + 2^{m(1)} + \cdots + 2^{m(p)}$, $m(0) < m(1) < \cdots < m(p)$.

Let $f \in L^2[0,1)$ and $\varphi \in \mathcal{S}$. Set

$$\Phi(\alpha_1, \alpha_2, \alpha_3, \ldots) = \|f - \varphi\|_2^2. \tag{35}$$

The necessary conditions for a minimum of Φ, according to (34), are

$$\frac{\partial \Phi}{\partial \alpha_k} = -2 \int_0^1 (f(x) - \varphi(x)) \, \varphi_{2^k} \, dx = -2 \int_0^1 f(x) \varphi_{2^k} \, dx = 0,$$

or

$$f_{2^{k-1}}^{(\wedge)}(\Lambda) = 0 \quad \text{for} \quad k = 1, 2, 3, \ldots. \tag{36}$$

If the parameters $\lambda_i = 1$ are fixed for $i > s$, one may calculate λ_i for $i = 1, 2, 3, \ldots, s$ from the conditions

$$f_{2^{k-1}}^{(\wedge)}(\Lambda) = 0 \quad \text{for} \quad k = 1, 2, 3, \ldots, s.$$

Call the starting function φ_f a **self function** of f if

$$\|f - \varphi_f\|_2 = \inf \left\{ \|f - \varphi\|_2 : \varphi \in \mathcal{S} \right\}.$$

If the sequence of the first 2^s Fourier coefficients of the function $f \in L^2[0,1)$, in respect to a self function is

$$c_0, c_1, c_2, \ldots c_{2^s-1},$$

then $c_{2^p-1} = 0$ for $p = 1, 2, 3, \ldots, s$. In the places of these coefficients we put the numbers $\epsilon_p = \lambda_p - 1$, where $\{\lambda_p\}_{p=1}^s$ are the parameters defining the self function ($\lambda_p = 1$ for $p > s$). In this way we preserve the number of coefficients needed for the presentation of f.

Question 2. What are the conditions for existence and for the uniqueness of the self function of $f \in L^2[0,1)$?

If f is continuous in $[0,1]$ and $f(x) = f(1 - x)$ for $x \in [0,1]$, then the self function of f is the constant 1.

Question 3. What are the necessary conditions for the function f to have self function equal to the constant 1 ?

Question 4. Let $\varphi \in \mathcal{S}$ be the self function of f. Is it true that

$$\hat{\omega}_\varphi(f; \delta)_2 \le \hat{\omega}_g(f; \delta)_2 \quad \text{for every} \quad g \in \mathcal{S} ?$$

5 Some properties of the Walsh-similar functions

The Walsh-similar functions have properties similar to these of the Walsh functions. It is interesting to study the limits of these similarities.

5.1 Relative dyadic derivative

There are various forms of dyadic differentiation, some of which interact with the Walsh functions similarly as the classical derivative does with the exponential functions, see B. L. Butzer and V. Engels [3] and the references there.

It is possible to define a relative dyadic derivative, which interacts with the Walsh-similar functions as the dyadic derivative does with the Walsh functions.

Definition 5.1 *Let* $\varphi(x) = \varphi_0(\Lambda; x) \in \mathcal{S}$ *and* $f \in L^2[0,1)$*. Define the pointwise* **relative dyadic derivative** *as follows*

$$f^{[1,\Lambda]}(x) = f^{[1]}(x) = \sum_{i=1}^{\infty} 2^{i-2}\lambda_i(\overline{x_i}) \left[\lambda_i(\overline{x_i})f(x) - \lambda_i(x_i)f(x\dot+2^{-i})\right]. \qquad (37)$$

Lemma 5.1 *Let* $\varphi(x) = \varphi(\Lambda; x) \in \mathcal{S}$ *and* $\{\varphi_m\}_{m=0}^{\infty}$ *are the WS functions produced from* φ*. Then*

$$\varphi_m^{[1,\Lambda]}(x) = \varphi_m^{[1]}(x) = m\varphi_m(x).$$

Proof.

$$\varphi_m^{[1]}(x) = \sum_{i=1}^{\infty} 2^{i-2}\lambda_i(\overline{x_i}) \left[\lambda_i(\overline{x_i})\varphi_m(x) - \lambda_i(x_i)\varphi_m(x\dot+2^{-i})\right] =$$

$$\sum_{i=1}^{\infty} 2^{i-2}\lambda_i(\overline{x_i})\varphi_m^{<i-1}(x)\varphi_m^{>i}(x)\theta_i(\Lambda; m, x),$$

where

$$\theta_i(\Lambda; m, x) = \left[(-1)^{x_i m_{1-i}}\lambda_i(\overline{x_i})\lambda_i(x_i\dot+m_{1-i}) - (-1)^{\overline{x_i}m_{1-i}}\lambda_i(x_i)\lambda_i(\overline{x_i}\dot+m_{1-i})\right].$$

It is directly verified that

$$\theta_i(\Lambda; m, x) = (-1)^{x_i m_{1-i}}2m_{1-i}.$$

Then

$$\varphi_m^{[1]}(x) = \sum_{i=1}^{\infty} 2^{i-2}(-1)^{x_i m_{1-i}}2m_{1-i}\lambda_i(x_i\dot+m_{1-i})\varphi_m^{<i-1}(x)\varphi_m^{>i}(x) =$$

$$\varphi_m(x) \sum_{i=1}^{\infty} 2^{i-1}m_{1-i} = m\varphi_m(x).$$

That completes the proof. ∎

5.2 Lebesgue constants

Now we estimate the Lebesgue's constants

$$L_n(\Lambda) = \sup_{x \in [0,1)} \int_0^1 |K_n(\Lambda; x, t)| \, dt, \qquad (38)$$

of the orthonormal system generated by the starting function $\varphi(x) = \varphi(\Lambda; x) \in \mathcal{S}$, generalizing the corresponding result in [6].

Lemma 5.2 *For every natural n, the inequality*

$$L_n(\Lambda) = \overline{\Lambda}^2 \log_2 n$$

holds.

Proof. Using (30), (1) and the relation

$$\lambda_i(x_i)\lambda_i(t_i) - (-1)^{x_i t_i} \lambda_i(\overline{x_i})\lambda_i(\overline{t_i}) = \begin{cases} 2 & \text{for} \quad x_i = t_i, \\ 0 & \text{for} \quad x_i \neq t_i, \end{cases}$$

we have

$$K_n(\Lambda; x, t) = \sum_{m=0}^{n-1} \varphi_m(x)\varphi_m(t) =$$

$$\varphi_m^{>n(p)}(x)\varphi_m^{>n(p)}(t) \sum_{m=0}^{n-1} \varphi_m^{<n(p)}(x)\varphi_m^{<n(p)}(t)$$

and

$$\left| \sum_{m=0}^{n-1} \varphi_m^{<n(p)}(x)\varphi_m^{<n(p)}(t) \right| \leq$$

$$\left| \sum_{m=0}^{2^{n(p)}-1} \varphi_m^{<n(p)}(x)\varphi_m^{<n(p)}(t) \right| + \left| \sum_{m=2^{n(p)}}^{n-1} \varphi_m^{<n(p)}(x)\varphi_m^{<n(p)}(t) \right| \leq$$

$$\Omega_{n(p)}(x, t) + \prod_{i=n(p-1)}^{n(p)-1} \overline{\lambda}_i^2 \left| \sum_{m=0}^{n-2^{n(p)}-1} \varphi_m^{<n(p-1)}(x)\varphi_m^{<n(p-1)}(t) \right|,$$

or

$$|K_n(\Lambda; x, t)| \leq \overline{\Lambda}^2 \left(\Omega_{n(p)}(x, t) + \Omega_{n(p-1)}(x, t) + \cdots + \Omega_{n(0)}(x, t) \right). \qquad (39)$$

The Lemma follows from (39) and (31). ∎

6   Generalized Walsh-similar functions

Set by definition

$$\lambda_{1-i} = \lambda_i; \quad i = 1, 2, 3, \dots \, . \tag{40}$$

Definition 6.1 *Let Λ be a normal sequence, then, according to (3.1), the function*

$$\phi_y(\Lambda; x) = (-1)^{x \circ y} \prod_{i=-\infty}^{\infty} \lambda_i(x_i \dot{+} y_{1-i})$$

*is defined for every $x, y \geq 0$. Call the function $\phi_y(\Lambda; x)$ **Generalized Walsh-similar** function, or **GWS** function with sequence Λ.*

The motivation for this name is the representation

$$\phi_y(x) = w_y(x) \prod_{i=-\infty}^{\infty} \lambda_i(x_i \dot{+} y_{1-i}),$$

where $w_y(x)$ is the Generalized Walsh function defined by N. J. Fine [7] and represented in the form (8) by G. R. Redinbo [14], and that $\phi_y(x) = w_y(x)$ if $\lambda_i = 1$ for $i = 1, 2, 3, \dots$.
Every GWS function is represented in the form

$$\phi_y(\Lambda; x) = \varphi_{[y]}(\Lambda; x)\varphi_{[x]}(\Lambda; y), \tag{41}$$

where

$$\varphi_{[y]}(\Lambda; x) = (-1)^{x \circ [y]} \prod_{i=1}^{\infty} \lambda_i(x_i \dot{+} y_{1-i}) = \prod_{i=1}^{\infty} (-1)^{x_i y_{1-i}} \lambda_i(x_i \dot{+} y_{1-i}). \tag{42}$$

The set $\{\varphi_k(x)\}_{k=0}^{\infty}$ is a closed orthonormal system in $L^2[01]$, see [15] . Every function $\varphi_k(x)$ is periodic with period 1.

Lemma 6.1 *Let $\phi_y(\Lambda; x)$; $x, y \geq 0$ be a GWF function. Then*

$$\sup_{x \in [0,\infty)} |\phi_y(\Lambda; x)| = \sup_{y \in [0,\infty)} |\phi_y(\Lambda; x)| = \sup_{x, y \in [0,\infty)} |\phi_y(\Lambda; x)| = \|\phi(\Lambda)\|.$$

The function $\phi_y(\Lambda; x)$, for a fixed y, as a function of x, is continuous in every non dyadic rational $x \in (0, \infty)$, and for a fixed x, as a function of y, is continuous in every non dyadic rational $y \in (0, \infty)$.

Proof. From Definition 6.1, according to (19), it follows that

$$\sup_{x \in [0,\infty)} |\phi_y(\Lambda; x)| = \sup_{y \in [0,\infty)} |\phi_y(\Lambda; x)| = \sup_{x, y \in [0,\infty)} |\phi_y(\Lambda; x)| = \prod_{i=1}^{\infty} \overline{\lambda_i}^2 = \|\phi(\Lambda)\| < \infty.$$

Let $x^* \in (0, \infty)$ be dyadic irrational and $\Delta_k(x^*, x) = [p2^{-k}, (p+1)2^{-k})$ be the smallest dyadic interval containing x^* and x. Obviously $|x^* - x| < 2^{-k}$. Let $\delta = 2^{-k} < 1$, then for every $|x - x^*| < \delta$

$$\phi_y(\Lambda; x) - \phi_y(\Lambda; x^*) =$$

$$(-1)^{xoy} \prod_{i=-\infty}^{k} \lambda_i(x_i \dot{+} y_{1-i}) \left(\prod_{i=k+1}^{\infty} \lambda_i(x_i \dot{+} y_{1-i}) - \prod_{i=k+1}^{\infty} \lambda_i(x_i^* \dot{+} y_{1-i}) \right),$$

or, according to (19),

$$|\phi_y(\Lambda; x) - \phi_y(\Lambda; x^*)| \le$$

$$\prod_{i=1}^{\infty} \overline{\lambda_i} \prod_{i=1}^{k} \overline{\lambda_i} \left(\prod_{i=k+1}^{\infty} \overline{\lambda_i} - \prod_{i=k+1}^{\infty} \underline{\lambda_i} \right) = \epsilon(\delta) \to 0 \quad \text{for} \quad k \to \infty.$$

The second part is proved in the same way. ∎

Corollary 6.1 *Every GWS function is locally Lebesgue integrable.*

From Definition 6.1 it follows that

$$\phi_y(\Lambda; x \dot{+} 2^{-k}) = \left(\frac{\lambda_k}{(2 - \lambda_k^2)^{1/2}} \right)^{(-1)^{x_k}} \phi_y(\Lambda; x). \tag{43}$$

The relation (43) shows a type of self-similarity of the GWS functions.

6.1 Riemann-Lebesgue Theorem

The corresponding Riemann-Lebesgue Theorem is proved following [7] and using Lemma 6.1.

Theorem 6.1 *If f is absolutely integrable on $[0, \infty)$, then for a fixed normal sequence Λ, $\phi_y(\Lambda; x) = \phi_y(x)$,*

$$\lim_{y \to \infty} \int_0^\infty f(x)\phi_y(x) \, dx = 0.$$

Proof. Write

$$\int_0^\infty f(x)\phi_y(x) \, dx = \int_0^n f(x)\phi_y(x) \, dx + \int_n^\infty f(x)\phi_y(x) \, dx = I_n + J_n. \tag{44}$$

We may choose n so that for all $y \ge 0$,

$$|J_n| \le \|\phi(\Lambda)\| \int_n^\infty |f(x)| \, dx < \epsilon/2.$$

Now, from (41),

$$I_n = \sum_{k=1}^{n-1} \int_k^{k+1} f(x)\varphi_{[y]}(x)\varphi_{[x]}(y)\ dx = \sum_{k=1}^{n-1} \varphi_k(y) \int_k^{k+1} f(x)\varphi_{[y]}(x)\ dx,$$

$$|I_n| \le \|\phi(\Lambda)\| \sum_{k=1}^{n-1} \left| \int_k^{k+1} f(x)\varphi_{[y]}(x)\ dx \right|. \tag{45}$$

On the right side of (45) we have a sum of a fixed number of Fourier coefficients of order $[y]$. We may choose y so large that this sum is less than $\epsilon/2$, so that $|I_n| + |J_n| < \epsilon$, which completes the proof. ∎

6.2 Fourier Integral Theorem

Theorem 6.2 *Let f be absolutely integrable on $[0,\infty)$ and continuous in every dyadic irrational $x \in (0,\infty)$. Then, for a fixed normal sequence Λ, $\phi_y(\Lambda; x) = \phi_y(x)$, the spectrum function*

$$F(y) = \int_0^\infty f(x)\phi_y(x)\ dx \tag{46}$$

exists and for every dyadic irrational $x \in (0,\infty)$, we have the inversion formula

$$f(x) = \int_0^\infty F(y)\phi_y(x)\ dy. \tag{47}$$

Proof. Let

$$c_{k,l} = \int_k^{k+1} f(x)\varphi_l(x)\ dx,$$

then

$$f(x) = \sum_{l=0}^\infty c_{k,l}\varphi_l(x) \tag{48}$$

for every dyadic irrational $x \in (k, k+1)$.

As f is absolutely integrable on $[0,\infty)$, $\phi_y(x)$ is uniformly, in respect to y, bounded as function of x on $[0,\infty)$ and $\phi_y(x)$, for every fixed $x \in [0,\infty)$, as a function of y, is continuous in every dyadic irrational $y \in (0,\infty)$, the spectrum function (46) exists and $F(y)$ is continuous in every dyadic irrational $y \in (0,\infty)$.

From (46) and (48) it follows

$$F(y) = \int_0^\infty f(x)\phi_y(x)\ dx = \sum_{k=0}^\infty \int_k^{k+1} f(x)\varphi_{[y]}(x)\varphi_{[x]}(y)\ dx$$

$$= \sum_{k=0}^\infty \varphi_k(y) \int_k^{k+1} f(x)\varphi_{[y]}(x)\ dx,$$

or

$$F(y) = \sum_{k=0}^{\infty} c_{k,l}\varphi_k(y) \tag{49}$$

for every dyadic irrational $y \in (l, l+1)$.

On the other hand

$$\int_0^{\infty} F(y)\phi_x(y) \, dy = \sum_{l=0}^{\infty} \int_l^{l+1} F(y)\varphi_{[x]}(y)\varphi_{[y]}(x) \, dy =$$

$$\sum_{l=0}^{\infty} \varphi_l(x) \int_l^{l+1} \left(\sum_{k=0}^{\infty} c_{k,l}\varphi_k(y) \right) \varphi_{[x]}(y) \, dy =$$

$$\sum_{l=0}^{\infty} \varphi_l(x) \left(\sum_{k=0}^{\infty} c_{k,l} \int_l^{l+1} \varphi_k(y)\varphi_{[x]}(y) \, dy \right) = \sum_{l=0}^{\infty} c_{m,l}\varphi_l(x)$$

for $x \in [m, m+1)$, or

$$\int_0^{\infty} F(y)\phi_y(x) \, dy = f(x)$$

for every dyadic irrational $x \in (0, \infty)$. ∎

7 Walsh-similar functions of higher rank

A Rademacher transformation (9) for $s > 0$ transforms every function $f \in L^2[0,1)$ in a function orthogonal to f,

$$\langle f, r_s f \rangle = 0 \quad \text{and} \quad \|r_s f\| = \|f\|.$$

There exist two more sets of transformations with this property.

7.1 Rademacher transformations of rank 2 and 3

Let ρ_j; $j = 0, 1, 2, \ldots, k-1$ be transformations in the set of k dimensional vectors $a = (a_0, a_1, a_2, \ldots, a_{k-1})$ of the form

$$\rho_j(a) = (\varepsilon_{j,0}a_j, \varepsilon_{j,1}a_{j+1}, \varepsilon_{j,2}a_{j+2}, \ldots, \varepsilon_{j,k-1}a_{j+(k-1)}), \tag{50}$$

where $\varepsilon_{j,l} = 1, -1$; $j, l = 0, 1, 2, \ldots, k-1$ and $\varepsilon_{0,l} = \varepsilon_{l,0} = 1$; $l = 0, 1, 2, \ldots, k-1$.

Call the transformations (50) Rademacher vector transformations and the matrix $\{\varepsilon_{j,l}\}$ - Rademacher ε-matrix if

$$\langle \rho_j(a), \rho_l(a) \rangle = \sum_{i=0}^{k-1} \varepsilon_{j,i}a_{j+i}\varepsilon_{l,i}a_{l+i} = 0 \quad \text{for} \quad j \neq l, \tag{51}$$

$j, l = 0, 1, 2, \ldots, k-1$, for an arbitrary vector a.

There are Rademacher vector transformations only for $k = 2, 2^2, 2^3$, corresponding to the complex numbers, the quaternions and the octets of Kelly respectively . Three Rademacher ε-matrices are the following

$$
\begin{vmatrix} 1 & 1 \\ 1 & -1 \end{vmatrix}, \quad
\begin{vmatrix}
1 & 1 & 1 & 1 \\
1 & -1 & 1 & -1 \\
1 & -1 & -1 & 1 \\
1 & 1 & -1 & -1
\end{vmatrix}, \tag{52}
$$

$$
\begin{vmatrix}
1 & 1 & 1 & 1 & 1 & 1 & 1 & 1 \\
1 & -1 & 1 & -1 & -1 & 1 & -1 & 1 \\
1 & -1 & -1 & 1 & -1 & 1 & 1 & -1 \\
1 & 1 & -1 & -1 & -1 & -1 & 1 & 1 \\
1 & 1 & 1 & 1 & -1 & -1 & -1 & -1 \\
1 & -1 & -1 & 1 & 1 & -1 & -1 & 1 \\
1 & 1 & -1 & -1 & 1 & 1 & -1 & -1 \\
1 & -1 & 1 & -1 & 1 & -1 & 1 & -1
\end{vmatrix}. \tag{53}
$$

Definition 7.1 *For a fixed natural number s, a set of 2^n; $n = 1, 2, 3$ transformations $r_{s,j} : L^2[0,1) \to L^2[0,1)$, $j = 0, 1, 2, \ldots, 2^n - 1$ is called:*
1) Rademacher set of transformations of rank 1 if

$$
r_{s,j}(f; x) = \varepsilon_{j,l} f(x \dotplus j 2^{-s}), \quad \text{for} \quad x_s = j; \quad j, l = 0, 1,
$$

where $\{\varepsilon_{j,l}\}$ is a $2^1 \times 2^1$ Rademacher ε-matrix.
2) Rademacher set of transformations of rank 2 if

$$
r_{s,j}(f; x) = \varepsilon_{j,l} f(x \dotplus j 2^{-s}), \quad \text{for} \quad 2x_{2s-1} + x_{2s} = j; \quad j, l = 0, 1, 2, 3,
$$

where $\{\varepsilon_{j,l}\}$ is a $2^2 \times 2^2$ Rademacher ε-matrix.
3) Rademacher set of transformations of rank 3 if

$$
r_{s,j}(f; x) = \varepsilon_{j,l} f(x \dotplus j 2^{-s}), \quad \text{for} \quad 2^2 x_{3s-2} + 2x_{3s-1} + x_{3s} = j; \quad j, l = 0, 1, 2, \ldots, 7,
$$

where $\{\varepsilon_{j,l}\}$ is a $2^2 \times 2^2$ Rademacher ε-matrix.

The transformation $r_{s,0}$ is the identity for $s = 1, 2, 3, \ldots$.
It is easy to see that every Rademacher transformation preserves the scalar product, or for every two functions $f, g \in L^2[0,1)$

$$
\langle r_{s,j}(f; .), r_{s,j}(g; .) \rangle = \langle f, g \rangle. \tag{54}
$$

Lemma 7.1 *Let $f \in L^2[0,1)$, s be a natural number and $r_{s,j}$ be a Rademacher transformation of rank $n = 1, 2, 3$, $j \neq 0$. Then*

$$
\langle r_{s,j}(f; .), f \rangle = 0 \quad \text{for} \quad j = 1, 2, \ldots, 2^n - 1.
$$

The proof follows directly from the definition of a Rademacher transformation.

We considered in this paper the Walsh-similar functions of rank 1, produced by the Rademacher transformations of rank 1. It is possible to define Walsh-similar functions of rank 2 and rank 3 using the Rademacher transformations of rank 2 and rank 3 respectively. This is done in [16].

References

[1] BARNSLEY, M. F. (1988), *Fractals Everywhere*, Boston.

[2] BARNSLEY, M. F. AND L. P. HURD (1993), *Fractal Image Compression*, AK Peters, Ltd. Wellesley, Massachusetts.

[3] BUTZER, P. L., AND V. ENGELS (1989), *Theory and Applications of Gibbs Derivatives*, Matematički Institut, Beograd.

[4] BRUDNYĬ, YU. A. (1994): *Adaptive approximation of functions with singularities*. Trudy Moskov. Mat. Obshch., **55**, 149 - 242.

[5] CHRESTENSON, H. E. (1955): *A Class of Generalized Walsh Functions*. Pacific J. Math., **5**, 17 - 31.

[6] FINE, N. J. (1949): *On the Walsh Functions*. Trans. Am. Math. Soc., **65**, 373 - 414.

[7] FINE, N. J. (1950): *The Generalized Walsh Functions*. Trans. Am. Math. Soc., **69**, 66 - 77.

[8] A. N. KOLMOGOROV (1936), *Über die beste annaherung von functionen einer functionsklasse*, Math. Ann., **37**, 107 - 111.

[9] LÉVY, P. (1944): *Sur une généralisation des fonctions orthogonales de M. Rademacher*. Comm. Math. Helv., **16**, 146 - 152.

[10] LARSEN, R. D. AND W. R. MADYCH (1976): *Walsh-like Expansions and Hadamard Matrices*. IEEE Trans. Acoust. Speech Signal Processing, **ASSP-24**(1), 71 - 75.

[11] MADYCH, W. R. (1978): *Generalized Walsh-like Expansions*. IEEE Midwest Symp. Circ. Syst., 21st, 378 - 382.

[12] MEYER, Y. (1993), *Wavelets*, SIAM, Philadelphia.

[13] PALEY, R. E. (1932): *A Remarkable Series of Orthogonal Functions*. Proc. London Math. Soc., **34**, 241 - 279.

[14] REDINBO, G. R. (1971): *A Note on the Construction of Generalized Walsh Functions.* SIAM J. Math. Anal., **2**(3), 166 - 167.

[15] BL. SENDOV (1997), *Multiresolution analysis of functions defined on the dyadic topological group.* East J. on Approx. **3**, n. 2, 225 - 239.

[16] BL. SENDOV (1998), *Adaptive Multiresolution Analysis on the Dyadic Topological Group.* J. of Approx. Theory (to appear).

[17] SUNOUCHI, G. (1964): *Strong Summability of Walsh Fourier Series.* Tôhoku Math. J., **16** , 228 - 237.

[18] WALSH, J. L. (1923): *A Closed Set of Normal Orthogonal Functions.* Amer. J. Math., **55**, 5 - 24.

[19] WATARI, C. (1956): *A Generalization of Haar Functions.* Tôhoku Math. J., **8**, 286 - 290.

[20] WATARI, C. (1957): *On Generalized Walsh Fourier Series, I.* Proc. Japan Acad., **33**, 435 - 438.

[21] WATARI, C. (1958): *On Generalized Walsh Fourier Series.* Tôhoku Math. J., **10** (2), 211 - 241.

[22] YANO, S. (1951): *On Walsh-Fourier Series.* Tôhoku Math. J., **3**, 223 - 242.

Center of Informatics and Computer Technology
Bulgarian Academy of Sciences
"Acad. G. Bonchev" street, Block 25A,
1113 Sofia, Bulgaria.
Email address: bsendov@argo.bas.bg

International Series of Numerical Mathematics
Vol. 132, © 1999 Birkhäuser Verlag Basel/Switzerland

Dual Recurrence and Christoffel-Darboux-Type Formulas for Orthogonal Polynomials

Michael - Ralf Skrzipek

Abstract

Let a sequence $\{\Phi_n\}_{n\geq 0}$ of polynomials be given which are orthogonal on the unit circle with respect to an inner product. By a k-shift of their reflection coefficients we obtain the associated polynomials $\{\Phi_n^{(k)}\}_{n\geq 0}$ of order $k \geq 0$, analogously to the associated polynomials on the real line. Using these polynomials we derive a dual recurrence formula for polynomials orthogonal ont the unit circle.

Modifying the associated polynomials by starting the Geronimus recurrence earlier we derive a Christoffel-Darboux-type formula for some classes of orthogonal polynomials. This formula expresses derivatives of orthogonal polynomials in terms of orthogonal polynomials and their (modified) associated polynomials.

1 Introduction

In this article we transfer some properties and techniques for orthogonal polynomials on the real axis to polynomials orthogonal on the unit circle. By using their defining recurrence relations we give in Sect. 3 a dual recurrence relation by the way of an example how techniques which are known for real orthogonal polynomials can be modified and used for the complex case. But for deriving an analogue Christoffel-Darboux-type formula which based on the dual recurrence relation we see that concluding for the complex case in analogy to the real case is not always suitable. Thus we modify in Sec. 4 the associated polynomials. Using these modified polynomials we derive a corresponding dual recurrence relation for them and obtain in Sect. 5 a Christoffel-Darboux-type formula. This formula is valid for orthogonal polynomials for which at least the first reflection coefficients are non zero (the rest of them may be zero). It expresses derivatives of orthogonal polynomials in terms of orthogonal polynomials and their (modified) associated polynomials. It corresponds to a mixed Christoffel-Darboux-type formula on the real line. But at first we start in Sect. 2 with some (known) results which describe the situation for orthogonal polynomials on the real line.

2   The Real Case

Let $\{P_n\}_{n\in\mathbf{N}_0}$ be a sequence of monic polynomials, orthogonal on the real line with respect to a measure ω i.e.

$$\int_{\mathbf{R}} P_n(x)P_m(x)\,d\omega(x) = d_n\delta_{m,n}, \quad d_n \neq 0.$$

where we assume that all moments $\int_{\mathbf{R}} x^n\,d\omega(x)$, $n \in \mathbf{N}_0$, are finite and (for simplicity) that ω has an infinite number of points of increase. These polynomials form a polynomial solution of the three term recurrence

$$u_{n+2}(x) = (x-\beta_{n+1})u_{n+1}(x)-\gamma_{n+1}u_n(x), \quad 0 \neq \gamma_{n+1}, \quad \beta_{n+1} \in \mathbf{R}, \quad n \geq 0, \quad (1)$$

with initial values $P_0(x) = 1$, $P_1(x) = x - \beta_0$ [3, 6]. Defining $P_{-1}(x) \equiv 0$ then (1) remains valid for $n \geq -1$ where γ_0 can be chosen arbitrarily. We choose $\gamma_0 := \int_{\mathbf{R}} d\omega(x)$. By an r-shift of the recurrence coefficients we obtain the r-associated polynomials $P_n^{(r)}$ as a polynomial solution of

$$u_{n+2}(x) \;=\; (x - \beta_{n+r+1})u_{n+1}(x) - \gamma_{n+r+1}u_n(x)\,, \quad n \geq 0, \quad (2)$$

with initial values

$$P_{-1}^{(r)}(x) = 0, \qquad P_0^{(r)} = 1\,. \quad (3)$$

Obviously we have $P_n^{(0)} = P_n$. By the theorem of Farvard [3, pp. 21] we see that associated polynomials are orthogonal polynomials, too. Since the elements of $\{P_n\}_{n\geq 0}$ and $\{P_{n-r}^{(r)}\}_{n\geq r}$ form (for $r \neq 0$) a pair of linearly independent solutions of (1) every solution of (1) can be written as a linear combination of them. This idea was used by W. Van Assche [7] (c. f. [1]) to prove the dual recurrence formula

$$P_{n+1}^{(r)}(x) = (x - \beta_r)P_n^{(r+1)}(x) - \gamma_{r+1}P_{n-1}^{(r+2)}(x)\,, \quad n \geq 0\,. \quad (4)$$

We see that in this recurrence relation the recurrence coefficients are independent of n. Using (4), a mixed type Christoffel-Darboux identity can be shown [7, Theorem 1]:

$$\sum_{j=0}^{n} P_{n-j}^{(j+1)}(x)P_j(y) \;=\; \left\{ \begin{array}{ll} \dfrac{P_{n+1}(x) - P_{n+1}(y)}{x - y}, & x \neq y\,, \\[2mm] P_{n+1}'(x), & x = y\,, \end{array} \right\}$$

$$=\; \sum_{j=0}^{n} P_{n-j}^{(j+1)}(y)P_j(x). \quad (5)$$

Now we want to derive the corresponding result for polynomials, orthogonal on the unit circle (i. e. Szegő polynomials).

3 Recurrence Relations for Szegő Polynomials

Monic polynomials Φ_n, $\deg \Phi_n = n$, orthogonal on the unit circle $\mathbf{T} := \{z \in \mathbf{C} : |z| = 1\}$ with respect to a measure μ are defined by

$$\int_{\mathbf{T}} \Phi_n(z) \overline{\Phi_m(z)} \, d\mu(\theta) = g_m \delta_{m,n} \quad z = \exp(i\theta) \,, \quad g_m \neq 0 \,, \quad m, n \in \mathbf{N}_0 \,,$$

where μ is a positive measure on $[-\pi, \pi)$ with infinite support. They satisfy a recurrence relation

$$u_{n+1}(z) = z u_n(z) - a_n u_n^*(z) \,, \quad n \geq 0 \,, \tag{6}$$

where $u_n^*(z) := z^n \bar{u}_n(1/z)$. Starting with $\Phi_0(z) := 1$, the sequence $\{\Phi_n\}_{n \geq 0}$ can be generated by using this recurrence. Obviously the reflection coefficients and the orthogonal polynomials are connected by $a_n = -\Phi_{n+1}(0)$. And it can be shown that $|a_n| < 1$ for all $n \in \mathbf{N}_0$ (see e. g. [4, p. 5 in connection with the remark on p. 3]. From (6) we see that the reciprocal polynomials Φ_n^* satisfy the recurrence

$$u_{n+1}^*(z) = u_n^*(z) - \bar{a}_n z u_n(z) \,, \quad n \geq 0, \tag{7}$$

with initial value $\Phi_0^*(z) := 1$. Using this recurrence relation, the reciprocal polynomials in (6) can be eliminated and we obtain [4, p. 4]

$$a_{n-1} u_{n+1}(z) = [z a_{n-1} + a_n] u_n(z) - a_n(1 - |a_{n-1}|^2) z u_{n-1}(z) \,, \quad n = 1, 2, \ldots \tag{8}$$

The Φ_n satisfy this recurrence with initial values $\Phi_0(z) := 1$, $\Phi_1(z) = z - a_0$. Since $a_{-1} = -1$, (8) remains formally true for $n = 0$, although u_{-1} is not defined. We use this sometimes for the purpose of simplification of some formulas.

Writing both mixed type recurrences (6), (7) in one vector valued recurrence we obtain

$$\begin{pmatrix} u_{n+1} \\ v_{n+1} \end{pmatrix} = A_n(z) \begin{pmatrix} u_n \\ v_n \end{pmatrix} \,, \quad A_n(z) = \begin{pmatrix} z & -a_n \\ -\bar{a}_n z & 1 \end{pmatrix} \,. \tag{9}$$

As in the real case the k-associated polynomials $\Phi_n^{(k)}$ are defined by a k-shift of the reflection coefficients in (6), i. e. they satisfy

$$\begin{pmatrix} u_{n+1} \\ v_{n+1} \end{pmatrix} = A_{n+k}(z) \begin{pmatrix} u_n \\ v_n \end{pmatrix} \,. \tag{10}$$

with initial value $(\Phi_0^{(k)}(z), \Phi_0^{(k)^*}(z))^T = (1, 1)^T$. Obviously, Φ_n and $\Phi_{n-k}^{(k)}$ satisfy for $n \geq k$ the same recurrence (6). The Wronskian $W(u_n, w_n)$ of two solutions of (9) is given as

$$W(u_n, w_n) = \begin{vmatrix} u_n & w_n \\ u_n^* & w_n^* \end{vmatrix} = u_n w_n^* - w_n u_n^* \,.$$

If this determinant is different from zero, u_n and v_n are linearly independent, and every solution of (6) can be written as a linear combination of them. Using a formula from [5, p. 178] we get

$$
\begin{aligned}
W(\Phi_n, \Phi_{n-k}^{(k)}) &= \Phi_n(\Phi_{n-k}^{(k)})^\star - \Phi_{n-k}^{(k)}\Phi_n^\star \\
&= \prod_{\nu=k}^{n}(1-|a_\nu|^2)z^{n-k}(\Phi_k^\star(z) - \Phi_k(z)) .
\end{aligned}
$$

Especially for $k=1$ we have $\Phi_1^\star(z) - \Phi_1(z) = -z(1+\bar{a}_0) + a_0 + 1$, which means that for $z \notin \{0, (a_0 + 1)/(\bar{a}_0 + 1)\}$ the polynomials Φ_n and $\Phi_{n-1}^{(1)}$ are linearly independent and every solution of (6) can be written as a linear combination of them.

Writing

$$
\begin{pmatrix} \Phi_{n-k}^{(k)} \\ (\Phi_{n-k}^{(k)})^\star \end{pmatrix} = \gamma \begin{pmatrix} \Phi_{n-1}^{(1)} \\ (\Phi_{n-1}^{(1)})^\star \end{pmatrix} + \delta \begin{pmatrix} \Phi_n \\ \Phi_n^\star \end{pmatrix} ,
$$

choosing $n = k$ and $n = k+1$, we obtain a system of linear equations for γ, δ and we get

$$
\gamma = \frac{\Phi_k^\star - \Phi_k}{\Phi_{k-1}^{(1)}\Phi_k^\star - \Phi_k(\Phi_{k-1}^{(1)})^\star} , \qquad \delta = \frac{(\Phi_{k-1}^{(1)})^\star - \Phi_{k-1}^{(1)}}{\Phi_{k-1}^{(1)}\Phi_k^\star - \Phi_k(\Phi_{k-1}^{(1)})^\star}
$$

and thus

$$
\begin{aligned}
z^{k-1}\prod_{\nu=1}^{k-1}(1-|a_\nu|^2)\ (\Phi_1^\star - \Phi_1)&\begin{pmatrix} \Phi_{n-k}^{(k)} \\ (\Phi_{n-k}^{(k)})^\star \end{pmatrix} \\
=(\Phi_k^\star - \Phi_k)&\begin{pmatrix} \Phi_{n-1}^{(1)} \\ (\Phi_{n-1}^{(1)})^\star \end{pmatrix} + ((\Phi_{k-1}^{(1)})^\star - \Phi_{k-1}^{(1)})\begin{pmatrix} \Phi_n \\ \Phi_n^\star \end{pmatrix}
\end{aligned}
\tag{11}
$$

which is valid for all $z \in \mathbf{C}$. Especially for $k = 2$ we get

$$
\begin{aligned}
z(1-|a_1|^2)\ (\Phi_1^\star - \Phi_1)&\begin{pmatrix} \Phi_{n-2}^{(2)} \\ (\Phi_{n-2}^{(2)})^\star \end{pmatrix} \\
=(\Phi_2^\star - \Phi_2)&\begin{pmatrix} \Phi_{n-1}^{(1)} \\ (\Phi_{n-1}^{(1)})^\star \end{pmatrix} + (\Phi_1^{(1)^\star} - \Phi_1^{(1)})\begin{pmatrix} \Phi_n \\ \Phi_n^\star \end{pmatrix} .
\end{aligned}
$$

If we consider the k-associated version of this formula we obtain

Theorem 3.1 (Dual recurrence formula) *The polynomials* Φ_n *and their associated polynomials satisfy for* $n \geq 2$

$$z(1 - |a_{k+1}|^2) \; ((\Phi_1^{(k)})^\star - \Phi_1^{(k)}) \begin{pmatrix} \Phi_{n-2}^{(k+2)} \\ (\Phi_{n-2}^{(k+2)})^\star \end{pmatrix}$$

$$= (\Phi_2^{(k)\star} - \Phi_2^{(k)}) \begin{pmatrix} \Phi_{n-1}^{(k+1)} \\ (\Phi_{n-1}^{(k+1)})^\star \end{pmatrix} + ((\Phi_1^{(k+1)})^\star - \Phi_1^{(k+1)}) \begin{pmatrix} \Phi_n^{(k)} \\ (\Phi_n^{(k)})^\star \end{pmatrix} . \tag{12}$$

The first component can be written as

$$(z(1 + \bar{a}_{k+1}) - (1 + a_{k+1})) \, \Phi_n^{(k)}(z) = (z^2(1 + \bar{a}_{k+1}) + z((1 + a_{k+1})\bar{a}_k$$
$$- (1 + \bar{a}_{k+1})a_k) - (1 + a_{k+1})) \, \Phi_{n-1}^{(k+1)}(z) \tag{13}$$
$$+ (1 - |a_{k+1}|^2)z((1 + a_k) - z(1 + \bar{a}_k))\Phi_{n-2}^{(k+2)}(z) , \quad n \geq 2 ,$$

with initial values $\Phi_0^{(k+2)}(z) = 1$, $\Phi_1^{(k+1)}(z) = z - a_{k+1}$. The reciprocal polynomials satisfy the same recurrence relation but with initial values $(\Phi_0^{(k+2)})^\star(z) = 1$, $(\Phi_1^{(k+1)})^\star(z) = 1 - \bar{a}_{k+1}z$. Furthermore the 'recurrence coefficient polynomials' in (13) are independent of n. This is analogous to the real case.

4   A Modification of Associated Polynomials

If we compare (4) with (13) we see that an advantage for the real case is that the appearance of the recurrence coefficients is simpler. To get a corresponding formula for Szegő polynomials it is necessary to start the recurrence earlier. This is not possible if we use (6) for $n = -1$ because of the presence of the reciprocal polynomial. But we can take (8) or its k-associated version

$$a_{n+k-1}u_{n+1}(z) = [za_{n+k-1} + a_{n+k}] u_n(z) - a_{n+k}(1 - |a_{n+k-1}|^2)zu_{n-1}(z) , \quad (14)$$

for $n \geq 1$ which is satisfied for $\{\Phi_n^{(k)}\}_{n \geq 0}$ with initial values $\Phi_0^{(k)}(z) := 1$, $\Phi_1^{(k)}(z) = z - a_k$, $k \geq 0$.

Obviously the recurrence (14) is only useful for determining u_{n+1} if $a_{n+k-1} \neq 0$. If $a_{n+k-1} = \ldots = a_{n+k-m} = 0 \neq a_{n+k-m-1}$, $n - m \geq 0$, we can use

$$u_n^{(k)}(z) = z^m u_{n-m}^{(k)}(z) , \tag{15}$$
$$(u_n^{(k)})^\star(z) = (u_{n-m}^{(k)})^\star(z) , \tag{16}$$

to modify the proof of (14): Using the subsequent formulas which can be derived directly from (6), (7), (15) and (16),

$$u_{n+1}^{(k)}(z) = zu_n^{(k)}(z) - a_{n+k}(u_n^{(k)})^\star(z) = z^{m+1}u_{n-m}^{(k)}(z) - a_{n+k}(u_{n-m}^{(k)})^\star(z) ,$$

$$a_{n+k-m-1}(u_{n-m}^{(k)})^{\star}(z) = a_{n+k-m-1}(u_{n-m-1}^{(k)})^{\star}(z) - |a_{n+k-m-1}|^2 z u_{n-m-1}^{(k)}(z) ,$$
$$a_{n+k-m-1}(u_{n-m-1}^{(k)})^{\star}(z) = z u_{n-m-1}^{(k)}(z) - u_{n-m}^{(k)}(z) ,$$

we obtain for $n \geq m+1$

$$
\begin{aligned}
&a_{n+k-m-1}u_{n+1}^{(k)}(z) \\
&=(a_{n+k-m-1}z^{m+1} + a_{n+k})u_{n-m}^{(k)}(z) - a_{n+k}(1 - |a_{n+k-m-1}|^2)z u_{n-m-1}^{(k)}(z)
\end{aligned}
\tag{17}
$$

instead of (14).

If we want to start (14) with $n = 0$ and $u_{-1} = 0$, $u_0 = 1$, we must assume $a_{k-1} \neq 0$. Otherwise we obtain $0 = a_{k-1}u_1(0) = a_k$. Increasing k we would have $a_\nu = 0$ for all $\nu \geq k - 1$ which is in general false (i. e. with the exception of Bernstein-Szegő polynomials for which $\Phi_{\nu+k} = z^{\nu+1}\Phi_{k-1}$, $\nu \geq 0$, hold). For a fixed k we assume in the subsequent $a_{k-1} \neq 0$. Starting (14) with $n = 0$, choosing $\widehat{\Phi}_{-1}^{(k)}(z) = 0$, $\widehat{\Phi}_0^{(k)}(z) = 1$ as initial values, we get a sequence $\{\widehat{\Phi}_n^{(k)}\}_{n \geq -1}$ which also satisfy (17) for $n \geq m$. From (14) we conclude

$$
\prod_{\nu=k-1}^{n+k-1} a_\nu \widehat{\Phi}_{n+1}^{(k)}(0) = \prod_{\nu=k}^{n+k} a_\nu \widehat{\Phi}_0^{(k)} \quad \text{or} \quad -\hat{a}_{n+k} := \widehat{\Phi}_{n+1}^{(k)}(0) = \frac{a_{n+k}}{a_{k-1}} .
$$

Theorem 4.1 *If $a_\nu \neq 0$, $\nu \geq k-1$, then the sequence of polynomials $\{\widehat{\Phi}_n^{(k)}\}_{n \in \mathbf{N}_0}$, $k \geq 1$, yields not a system of Szegő polynomials.*

Proof. We assume that they were orthogonal polynomials. Then they satisfy a Szegő recurrence

$$\widehat{\Phi}_{n+1}^{(k)}(z) = z\widehat{\Phi}_n^{(k)}(z) - \hat{a}_{n+k}(\widehat{\Phi}_n^{(k)})^{\star}(z) , \quad n \geq 0 .$$

Using this reflection coefficients for the corresponding Geronimus type recurrence we get

$$a_{n+k-1}\widehat{\Phi}_{n+1}^{(k)}(z) = [za_{n+k-1} + a_{n+k}]\widehat{\Phi}_n^{(k)}(z) - a_{n+k}(1 - |\frac{a_{n+k-1}}{a_{k-1}}|^2)z\widehat{\Phi}_{n-1}^{(k)}(z)$$

for $n \geq 1$. This contradicts our assumption that $\widehat{\Phi}_n^{(k)}$ is a solution of (14) since $|a_{k-1}| < 1$ for $k \geq 1$ and $a_{n+k-1} \neq 0$. ∎

5   A Christoffel-Darboux-Type Formula

Now we want to derive an analogon to the mixed Christoffel-Darboux-type formula (5) for the complex case. The simplicity of formula (5) is based on the simplicity of the recurrences (1) and (4). Considering (13) we see that this is not so for the complex case if we use the associated polynomials. Thus we make a compromise. We want to get difference quotients and derivatives of orthogonal polynomials as a sum of products of polynomials analogous to the real case. To preserve the 'simplicity' of this formula we cannot expect that this formula hold for arbitrary sequences of Szegő polynomials. In this section we show that we obtain a corresponding result if we use the modified associated polynomials. To obtain it we need a dual recurrence formula for the $\widehat{\Phi}_n^{(k)}$. This can be obtained in a similar manner as the dual recurrence relation for the $\Phi_n^{(k)}$. At first we need the following

Lemma 5.1 *If $a_0 \neq 0$, $l \geq 1$, then we have*

$$\widehat{\Phi}_{l-1}^{(1)}(z)\Phi_{l-1}(z) - \widehat{\Phi}_{l-2}^{(1)}(z)\Phi_l(z) = \frac{a_{l-1}}{a_0}z^{l-1}\prod_{\nu=0}^{l-2}\left(1 - |a_\nu|^2\right). \tag{18}$$

Proof. The case $l = 1$ is trivial. Let $l \geq 2$. At first we assume that $a_0, \ldots, a_{l-1} \neq 0$. By using the recurrence (14) for Φ_l and $z\widehat{\Phi}_{l-2}^{(1)}(z)$ we have

$$a_{l-2}\left(\widehat{\Phi}_{l-1}^{(1)}(z)\Phi_{l-1}(z) - \widehat{\Phi}_{l-2}^{(1)}(z)\Phi_l(z)\right)$$
$$= a_{l-1}(1 - |a_{l-2}|^2)z\left(\widehat{\Phi}_{l-2}^{(1)}(z)\Phi_{l-2}(z) - \widehat{\Phi}_{l-3}^{(1)}(z)\Phi_{l-1}(z)\right).$$

Recursively we obtain

$$a_1\left(\widehat{\Phi}_{l-1}^{(1)}(z)\Phi_{l-1}(z) - \widehat{\Phi}_{l-2}^{(1)}(z)\Phi_l(z)\right)$$
$$= a_{l-1}z^{l-2}\prod_{\nu=1}^{l-2}\left(1 - |a_\nu|^2\right)\left(\widehat{\Phi}_{1}^{(1)}(z)\Phi_1(z) - \widehat{\Phi}_{0}^{(1)}(z)\Phi_2(z)\right).$$

Using $\widehat{\Phi}_{1}^{(1)}(z)\Phi_1(z) - \widehat{\Phi}_{0}^{(1)}(z)\Phi_2(z) = z\frac{a_1}{a_0}(1 - |a_0|^2)$, (18) is proved for non vanishing reflection coefficients. Since the a_ν, $\nu = l - 2, \ldots, 1$, are canceled during the recursion (independent of their values), the proof can be modified in a simple manner if some of the appearing reflection coefficients are zero which shows the validity of (18) in this case, too. ∎

If $a_0 \neq 0$ then by Lemma 5.1 the Wronskian $\widetilde{W}(\Phi_n, \widehat{\Phi}_{n-1}^{(1)})$ of two solutions Φ_n, $\widehat{\Phi}_{n-1}^{(1)}$ of (14) can be written as

$$\widetilde{W}(\Phi_n, \widehat{\Phi}_{n-1}^{(1)}) = \begin{vmatrix} \Phi_n & \widehat{\Phi}_{n-1}^{(1)} \\ \Phi_{n+1} & \widehat{\Phi}_{n}^{(1)} \end{vmatrix} = \begin{cases} 1, & n = 0, \\ \frac{a_n}{a_0}z^n\prod_{\nu=0}^{n-1}(1 - |a_\nu|^2), & a_0 \neq 0, \ n \geq 1, \end{cases}$$

which is nonzero for $z \neq 0 \neq a_n$. For this case we can write every solution of (14) as a linear combination of $\Phi_n, \widehat{\Phi}_{n-1}^{(1)}$:

$$\widehat{\Phi}_{n-l}^{(l)} = A\Phi_n + B\widehat{\Phi}_{n-1}^{(1)}, \quad n \geq l - 1.$$

By setting $n = l$ resp. $n = l - 1$ we determine A and B and get formally

$$\widehat{\Phi}_{n-l}^{(l)} = \frac{\Phi_{l-1}\widehat{\Phi}_{n-1}^{(1)} - \widehat{\Phi}_{l-2}^{(1)}\Phi_n}{\Phi_{l-1}\widehat{\Phi}_{l-1}^{(1)} - \widehat{\Phi}_{l-2}^{(1)}\Phi_l} .$$

Setting $l = 2$ we obtain (c. f. Lemma 5.1)

$$\Phi_n(z) = \Phi_1(z)\widehat{\Phi}_{n-1}^{(1)}(z) - \frac{a_1}{a_0}(1 - |a_0|^2)z\widehat{\Phi}_{n-2}^{(2)}(z) , \quad n \geq 1,$$

where we require $a_0 \neq 0$. Using $\Phi_\nu = \widehat{\Phi}_\nu$, $\nu \in \mathbf{N}_0$, we get

$$\widehat{\Phi}_n(z) = \widehat{\Phi}_1(z)\widehat{\Phi}_{n-1}^{(1)}(z) - \frac{a_1}{a_0}(1 - |a_0|^2)z\widehat{\Phi}_{n-2}^{(2)}(z) , \quad n \geq 1, \quad a_0 \neq 0 .$$

Formulating the k-associated version of these formulas, using that we can only use $\widehat{\Phi}_n^{(k)}$ if $a_{k-1} \neq 0$ we get

Theorem 5.2 *If $a_k \neq 0$ then we have*

$$\Phi_n^{(k)}(z) = \Phi_1^{(k)}(z)\widehat{\Phi}_{n-1}^{(k+1)}(z) - \frac{a_{k+1}}{a_k}(1 - |a_k|^2)z\widehat{\Phi}_{n-2}^{(k+2)}(z) , \quad n \geq 1 .$$

If additionally $a_{k-1} \neq 0$ then for $\widehat{\Phi}_n^{(k)}$ we have a dual recurrence relation

$$\widehat{\Phi}_n^{(k)}(z) = \widehat{\Phi}_1^{(k)}(z)\widehat{\Phi}_{n-1}^{(k+1)}(z) - \frac{a_{k+1}}{a_k}(1 - |a_k|^2)z\widehat{\Phi}_{n-2}^{(k+2)}(z) , \quad n \geq 1 . \quad (19)$$

Using this result we now formulate a Christoffel-Darboux-type formula:

Theorem 5.3 *1) If $a_j \neq 0$ for $j = -1, \ldots, n$, then*

$$\sum_{k=0}^{n} \left(\Phi_k(w) - \frac{a_k}{a_{k-1}}(1 - |a_{k-1}|^2)\Phi_{k-1}(w) \right) \widehat{\Phi}_{n-k}^{(k+1)}(z)$$

$$= \begin{cases} \dfrac{\Phi_{n+1}(w) - \Phi_{n+1}(z)}{w - z} & \text{for } w \neq z , \\ \dfrac{d}{dw}\Phi_{n+1}(w) & \text{for } w = z . \end{cases} \quad (20)$$

2) If $a_{-1}, \ldots, a_j \neq 0$, $a_{j+1} = \ldots = a_n = 0$, for a j, $-1 \leq j \leq n$, then

$$\sum_{k=0}^{j} \left(\Phi_k(w) - \frac{a_k}{a_{k-1}}(1 - |a_{k-1}|^2)\Phi_{k-1}(w) \right) \widehat{\Phi}_{n-k}^{(k+1)}(z) + \sum_{k=j+1}^{n} \Phi_k(w)\Phi_{n-k}^{(k+1)}(z)$$

$$= \begin{cases} \dfrac{\Phi_{n+1}(w) - \Phi_{n+1}(z)}{w - z} & \text{for } w \neq z , \\ \dfrac{d}{dw}\Phi_{n+1}(w) & \text{for } w = z . \end{cases} \quad (21)$$

(Notice that on the left side there appear the associated and the modified associated polynomials).

Proof. From (8) we have for $n = k \geq 1$

$$z a_{k-1} \Phi_k(z) = a_{k-1} \Phi_{k+1}(z) - a_k \Phi_k(z) + a_k (1 - |a_{k-1}|^2) z \Phi_{k-1}(z) ,$$

which also holds for $k = 0$ since $a_{-1} = -1$. From (19), if $a_k \neq 0$, replacing n by $n - k + 1$ we get

$$w a_{k-1} \widehat{\Phi}_{n-k}^{(k+1)}(w)$$
$$= a_{k-1} \widehat{\Phi}_{n-k+1}^{(k)}(w) - a_k \widehat{\Phi}_{n-k}^{(k+1)}(w) + a_{k-1} \frac{a_{k+1}}{a_k} (1 - |a_k|^2) w \widehat{\Phi}_{n-k-1}^{(k+2)}(w) .$$

From these formulas we obtain for $a_{k-1} a_k \neq 0$

$$
\begin{aligned}
(z - w) \Phi_k(z) \widehat{\Phi}_{n-k}^{(k+1)}(w) &= \Phi_{k+1}(z) \widehat{\Phi}_{n-k}^{(k+1)}(w) - \Phi_k(z) \widehat{\Phi}_{n-(k-1)}^{(k)}(w) \\
&+ \frac{a_k}{a_{k-1}} (1 - |a_{k-1}|^2) z \Phi_{k-1}(z) \widehat{\Phi}_{n-k}^{(k+1)}(w) \\
&- \frac{a_{k+1}}{a_k} (1 - |a_k|^2) w \Phi_k(z) \widehat{\Phi}_{n-(k+1)}^{(k+2)}(w) .
\end{aligned}
$$

If $a_0, \ldots, a_n \neq 0$ we get by using $-a_{-1} = \Phi_0(0) = 1$, $\widehat{\Phi}_{-1}^{(k+1)}(z) = 0$, $\Phi_\nu^{(0)} = \Phi_\nu$, $\nu \in \mathbf{N}_0$,

$$
\begin{aligned}
(z - w) \sum_{k=0}^{n} \Phi_k(z) \widehat{\Phi}_{n-k}^{(k+1)}(w) &= \sum_{k=0}^{n} \left(\Phi_{k+1}(z) \widehat{\Phi}_{n-k}^{(k+1)}(w) - \Phi_k(z) \widehat{\Phi}_{n-(k-1)}^{(k)}(w) \right) \\
&+ \sum_{k=0}^{n} \frac{a_k}{a_{k-1}} (1 - |a_{k-1}|^2) z \Phi_{k-1}(z) \widehat{\Phi}_{n-k}^{(k+1)}(w) \\
&- \sum_{k=0}^{n} \frac{a_{k+1}}{a_k} (1 - |a_k|^2) w \Phi_k(z) \widehat{\Phi}_{n-(k+1)}^{(k+2)}(w) \\
&= \Phi_{n+1}(z) - \Phi_{n+1}(w) \\
&+ (z - w) \sum_{k=0}^{n} \frac{a_k}{a_{k-1}} (1 - |a_{k-1}|^2) \Phi_{k-1}(z) \widehat{\Phi}_{n-k}^{(k+1)}(w) .
\end{aligned}
$$

Thus we have

$$(z - w) \sum_{k=0}^{n} \left[\Phi_k(z) - \frac{a_k}{a_{k-1}} (1 - |a_{k-1}|^2) \Phi_{k-1}(z) \right] \widehat{\Phi}_{n-k}^{(k+1)}(w) = \Phi_{n+1}(z) - \Phi_{n+1}(w)$$

from which (21) follows.

Now, let $a_{-1}, \ldots, a_j \neq 0$, $a_{j+1} = \ldots = a_n = 0$, where $-1 \leq j \leq n$. Then we have

$$
\begin{aligned}
\Phi_{j+1+l}(z) &= z^l \Phi_{j+1}(z) , & 0 \leq l \leq n - j - 1, & \quad (22) \\
\Phi_{n-k}^{(k+1)}(z) &= z^{n-k} & \text{for } k = j + 1, \ldots, n , & \quad (23) \\
\widehat{\Phi}_{n-k}^{(k+1)}(z) &= z^{n-j} \widehat{\Phi}_{j-k}^{(k+1)}(z) & \text{for } n, j \geq k . & \quad (24)
\end{aligned}
$$

Using (22), (23) we obtain

$$\sum_{k=j+1}^{n} \Phi_k(z)\Phi_{n-k}^{(k+1)}(w) = \Phi_{j+1}(z)\sum_{k=j+1}^{n} z^{k-j-1}\Phi_{n-k}^{(k+1)}(w)$$

$$= \Phi_{j+1}(z)\sum_{k=j+1}^{n} z^{k-j-1}w^{n-k} .$$

For $z \neq w$, using (24), we have

$$\sum_{k=0}^{j}\left(\Phi_k(z) - \frac{a_k}{a_{k-1}}(1-|a_{k-1}|^2)\Phi_{k-1}(z)\right)\widehat{\Phi}_{n-k}^{(k+1)}(w) + \sum_{k=j+1}^{n}\Phi_k(z)\Phi_{n-k}^{(k+1)}(w)$$

$$=w^{n-j}\frac{\Phi_{j+1}(z) - \Phi_{j+1}(w)}{z - w} + \Phi_{j+1}(z)z^{-j-1}w^n\sum_{k=j+1}^{n}\left(\frac{z}{w}\right)^k$$

$$=w^{n-j}\frac{\Phi_{j+1}(z) - \Phi_{j+1}(w)}{z - w} + \Phi_{j+1}(z)z^{-j-1}w^{n+1}\frac{\left(\frac{z}{w}\right)^{n+1} - \left(\frac{z}{w}\right)^{j+1}}{z - w}$$

$$=\frac{1}{z - w}\left(w^{n-j}\Phi_{j+1}(z) - w^{n-j}\Phi_{j+1}(w) + \Phi_{j+1}(z)(z^{n-j} - w^{n-j})\right)$$

$$=\frac{z^{n-j}\Phi_{j+1}(z) - w^{n-j}\Phi_{j+1}(w)}{z - w}$$

$$=\frac{\Phi_{n+1}(z) - \Phi_{n+1}(w)}{z - w} .$$

For $z = w$ we get

$$\sum_{k=0}^{j}\left(\Phi_k(z) - \frac{a_k}{a_{k-1}}(1-|a_{k-1}|^2)\Phi_{k-1}(z)\right)\widehat{\Phi}_{n-k}^{(k+1)}(z) + \sum_{k=j+1}^{n}\Phi_k(z)\Phi_{n-k}^{(k+1)}(z)$$

$$=z^{n-j}\Phi_{j+1}'(z) + \Phi_{j+1}(z)(n - j)z^{n-j-1} = z^{n-j}\Phi_{j+1}'(z) + \frac{d}{dz}\left(z^{n-j}\right)\Phi_{j+1}(z)$$

$$=\frac{d}{dz}\left(z^{n-j}\Phi_{j+1}(z)\right) = \Phi_{n+1}'(z) ,$$

which proves the Theorem.   ∎

Remark 5.4 *1) The usage of the $\widehat{\Phi}_n^{(l)}$ was the reason for the restrictions under which the results given in Theorem 5.3 hold. This was the price for our aim to derive a mixed Christoffel-Darboux-type formula whose structure is comparable in its simplicity to the corresponding result (5) for polynomials orthogonal on the real line.*
2) If $a_k \neq 0$, $k = 0,\ldots,j$, then the polynomials

$$\Lambda_k^L := \Phi_k - \frac{a_k}{a_{k-1}}(1-|a_{k-1}|^2)\Phi_{k-1}, \quad 0 \le k \le j ,$$

which appear in Theorem 5.3 have an interesting property: From (8) we have

$$z\Lambda_k^L = \Phi_{k+1} - \frac{a_k}{a_{k-1}}\Phi_k, \ 0 \le k \le j \ .$$

Thus $\{z\Lambda_k^L(z)\}_{0\le k\le j}$ *are quasi orthogonal of order one on* $\mathbf{T}$ *with respect to* $d\mu$, *resp.* $\{\Lambda_k^L\}_{0\le k\le j}$ *are left orthogonal monic polynomials with respect to* $z\,d\mu(\theta)$, $z = \exp(i\theta)$ *on* $\mathbf{T}$ *(the underlying linear functional which describes the orthogonality isn't Hermitian [2, Theorem 1]).*

References

[1] S. BELMEHDI, On the associated orthogonal polynomials, *J. Comput. Appl. Math.* **32** (1990) 311–319.

[2] M. A. CACHAFEIRO, Y. M. SUAREZ, Kernels on the unit circle. Orthogonality, in: M. ALFARO, A. GARCÍA, C. JAGELS, F. MARCELLÁN (eds.), "Orthogonal Polynomials on the unit circle: Theory and Applications", (Proceedings of a workshop), Universidad Carlos III de Madrid (1994) 43 – 57.

[3] T. S. CHIHARA, "An Introduction to Orthogonal Polynomials", (Gordon and Breach, New York, 1978).

[4] YA. L. GERONIMUS, Polynomials orthogonal on a circle and their applications, *Amer. Math. Soc. Transl.* **3** (1954) 1 - 78.

[5] F. PEHERSTORFER, A special class of polynomials orthogonal on the unit circle including the associated polynomials, *Constructive Approximation*, **12** (1996) 161 - 185.

[6] G. SZEGŐ, "Orthogonal Polynomials", Amer. Math. Colloq. Publ., Amer. Math. Soc., Providence, R. I., 4th ed., 1975 .

[7] W. VAN ASSCHE, Orthogonal polynomials, associated polynomials and functions of the second kind, *J. Comput. Appl. Math.* **37** (1991) 237–249.

FB Mathematik
Fernuniversität -GHS- in Hagen
Postfach 940
D-58084 Hagen, Germany
Email address: michael.skrzipek@fernuni-hagen.de

International Series of Numerical Mathematics
Vol. 132, © 1999 Birkhäuser Verlag Basel/Switzerland

On Some Problems of Weighted Polynomial Approximation and Interpolation

József Szabados

Abstract

This is a survey on some recent results in the theory of weighted polynomial approximation and interpolation. Some proofs are only sketched, other are presented in details (cf. [4] and [7]).

1

Recently, there has been a considerable interest in different aspects of polynomial approximation (orthogonal polynomials, interpolation) with respect to Freud and Erdős weights on the real line. First we extend some of these results for a more general class of weights. We will consider weights which have finitely many zeros on the real line, and state density theorems for polynomial approximation in the corresponding space of functions. Also, we will construct systems of nodes of interpolation where the weighted Lebesgue constant is of optimal order. Allowing roots for the weight opens the possibility of considering spaces of piecewise continuous (unbounded) functions. As far as we know, it was D. S. Lubinsky and E. B. Saff ([5], Theorems 3.4–3.5) who considered such weights (from different aspects).

Our starting point is the following result, attributed to Akhiezer, Babenko, Carleson and Dzrbasjan (see D. S. Lubinsky [6]).

Theorem 1.1 *Let $w = e^{-Q}$ where Q is even on $\mathbf{R}$, $Q(e^x)$ is convex on $(0, \infty)$, and let*

$$C_w(\mathbf{R}) := \{f \mid f \in C(\mathbf{R}), \lim_{|x| \to \infty} (f(x)w(x)) = 0\}. \tag{1}$$

For an $f \in C_w(\mathbf{R})$ define the best polynomial approximation

$$E_n(f)_w := \inf_{p \in \Pi_n} \|w(f - p)\|$$

where $\| \cdot \|$ is the supremum norm over $\mathbf{R}$, and Π_n is the set of polynomials of degree at most n. Then

$$\lim_{n \to \infty} E_n(f)_w = 0 \quad for \ all \quad f \in C_w(\mathbf{R}) \tag{2}$$

if and only if

$$\int_0^\infty \frac{Q(x)}{1+x^2}\,dx = \infty.$$

Our first result generalizes the "if" part of this theorem for a wider class of weights defined below.

Definition 1.2 *The set of weight-functions $w(x) = e^{-Q(x)} \in \mathcal{W}_1$ is defined by the following conditions. Let $-\infty < t_1 < \ldots < t_s < \infty$ be arbitrary fixed real numbers, and let $Q(x)$ satisfy the following properties:*
   (i) $0 < Q \in C(\mathbf{R} \setminus \cup_{i=1}^s \{t_i\})$, $\lim_{x \to t_i} Q(x) = \infty$ $(i = 1, \ldots, s)$,
   (ii) $\limsup\limits_{x \to \infty} |Q(x) - Q(-x)| < \infty$,
   (iii) $Q(e^x)$ is convex for x large, and
   (iv) $\int\limits_{t_s+1}^{\infty} \frac{Q(x)dx}{1+x^2} = \infty$.

This class of weights $\mathcal{W}_1$ is more general than those considered in the above cited theorem, since $w \in \mathcal{W}_1$ vanishes at t_i $(i = 1, \ldots, s)$. Also, condition (ii) permits a certain asymmetry of the weight at $\pm\infty$. Finally, (iii) requires convexity only for large x.

Here are two characteristic examples for weights in $\mathcal{W}_1$:

$$w(x) = e^{-|x|^\alpha} \prod_{i=1}^s |x - t_i|^{\alpha_i} |\log|x - t_i||^{\beta_i} \tag{3}$$

$$(\alpha \geq 1, \alpha_i \geq 0, \beta_i \in \mathbf{R}, \beta_i < 0 \text{ if } \alpha_i = 0, i = 1, \ldots, s),$$

and

$$w(x) = \exp\left(-|x|^\alpha - \sum_{i=1}^s \frac{b_i}{|x - t_i|^{\alpha_i}}\right) \quad (\alpha \geq 1, b_i, \alpha_i > 0, i = 1, \ldots, s). \tag{4}$$

Now let (compare (1))

$$C_w(\mathbf{R}) := \{f \mid f \in C(\mathbf{R} \setminus \cup_{i=1}^s \{t_i\}), \lim_{x \to t_i} (w(x)f(x)) = 0, i = 0, 1, \ldots, s + 1\}, \tag{5}$$

where $t_0 = -t_{s+1} = -\infty$. Hence $C_w(\mathbf{R})$ contains functions which are unbounded at the t_i's.

Theorem 1.3 *We have (2) for all $w \in \mathcal{W}_1$ and $f \in C_w(\mathbf{R})$.*

This generalization of Theorem 1.1 above can be proved by approximating f by a continuous function and then using Theorem A itself.

Now we define a subset $\mathcal{W}_2$ of $\mathcal{W}_1$.

Definition 1.4 *We shall say that* $w(x) = v(x)e^{-Q(x)} \in W_2$ *if the following conditions hold:*

(a) *Q is even, continuous in* $\mathbf{R}$, $0 < Q' \in C(0, \infty)$, *and there exist* $1 < A \leq B < \infty$ *such that*

$$A \leq \frac{(xQ')'}{Q'} \leq B \qquad (x \geq 0);$$

(b) $v(x) \geq 0$ *is continuous in* $\mathbf{R}$, $v(x) > 0$ *if* $x \in \mathbf{R} \setminus \cup_{i=1}^{s}\{t_i\}$, *and there exist
integers* $m_i \geq 0$ $(i = 1, \ldots, s)$ *and constants* $c_1, c_2, c_3 \geq 0$ *such that*

$$c_1 \left|\frac{x - t_i}{y - t_i}\right|^{m_i+1} \leq \frac{v(x)}{v(y)} \leq c_2 \left|\frac{x - t_i}{y - t_i}\right|^{m_i} \tag{6}$$

$$(|x - t_i| \leq |y - t_i| \leq c_3, \ i = 1, \ldots, s).$$

*(In what follows, $c_1, \ldots$ will denote positive constants possibly depending on the
weights but independent of n.)*

(c) $v(x)$ *is twice differentiable for large* $|x|$, *and (∼ means that the ratio of
the two sides remains between two positive constants as* $x \to \infty$)

$$v(x) \sim v(-x) \qquad (x \to \infty),$$

$$\left|\left(\frac{v'(x)}{v(x)}\right)'\right| = o(|x|^{A-2}) \qquad (|x| \to \infty).$$

It is easy to see that $W_1 \subset W_2$. A characteristic example for weight in W_2 is
the function (3). The only difficulty in checking this is to choose the m_i's in (6):

$$m_i = \begin{cases} [\alpha_i] & \text{if } \alpha_i \text{ is not an integer,} \\ \alpha_i & \text{if } \alpha_i \text{ is an integer and } \beta_i < 0, \\ \alpha_i - 1 & \text{if } \alpha_i \text{ is an integer and } \beta_i \geq 0. \end{cases}$$

However, it is easy to see that the function in (4) is not in W_2, since (6) does
not hold because of the non-polynomial decrease of the weight near the singulari-
ties. The weights with $Q(x) = |x|^\alpha$ and

$$v(x) = \begin{cases} -x & \text{if } -1 \leq x \leq 0, \\ x^2 & \text{if } 0 \leq x \leq 1, \\ 1 & \text{if } |x| \geq 1 \end{cases} \qquad \text{or} \qquad v(x) = |x|\left(|x| + \left|\sin\frac{1}{x}\right|\right)$$

are also not in W_2; the first because of the asymmetry and the second because of
the oscillation at the singularity (again, the critical condition (6) does not fulfil).
Nevertheless, these weights are easily seen to be in W_1.

Since Theorem 1.3 ensures the density of polynomials for weights in the class
W_∞, it makes sense to look for systems of nodes of interpolation for which the
weighted Lebesgue constant

$$\Lambda_n(w) := \left\|w(x)\sum_{k=1}^{n}\frac{|l_k(x)|}{w(x_k)}\right\|$$

is optimal in order for $w \in \mathcal{W}_2$. (Here $l_k(x)$ are the fundamental polynomials of Lagrange interpolation based on the nodes $-\infty < x_1 < \ldots < x_n < \infty$.) This quantity $\lambda_n(w)$ plays an essential role in the weighted error estimate of the Lagrange interpolation polynomial $L_n(f, x)$, namely

$$\|w(f - L_n(f))\| \leq (1 + \lambda_n(w)) E_n(f)_w.$$

Theorem 1.5 *For any $w \in \mathcal{W}_2$, there exists a system of nodes $\{x_k\}_{k=1}^n \subset \mathbf{R}$ such that*

$$\lambda_n(w) = O(\log n).$$

This order of magnitude of the Lebesgue constant is probably optimal, but we do not address this problem here. Theorem 1.5 is a generalization of Theorem 1, (7) from [10]. The construction of the system of nodes is the following. Let $a > \max_{1 \leq i \leq s} |t_i|$ and

$$V(x) := \begin{cases} \sqrt{\frac{v(x)v(-x)}{u(x)u(-x)}} & \text{if } |x| \geq a, \\ V(a)e^{\alpha(|x|^B - a^B) + \beta(|x|^B - a^B)^2} & \text{if } |x| < a, \end{cases} \tag{7}$$

where

$$u(x) := \prod_{i=1}^{s} |x - t_i|^{m_i}$$

and

$$\alpha := \frac{1}{Ba^{B-1}} \frac{V'}{V}(a), \quad \beta := \frac{1}{2B^2 a^{2B-2}} \left(\frac{V'}{V}\right)'(a) - \frac{B-1}{2B^2 a^{2B-1}} \frac{V'}{V}(a).$$

(Here $V'(a)$ and $V''(a)$ are meant to be left derivatives calculated from the first part of the definition of $V(x)$ in (7).)

It can be shown that $w_1(x) := e^{-Q_1(x)}$, where

$$Q_1(x) := Q(x) - \log V(x),$$

is a Freud weight, i.e. it satisfies (a) of Definition 1.4 (with Q_1 instead of Q). Then the point system realizing the optimal order of magnitude of the Lebesgue constant is the following. Let

$$r := \sum_{i=1}^{s} m_i,$$

and consider the roots of the polynomial $p_{n+r-2}(x)$ of degree $n + r + 2$ orthogonal with respect to the Freud weight $w_1(x)^2$. Let n be sufficiently large, and for each $1 \leq i \leq s$, let $y_{i,1}, y_{i,2}, \ldots, y_{i,m_i+2}$ be the $m_i + 2$ roots of this polynomial nearest to t_i, in such an order that

$$|t_i - y_{i,1}| \leq |t_i - y_{i,2}| \leq \ldots \leq |t_i - y_{i,m_i+2}| \qquad (i = 1, \ldots, s).$$

We drop the first $m_i + 1$ of these roots from, and add

$$z_i := \begin{cases} \frac{\lambda y_{i,1} + y_{i,2}}{\lambda + 1} & \text{if sgn}\,(t_i - y_{i,1}) = \text{sgn}\,(t_i - y_{i,2}), \\ \frac{\lambda t_i + y_{i,2}}{\lambda + 1} & \text{otherwise} \end{cases} \qquad (i = 1, \dots, s)$$

to the set of roots of p_{n+r-2}, where $\lambda > 0$ is a constant to be chosen later. In this way we get $n - 2$ roots. Further let $z_0 > 0$ be a point where the norm $\|\bar{w} p_{n+r-2}\|$ is attained; we add $\pm z_0$ to the previous system of nodes. These n nodes $x_1, \dots, x_n$ will be our system. In other words, these are the roots of the polynomial

$$\omega_n(x) := p_{n+r-2}(x)(x^2 - z_0^2) \prod_{i=1}^{s} \frac{x - z_i}{\prod_{j=1}^{m_i+1}(x - y_{i,j})}$$

of degree n.

Now let a_n be the Mhaskar–Rahmanov–Saff number belonging to the Freud weight $w_1 = e^{-Q_1}$ (i.e. Q_1 satisfies (a) of Definition 1.4), that is for any polynomial p of degree at most n we have

$$\|w_1 p\| = \max_{|x| \le a_n} w_1(x)|p(x)|$$

(cf. e.g. Mhaskar and Saff [9]). These numbers a_n possess some interesting properties which are listed in the following lemmas and their corollaries:

Lemma 1.6 *Given a Freud weight $w_1 = e^{-Q_1}$, there exist constants $0 < c_4 < 1 < c_5$ such that for any polynomial of degree at most n we have*

$$w_1(x)|p(x)| \le \|w_1 p\| c_4^n \qquad (|x| \ge c_5 a_n).$$

Lemma 1.7 *Let $x \in \mathbf{R}$ and*

$$|x - x_j| := \min_{1 \le k \le n} |x - x_k|. \tag{8}$$

Then we have

$$w(x) \left| \frac{\omega_n(x)}{x - x_j} \right| = O\left(\frac{v(x) n a_n^{1/2} \psi_n^{5/4}(x)}{u(x) V(x)} \right) \qquad (x \in \mathbf{R}), \tag{9}$$

where

$$\psi_n(x) := \max\{n^{-2/3}, 1 - |x|/a_n\}. \tag{10}$$

Here in case

$$|x - x_j| \le \frac{\eta a_n}{n \psi_n(x)^{1/2}}$$

with a small enough $\eta > 0$ the estimate is sharp in the sense of the order of magnitude.

Corollary 1.8 *We have*

$$w(x)|\omega_n(x)| = O\left(\frac{v(x)a_n^{3/2}\psi_n^{3/4}(x)}{u(x)V(x)}\right) \qquad (x \in \mathbf{R}).$$

This follows from (9) by taking into account that

$$|x - x_j| = O\left(\frac{a_n}{n\psi_n(x)^{1/2}}\right),$$

which is a consequence of the root distance relation

$$\Delta x_i := x_i - x_{i+1} \sim \frac{a_n}{n\psi_n(x_i)^{1/2}} \qquad (i = 1, \ldots, n-1) \tag{11}$$

(cf. [2], Lemma 4.4). The applicability of the last relations for x_i's instead of the original roots of the polynomial p_{n+r-2} follows from the construction of these nodes.

Corollary 1.9 *We have*

$$w(x_j)|\omega_n'(x_j)| \sim \frac{v(x_j)\psi_n(x_j)^{5/4}}{u(x_j)V(x_j)}na_n^{1/2} \qquad (j = 1, \ldots, n).$$

This follows again from (9) by letting $x \to x_j$ and using the sharpness of the estimate. We now return to the proof of Theorem 1.5. Using the notation (8), Lemma 1.7 and Corollaries 1.8 and 1.9 we get

$$w(x)\sum_{k=1}^{n}\frac{|l_k(x)|}{w(x_k)} = O\left(\frac{v(x)}{u(x)}\frac{u(x_j)}{v(x_j)}\frac{V(x_j)}{V(x)}\left(\frac{\psi_n(x)}{\psi_n(x_j)}\right)^{5/4}\right. \tag{12}$$

$$\left. + \frac{v(x)\psi_n(x)^{3/4}a_n}{u(x)V(x)n}\sum_{\substack{k=1\\k\neq j}}^{n}\frac{u(x_k)V(x_k)}{v(x_k)\psi_n(x_k)^{5/4}|x - x_k|}\right).$$

Here in the first term $\psi_n(x) = O(\psi_n(x_j))$, and in case $|x| \geq a$ we have $\frac{v(x)}{u(x)V(x)} = O(1)$ by (7). If $|x| < a$ then again by (7)

$$\frac{v(x)}{v(x_j)} = O\left(\frac{u(x)}{u(x_j)}\left(1 + \left|\frac{x - t_i}{x_j - t_i}\right|\right)\right) = O\left(\frac{u(x)}{u(x_j)}\right),$$

where t_i is the nearest to x, and by (7) $V(x) \sim 1$, $V(x_j) \sim 1$. This shows that the first term in (12) is $O(1)$.

For the rest of the right-hand side in (12), applying (11) we get

$$O\left(\sum_{k\neq j} \frac{v(x)}{u(x)V(x)} \frac{u(x_k)}{v(x_k)V(x_k)} \left(\frac{\psi_n(x)}{\psi_n(x_k)}\right)^{3/4} \frac{\Delta x_k}{|x-x_k|}\right)$$

$$\leq \sum_{i=1}^{s} \sum_{\substack{|x-t_i|\leq|x_k-t_i|\leq c_3 \\ k\neq j}} + \sum_{i=1}^{s} \sum_{\substack{|x_k-t_i|<|x-t_i|\leq c_3 \\ k\neq j}} + \sum_{\substack{|x_k-t_i|\geq c_3,\frac{1}{2}(t_{i-1}+t_i)\leq x_k\leq\frac{1}{2}(t_i+t_{i+1}) \\ k\neq j}}$$

$$+ \sum_{\substack{x_k<\frac{1}{2}(t_0+t_1)\,or\,x_k>\frac{1}{2}(t_s+t_{s+1}) \\ k\neq j}} = A_1 + A_2 + A_3 + A_4,$$

where now $t_0 = -a = t_{s+1}$.

For A_1, we get from (6), (7) and (10)

$$\frac{v(x)}{v(x_k)} = O\left(\frac{u(x)}{u(x_k)}\right),$$

$$V(x) \sim 1, \quad V(x_k) \sim 1 \tag{13}$$

and

$$\frac{\psi_n(x)}{\psi_n(x_k)} = O(1), \tag{14}$$

whence

$$A_1 = O\left(\sum_{k\neq j} \frac{\Delta x_k}{|x-x_k|}\right) = O(\log n) \tag{15}$$

(cf.[10], Lemma 6). For A_2, (13) and (14) still hold, and again by (6)

$$\frac{v(x)}{v(x_k)} = O\left(\frac{u(x)}{u(x_k)}\left|\frac{x-t_i}{x_k-t_i}\right|\right) = O\left(\frac{u(x)}{u(x_k)}\left(1+\left|\frac{x-t_i}{x_k-t_i}\right|\right)\right),$$

whence

$$A_2 = O(A_1) + O\left(\sum_{k\neq j} \frac{\Delta x_k}{|t_i-x_k|}\right) = O(\log n),$$

by Lemma 6 in [2].

For A_3, by (6) and (7)

$$v(x) \sim u(x)V(x), \quad u(x_k) \sim v(x_k)V(x_k),$$

and (14) still holds. Therefore A_3 has the same estimate as A_1 in (15).

Finally, A_4 is estimated the same way as A_3. Theorem 1.5 is completely proved.

2

We now turn our attention to weighted polynomial approximation on a finite interval. Here we go beyond the usual and widely investigated, so called generalized Jacobi weights. We will allow *non-symmetric* weights, which means that they may have different behavior to the left and to the right of the zeros. This is a natural extension of the generalized Jacobi weights, and it would be desirable to exploit the theory of the associated orthogonal polynomials (estimates, asymptotics, root distance, etc.). We will see that for the weighted approximation, truncated Jackson integrals serve as a good approximation tool.

First we define the weight function we are dealing with. Let

$$v_{\alpha,\beta}(x) := \begin{cases} (-x)^\alpha & \text{if } x \le 0, \\ x^\beta & \text{if } x \ge 0 \end{cases} \qquad (\alpha, \beta > 0).$$

Further let $s \ge 1$ be an integer, and

$$-1 < t_1 < \ldots < t_s < 1$$

a finite partition of the interval $[-1, 1]$. With α_i, $\beta_i > 0$ $(i = 1, \ldots, s)$ real numbers we define our weight function as

$$w(x) := \prod_{i=1}^{s} v_{\alpha_i,\beta_i}(x - t_i) \qquad (|x| \le 1). \tag{16}$$

Next, we define the class of functions to be approximated, associated with the weight w:

$$C_w := \{f | f \in C([-1, 1] \setminus \cup_{i=1}^{s}\{t_i\}), \lim_{x \to t_i} (wf)(x) = 0, i = 1, \ldots, s\}.$$

(Compare (5).) Then the weighted best approximation is defined as

$$E_n(f)_w := \inf_{p \in \Pi_n} \|w(f - p)\| \qquad (f \in C_w) \tag{17}$$

where Π_n is the set of polynomials of degree at most n, and $\| \cdot \|$ is the supremum norm over the interval $[-1, 1]$.

Introducing the notations

$$\gamma_i := \min(\alpha_i, \beta_i), \quad \Gamma_i := \max(\alpha_i, \beta_i) \qquad (i = 1, \ldots, s),$$

$$\lambda := \min\left(\min_{\Gamma_i \le 1} \gamma_i, \min_{\Gamma_i > 1} \gamma_i/\Gamma_i\right),$$

our first result is a Jackson-type inequality for this weighted best approximation.

Theorem 2.1 *For any $f \in C_w$ we have*

$$E_n(f)_w = O\left(\omega^\varphi\left(wf, \frac{1}{n^\lambda}\right)\right)$$

where ω^φ is the Ditzian–Totik modulus of continuity of the corresponding function (cf. Ditzian–Totik [3], p. 8).

This result shows that the closer α_i and β_i are, the better the order of approximation is. Also, the estimate is optimal if $\alpha_i = \beta_i \geq 1$ $(i = 1, \ldots, s)$: in this case we have

$$E_n(f)_w = O\left(\omega^\varphi\left(wf, \frac{1}{n}\right)\right).$$

In general, Theorem 2.1 is not sharp, since G. Mastroianni and V. Totik prove in a forthcoming paper that the previous estimate holds whenever $\alpha_i = \beta_i$ $(i = 1, \ldots, s)$. The proof is based on the so-called truncated Jackson integrals. Let

$$l \geq \frac{1}{2} \max_{1 \leq i \leq s} \Gamma_i + \frac{3}{2}$$

be a fixed integer, and consider the trigonometric polynomial

$$T_n(u) := c_n \left(\frac{\sin \frac{2n+1}{2} u}{\sin \frac{u}{2}}\right)^{2l}$$

of order $2ln$, where

$$c_n := \left(\int_{-\pi}^{\pi} \left(\frac{\sin \frac{2n+1}{2} u}{\sin \frac{u}{2}}\right)^{2l} du\right)^{-1}.$$

For sufficiently large n let

$$f_n(x) := \begin{cases} f(x) & \text{if } |x - t_i| \geq n^{-\lambda}, \ i = 1 \ldots, s, \\ 0 & \text{otherwise} \end{cases} \qquad (|x| \leq 1),$$

and consider the Jackson integral

$$J_n(f_n, x) = \int_{\substack{|\cos(\arccos x - u) - t_i| \geq n^{-\lambda} \\ i=1,\ldots,s}} f(\cos(\arccos x - u)) T_n(u) \, du \qquad (18)$$

of the function f_n. It is known that (18) is a polynomial of degree at most $2ln$. The order of approximation stated in Theorem 2.1 is attained by this operator. We omit the details.

In Theorem 2.1, we excluded endpoint singularities. Now, with the notation (16), consider the weight

$$W(x) := (1-x)^\alpha (1+x)^\beta w(x) \qquad (\alpha, \beta \geq 0, |x| \leq 1).$$

The corresponding function class is defined as

$$C_W := \{f | f \in C([-1, 1] \setminus \cup_{i=0}^{s+1} \{t_i\}), \lim_{x \to t_i} (Wf)(x) = 0, \ i = 0, \ldots, s+1\},$$

where $t_0 := -t_{s+1} := -1$. Let

$$\mu := \min{}^* (\alpha, \beta, \lambda),$$

where the star indicates that in case $\alpha = 0$ or $\beta = 0$, these zeros are omitted when forming the minimum, and if there are no inner singularities, then we define $\lambda := 1$. Analogously to (17), let

$$E_n(f)_W := \inf_{p \in \Pi_n} \|W(f - p)\| \qquad (f \in C_W).$$

We state

Theorem 2.2 *For any $f \in C_W$ we have*

$$E_n(f)_W = O\left(\omega^\varphi\left(Wf, \frac{1}{n^\mu}\right)\right).$$

The proof can be carried out by a homogeneous transformation of the variable, and making use of the previous theorem.

3

Although we know that the above results, in general, are not sharp, in any case, in order to establish some converse theorems, one needs weighted Bernstein–Markov type inequalities for nonsymmetric weights. Let

$$\Delta_n(x) := \frac{\sqrt{1 - x^2}}{n} + \frac{1}{n^2},$$

$$\varphi_n(x) := \begin{cases} |x|^\alpha + n^{-\alpha}, & \text{if } -1 \leq x \leq \left(\frac{\log n}{n^{1+\alpha}}\right)^{\frac{1}{1+\beta}}, \\ \frac{nx^\beta}{n + \frac{1}{x}\log\frac{1}{x}}, & \text{if } \left(\frac{\log n}{n^{1+\alpha}}\right)^{\frac{1}{1+\beta}} < x \leq 1 \end{cases} \qquad (\beta < \alpha \leq 2\beta + 1),$$

and

$$\varphi_n(x) := \begin{cases} |x|^\alpha + n^{-\alpha}, & \text{if } -1 \leq x < -\frac{1}{n^2}, \\ |x|^{\beta + \frac{1}{2}} + n^{-2\beta - 1}, & \text{if } -\frac{1}{n^2} < x < \left(\frac{\log n}{n}\right)^2, \\ \frac{nx^\beta}{n + \frac{1}{x}\log\frac{1}{x}}, & \text{if } \left(\frac{\log n}{n}\right)^2 < x \leq 1 \end{cases} \qquad (2\beta + 1 < \alpha).$$

With the notation of Section 2, we prove the following Bernstein–Markov type inequality:

Theorem 3.1 *For any polynomial $p_n(x)$ of degree at most n we have*

$$\|\Delta_n(x)\varphi_n(x)p'_n(x)\| \leq \|v_{\alpha,\beta}p_n\|.$$

Proof. We start with the Markov–Bernstein inequality for generalized Jacobi weights

$$\|\Delta_n(x)|x|^\alpha p'_n(x)\| \leq C\||x|^\alpha p_n(x)\| \leq C\|v_{\alpha,\beta}p_n\| \tag{19}$$

(cf. Mastroianni–Totik [8], (7.28) and (7.30)). If $-1 \leq x \leq -1/n$ or $1/2 \leq x \leq 1$, then this proves our statement. If $-1/n \leq x \leq \left(\frac{\log n}{n^{1+\alpha}}\right)^{\frac{1}{1+\beta}}$ then our statement says

$$|p_n'(x)| \leq Cn^{1+\alpha}\|v_{\alpha,\beta}p_n\|. \tag{20}$$

But this follows again from (19), since (19) implies (20) for $1/n \leq |x| \leq 1/2$, and Remez's inequality extends it to $|x| \leq 1/2$.

Thus we have to prove our statement only for $\left(\frac{\log n}{n^{1+\alpha}}\right)^{\frac{1}{1+\beta}} < x \leq 1/2$. The basic tool is the so-called Nevanlinna inequality

$$\log|p_n(x+iy)| \leq \frac{y}{\pi}\int_{-\infty}^{\infty}\frac{\log|p_n(t)|}{(x-t)^2+y^2}\,dt \qquad (x,y \text{ real})$$

valid for all polynomials (cf. Boas [1], Theorem 6.5.5). With the substitution $t = x - uy$ we get

$$\log|p_n(x+iy)| \leq \frac{1}{\pi}\int_{-\infty}^{\infty}\frac{\log|p_n(x-uy)|}{u^2+1}\,du \qquad (x,y; \text{real}).$$

Here we split the integral in three parts:

$$\log|p_n(x+iy)| \leq \frac{1}{\pi}\left(\int_{|x-uy|\geq 1} + \int_{\frac{x-1}{y}}^{\frac{x}{y}} + \int_{\frac{x}{y}}^{\frac{x+1}{y}}\right) := \frac{1}{\pi}(I_1 + I_2 + I_3).$$

First we estimate I_1. Using Chebyshev's inequality and a Schur type inequality from [8] (cf. (7.33)) we obtain

$$|p_n(t)| \leq (2|t|)^n\|p_n\| \leq C(2|t|)^n n^\alpha \||x|^\alpha p_n(x)\| \leq C(3|t|)^n\|v_{\alpha,\beta}p_n\| \qquad (|t| \geq 1).$$

Thus

$$|I_1| \leq \log\|v_{\alpha,\beta}p_n\|\int_{|x-uy|\geq 1}\frac{du}{u^2+1} + n\int_{|x-uy|\geq 1}\frac{\log(3|x-uy|)}{u^2+1}\,du + O(1).$$

Here integrating by parts

$$\int_{|x-uy|\geq 1}\frac{\log(3|x-uy|)}{u^2+1}\,du \leq \int_{|x-uy|\geq 1}\frac{\log(3|x-uy|)}{u^2}\,du =$$

$$= \int_{-\infty}^{\frac{x-1}{y}}\frac{\log[3(x-uy)]}{u^2}\,du + \int_{\frac{x+1}{y}}^{\infty}\frac{\log[3(uy-x)]}{u^2}\,du = = -\frac{\log[3(x-uy)]}{u}\Big|_{-\infty}^{\frac{x-1}{y}} -$$

$$-y\int_{-\infty}^{\frac{x-1}{y}}\frac{du}{u(x-uy)} - \frac{\log[3(uy-x)]}{u}\Big|_{\frac{x+1}{y}}^{\infty} + y\int_{\frac{x+1}{y}}^{\infty}\frac{du}{u(uy-x)} \leq$$

$$\leq \frac{2\log 3}{1-x^2}y + y^2\int_{\frac{x+1}{y}}^{\infty}\frac{du}{(uy-x)^2} = O(y) \qquad (0 < x \leq 3/4).$$

Hence

$$|I_1| \leq \log \|v_{\alpha,\beta} p_n\| \int_{|x-uy|\geq 1} \frac{du}{u^2+1} + O(ny+1) \qquad (0 < x \leq 3/4).$$

We now turn to estimating I_2 and I_3:

$$|I_2| \leq \log \|v_{\alpha,\beta} p_n\| \int_{\frac{x-1}{y}}^{\frac{x}{y}} \frac{du}{u^2+1} - \beta \int_{\frac{x-1}{y}}^{\frac{x}{y}} \frac{\log(x-uy)}{u^2+1} du,$$

and

$$|I_3| \leq \log \|v_{\alpha,\beta} p_n\| \int_{\frac{x}{y}}^{\frac{x+1}{y}} \frac{du}{u^2+1} - \alpha \int_{\frac{x}{y}}^{\frac{x+1}{y}} \frac{\log(uy-x)}{u^2+1} du,$$

whence

$$|I_2 + I_3| \leq \log \|v_{\alpha,\beta} p_n\| \int_{|x-uy|\leq 1} \frac{du}{u^2+1} - \beta \int_{\frac{x-1}{y}}^{\frac{x}{y}} \frac{\log(x-uy)}{u^2+1} du -$$

$$- \alpha \int_{\frac{x}{y}}^{\frac{x+1}{y}} \frac{\log(uy-x)}{u^2+1} du = \log \|v_{\alpha,\beta} p_n\| \int_{|x-uy|\leq 1} \frac{du}{u^2+1} +$$

$$+ \beta \log \frac{1}{x} \int_{|x-uy|\leq 1} \frac{du}{u^2+1} +$$

$$+ \beta \int_{|x-uy|\leq 1} \frac{\log \frac{x}{|x-uy|}}{u^2+1} du + (\alpha - \beta) \int_{\frac{x}{y}}^{\frac{x+1}{y}} \frac{\log \frac{1}{uy-x}}{u^2+1} du.$$

Here

$$\int_{|x-uy|\leq 1} \frac{du}{u^2+1} < \pi,$$

and integrating by parts again

$$\int_{|x-uy|\leq 1} \frac{\log \frac{x}{|x-uy|}}{u^2+1} du \leq \int_{|x-uy|\leq x/2} + \int_{x/2\leq|x-uy|\leq 1} \leq$$

$$\leq \frac{4y^2}{x^2} \int_{\frac{x}{2y}}^{\frac{3x}{2y}} \log \frac{x}{|x-uy|} du + \pi \log 2 = \frac{4y^2}{x^2} \left(\int_{\frac{x}{2y}}^{\frac{x}{y}} + \int_{\frac{x}{y}}^{\frac{3x}{2y}} \right) \log \frac{x}{|x-uy|} du + O(1) =$$

$$= \frac{4y}{x^2} \left\{ \left[(uy-x) \log \frac{x}{x-uy} \right] \Big|_{\frac{x}{2y}}^{\frac{x}{y}} - -y \int_{\frac{x}{2y}}^{\frac{x}{y}} du + \left[(uy-x) \log \frac{x}{uy-x} \right] \Big|_{\frac{x}{y}}^{\frac{3x}{2y}} \right. +$$

$$\left. +y \int_{\frac{x}{y}}^{\frac{3x}{2y}} du \right\} + O(1) = O\left(\frac{y}{x} + 1 \right).$$

Finally,

$$\int_{\frac{x}{y}}^{\frac{x+1}{y}} \frac{\log \frac{1}{uy-x}}{u^2+1}\, du \le -\frac{y^2}{x^2} \int_{\frac{x}{y}}^{\frac{2x}{y}} \log(uy-x)\, du - \int_{\frac{2x}{y}}^{\frac{x+1}{y}} \frac{\log(uy-x)}{u^2}\, du =$$

$$= O\left(\frac{y}{x}\log\frac{1}{x}\right) - \frac{1}{u}\log\frac{1}{uy-x}\Big|_{\frac{2x}{y}}^{\frac{x+1}{y}} - y\int_{\frac{2x}{y}}^{\frac{x+1}{y}} \frac{du}{u(uy-x)} = O\left(\frac{y}{x}\log\frac{1}{x}\right).$$

Collecting these estimates,

$$|I_2 + I_3| \le \log\|v_{\alpha,\beta}\| \int_{|x-uy|\le 1} \frac{du}{u^2+1} + \pi\beta\log\frac{1}{x} + O\left(1 + \frac{y}{x}\log\frac{1}{x}\right).$$

Thus we obtain

$$\log|p_n(x+iy)| \le \log\frac{\|wp_n\|}{x^\beta} + O\left(1 + ny + \frac{y}{x}\log\frac{1}{x}\right) \qquad (0 < x \le 3/4, y > 0).$$

By symmetry, this easily extends to

$$\log|p_n(x+iy)| \le \log\frac{\|v_{\alpha,\beta}p_n\|}{x^\beta} + O\left(1 + n|y| + \frac{|y|}{x}\log\frac{1}{x}\right) \qquad (0 < x \le 3/4, y \text{ real}).$$

$$(21)$$

Now fix an x, $\left(\frac{\log n}{n^{1+\alpha}}\right)^{\frac{1}{1+\beta}} < x \le 1/2$, and let

$$r := \frac{1}{n + \frac{2}{x\log 2}\log\frac{1}{x}} < \frac{x}{2}.$$

Thus if $|z - x| = r$ then $0 < \frac{x}{2} \le \mathrm{Re} \le \frac{3}{2}x \le \frac{3}{4}$, $z \le 1/2$, $|\mathrm{Im}\, z| \le r$ and from Cauchy's integral formula and (21) we get

$$|p_n'(x)| = \frac{1}{2\pi}\left|\oint_{|z-x|=r} \frac{p_n(z)}{(z-x)^2}\, dz\right| \le \frac{1}{r}\max_{|z-x|=r}|p_n(z)|$$

$$= O\left(\frac{n + \frac{1}{x}\log\frac{1}{x}}{x^\beta}\right)\|v_{\alpha,\beta}p_n\|.$$

Hence we get the statement of the theorem in case $\beta < \alpha \le 2\beta + 1$. In order to get it in case $\alpha > 2\beta + 1$, we apply (...) from [8] with the weight $w(x) := |x|^\beta$ on the interval $[0,1]$ to get

$$\left(\frac{\sqrt{x(1-x)}}{n} + \frac{1}{n^2}\right)(x^\beta + n^{-2\beta})|p_n'(x)| \le C\max_{0\le x\le 1}|x|^\beta|p_n(x)| \le \|v_{\alpha,\beta}p_n\|$$

$$\left(-\frac{1}{n^2} \le x \le 1\right).$$

(Here we applied a Remez inequality to extend the inequality from [0,1].) Examining the corresponding subintervals, in this case we can improve the estimate to the second definition of $\varphi_n(x)$.   ∎

References

[1] R. P. Boas, *Entire Functions*, Academic Press (New York, 1954).

[2] G. Criscuolo, B. DellaVecchia, D. S. Lubinsky and G. Mastroianni, *Functions of the second kind for Freud weights and series expansions of Hilbert transforms,* Journal of Math. Anal. and Appl., **189** (1995), 256-296.

[3] Z. Ditzian and V. Totik, *Moduli of Smoothness*, Springer Verlag (New York, 1987).

[4] Á. Horváth and J. Szabados, *Polynomial approximation and interpolation on the real line with respect to general classes of weights,* Results in Mathematics, **34** (1998), 120-131.

[5] D. S. Lubinsky and E. B. Saff, *Strong Asymptotics for Extremal Polynomials Associated with Weights on* **R**, Lecture Notes in Mathematics, Vol. 1305, Springer Verlag (Berlin, 1988).

[6] D. S. Lubinsky, *Jackson and Bernstein theorems for exponential weights,* in Approximation Theory and Function Series, Bolyai Society Mathematical Studies, Vol. 5 (Budapest, 1996), pp. 85-115.

[7] G. Mastroianni and J. Szabados, *Jackson-type theorems on a finite interval with weights having non-symmetric inner singularities,* Acta Math. Hungar., **83** (1999) (to appear).

[8] G. Mastroianni and V. Totik, *Weighted polynomial inequalities with doubling and A_∞ weights,* Constructive Approximation, submitted.

[9] H. N. Mhaskar and E. B. Saff, *Where does the sup-norm of a weighted polynomial live?,* Constr. Approx., **1** (1985), 71-91.

[10] J. Szabados, *Weighted Lagrange and Hermite–Fejér interpolation on the real line,* Journal Inequ. Appl., **1** (1997), 99-123.

Mathematical Institute
of the Hungarian Academy of Sciences
H-1364 Budapest, POB 127
Email address: szabados@math-inst.hu

International Series of Numerical Mathematics
Vol. 132, © 1999 Birkhäuser Verlag Basel/Switzerland

Asymptotics of derivatives of orthogonal polynomials based on generalized Jacobi weights. Some new theorems and applications

P. Vértesi*

1 Introduction. Preliminary result

1.1

In this paper we state uniform asymptotic formulae for $n \to \infty$ valid on the unit circle line for derivatives of (complex) orthogonal polynomials based on Jacobi-type weights with a finite number of power type singularities. By these results we settle the corresponding problems for generalized Jacobi polynomials on $[-1, 1]$, i.e. for the (real) orthogonal polynomials based on power type weights with inner singularities.

We mention several applications,too.

The results are based on the works of G. Szegő [1] and V. M. Badkov [2] where the corresponding relations were obtained for the polynomials themselves.The detailed proofs will appear in the near future.

1.2

Everywhere below $\mathbb{C}$ is the complex plane, $\mathbb{R} = (-\infty, \infty)$, $\mathbb{N} = \{1, 2, \ldots\}$, $Z_+ = \{0, 1, 2, \ldots\}$.

Throughout this paper $c, c_1, c_2 \ldots$ denote positive constants; they may take different values even in subsequent formulae. It will always be clear what variables and indices the constants are independent of.

If r and s are two expressions depending on some variables then we write

$$r \sim s \quad \text{iff} \quad |r\,s^{-1}| \le c_1 \quad \text{and} \quad |r^{-1}s| \le c_2$$

uniformly for the variables in consideration.

[1]Research supported by Hungarian National Science Foundation Grant Ns. T7570, T22943, T17425 and by the Hungarian Academy of Sciences Grant 96-328/11. Version completed: July 19, 1998.

Let $L^p[a, b]$ denote the space of Lebesgue measurable complex valued functions f where

$$
\left.
\begin{aligned}
\|f\|_{L^p[a,b]} &:= \left\{ \int_a^b |f(t)|^p dt \right\}^{1/p}, & 1 \le p < \infty, \\
\|f\|_{L^\infty[a,b]} &:= \underset{a \le t \le b}{\text{ess sup}} \ |f(t)|, & p = \infty
\end{aligned}
\right\}
\tag{1}
$$

is finite. If $[a, b] = [-1, 1]$, we use the short notations L^p, $\|f\|_{L^p}$, a.s.o.; if $f \in C[a, b]$ (= the space of continuous functions on $[a, b]$) then $\|f\|_{[a,b]}$ stands for $\|f\|_{L^\infty[a,b]}$; if $[a, b] = [-1, 1]$ we write $\|f\|$. If g is of 2π-periodic, then we use notations L_p, L_∞, $\|g\|_p$ and $\|g\|_\infty$, respectively; finally if $g \in \tilde{C}$, i.e. g is continuous and of 2π-periodic, we write $\|g\|$ for $\|g\|_\infty$.

1.3

A function $F(x) \in L^1$ is an *(algebraic) weight* $(F \in \mathcal{AW})$ iff $F(x) \ge 0$, $F(x) \not\equiv 0$ $(x \in [-1, 1])$. The unique system of algebraic polynomials $\{F_n(F, x)\}_{n=0}^\infty$, $x \in [-1, 1]$, defined by

$$
\left.
\begin{aligned}
F_n(F, x) &= \gamma_n(F)x^n + \text{ lower degree terms (l.d.t.)}, \ \gamma_n(F) > 0, \ n \in \mathbb{Z}_+, \\
\int_{-1}^1 F_n(F, x)F_m(F, x)F(x)dx &= \delta_{nm},
\end{aligned}
\right\} \quad n, m \in \mathbb{Z}_+,
\tag{2}
$$

forms the *orthonormal polynomials* (ONP) *with respect to* $F \in \mathcal{AW}$.

A point x_0 is *regular* for $F \in \mathcal{AW}$ iff for some $\varepsilon > 0$ the essential suprema of F and $1/F$ on $[-1, 1] \cap [x_0 - \varepsilon, x_0 + \varepsilon]$ are finite. Otherwise, F has a *singularity* at x_0.

Similarly, $\Phi(\vartheta) \in \mathcal{TW}$ (trigonometric weight) iff $\Phi \in L_1$, $\Phi(\vartheta) \ge 0$, $\Phi(\vartheta) \not\equiv 0$ $(\vartheta \in \mathbb{R})$. The corresponding unique ONP $\{\Phi_n(\Phi, z)\}_{n=0}^\infty$ with respect to $\Phi \in \mathcal{TW}$ on the unit circle line $\partial \mathcal{U}$ are defined by relations

$$
\left.
\begin{aligned}
\Phi_n(\Phi, z) &= \chi_n(\Phi)z^n + \text{l.d.t.}, \ \chi_n(\Phi) > 0, \ z \in \mathbb{C}, \ n \in \mathbb{Z}_+, \\
\tfrac{1}{2\pi} \int_0^{2\pi} \Phi_n(\Phi, z) \overline{\Phi_m(\Phi, z)} \, \Phi(\vartheta)d\vartheta &= \delta_{nm}; \ z = e^{i\vartheta}, \ n, m \in \mathbb{Z}_+.
\end{aligned}
\right\}
\tag{3}
$$

We define the singular and regular points of Φ as we did for F.

G. Szegő [1, 11.5] established a close connection between $\{F_n(F)\}$ and $\{\Phi_n(\Phi)\}$. For example if $F \in \mathcal{AW}$ and with $x = \cos\vartheta$, $0 \le \vartheta \le \pi$,

$$
\Phi(\vartheta) := F(\cos\vartheta)|\sin\vartheta|,
\tag{4}
$$

then $\Phi \in \mathcal{TW}$, moreover if $a_{n-1}(\Phi) := -\Phi_n(\Phi, 0)/\chi_n(\Phi)$, $v(x) = \sqrt{1-x^2}$ and

$$
P_n^*(z) := z^n \overline{P_n}(z^{-1}), \qquad P_n \in \mathcal{P}_n \setminus \mathcal{P}_{n-1}, \ z \in \mathbb{C},
\tag{5}
$$

(where $\overline{P_n}$ denotes the polynomial whose coefficients are the complex conjugates of the corresponding coefficients of P_n), the relations

$$\left.\begin{array}{l} F_n\left(F, \frac{1}{2}(z+\frac{1}{z})\right) = \frac{\Phi_{2n}(\Phi,z)+\Phi^*_{2n}(\Phi,z)}{\sqrt{2\pi(1-a_{2n-1}(\Phi))}} \cdot z^{-n}, \\[2mm] F_{n-1}\left(Fv^2, \frac{1}{2}(z+\frac{1}{z})\right) = \frac{\Phi_{2n}(\Phi,z)-\Phi^*_{2n}(\Phi,z)}{\sqrt{2\pi(1+a_{2n-1}(\Phi))}} \cdot \frac{z^{-n}}{\frac{1}{2}(z-\frac{1}{z})} \end{array}\right\} \tag{6}$$

hold true for $z \in \mathbb{C}$, $z \neq 0$ and $n \in \mathbb{N}$. Notice that if $z = e^{i\vartheta}$ then $2^{-1}(z+z^{-1}) = \cos\vartheta = x \in [-1,1]$ and $2^{-1}(z-z^{-1}) = i\sin\vartheta$ (cf. [1, (11.5.2)] for similar formulae).

1.4

In applications getting asymptotic expressions for the orthonormal polynomials $F_n(F)$ or $\Phi_n(\Phi)$ when $n \to \infty$ is of fundamental importance.

From the extensive literature we refer to G. Szegő [1], Ya. L. Geronimus [4] and G. Freud [5]. However, the problem of finding asymptotics for the derivatives seems to be more difficult (see e.g. P. G. Nevai [6] and the references therein).

The present paper is concerned with the latter problem on the unit circle line ∂U and on the interval $[-1,1]$ for ONP based on generalized Jacobi weights (see (7)).

Before stating our results we give some further definitions and preliminary results with some hints of their proofs.

Definition 1.1 $F \in GJ$ *(generalized Jacobi weights) iff for $x \in [-1,1]$*

$$F(x) \equiv r(H, \mathbf{\Gamma}, t; x) := H(x)(1-x)^{\Gamma_0}(1+x)^{\Gamma_{m+1}} \prod_{k=1}^{m} |x-t_k|^{\Gamma_k}; \tag{7}$$

(i) $-1 \equiv t_{m+1} < t_m < \ldots < t_1 < t_0 \equiv 1$, $\Gamma_0, \Gamma_1, \ldots \Gamma_{m+1} > -1$;

(ii) $H \in C, H(x) > 0$ *on* $[-1,1]$ *and with* $h(\vartheta) := H(\cos\vartheta)$, $\omega(h,\delta)_{[0,2\pi]}\delta^{-1} \in L_1$.

Here $\omega(f,\delta) := \omega(f,\delta)_{[0,2\pi]}$ is the usual modulus of continuity of f on $[0,2\pi]$, that means,

$$\omega(f,\delta)_{[a,b]} := \sup\{|f(x)-f(y)|; \; |x-y| \leq \delta; \; x,y \in [a,b]\}.$$

If $H \equiv 1$, and $\Gamma_k = 0$ when $1 \leq k \leq m$ (no inner singularity!), then we use the notation

$$p^{(\alpha,\beta)}(x) = (1-x)^\alpha(1+x)^\beta, \qquad \alpha, \beta > -1, \tag{8}$$

for the classical Jacobi weights J. Clearly $J \subset GJ \subset AW$. Further, in what follows, $\{r_n(H,\mathbf{\Gamma},t;x)\}$ and $\{p_n^{(\alpha,\beta)}(x)\}$ will denote the corresponding ONP, respectively.

Similarly, we have the following

Definition 1.2 $\Phi \in TGJ$ *(trigonometric generalized Jacobi weight) iff for* $\vartheta \in \mathbb{R}$

$$\Phi(\vartheta) \equiv \varphi(h, \boldsymbol{\Delta}, \boldsymbol{\tau}; \vartheta) := h(\vartheta) \prod_{k=1}^{\mu} \left(\sin \frac{|\vartheta - \tau_k|}{2} \right)^{\Delta_k}; \qquad (9)$$

(i*) $-\pi < \tau_1 < \tau_2 < \ldots < \tau_\mu \leq \pi, \quad \Delta_1, \Delta_2, \ldots, \Delta_\mu > -1;$

(ii*) $h \in \widetilde{C}, \; h > 0$ *on* $\mathbb{R}$ *and* $\omega(h, \delta)_{[0,2\pi]}\delta^{-1} \in L_1.$

The special case

$$\begin{aligned}
\Phi(\vartheta) &= \left(2\sin^2 \frac{\vartheta}{2} \right)^{\alpha+1/2} \left(2\sin^2 \frac{\vartheta - \pi}{2} \right)^{\beta+1/2} \\
&= (1 - \cos\vartheta)^{\alpha+1/2}(1 + \cos\vartheta)^{\beta+1/2} \quad (\alpha, \beta > -1)
\end{aligned}$$

will be denoted by $\psi^{(\alpha,\beta)}(\vartheta)$, while the corresponding ONP are $\varphi_n(h, \boldsymbol{\Delta}, \boldsymbol{\tau}; z)$ and $\psi_n^{(\alpha,\beta)}(z)$, respectively. Again, $TJ \subset TGJ \subset TW$ (where TJ is the set of the weights $\psi^{(\alpha,\beta)}; \; \alpha, \beta > -1$).

1.5

In 1933 Gábor Szegő essentially proved the following asymptotic formula (cf. [1, (8.21.18)] and [2, (2.34)]). With $x = \cos\vartheta$

$$p_n^{(\alpha,\beta)}(x) = \sqrt{\frac{2}{\pi}} \frac{\cos(N\vartheta + \gamma) + O\left(\frac{1}{n\sin\vartheta}\right)}{p^{\left(\frac{\alpha}{2}+\frac{1}{4}, \frac{\beta}{2}+\frac{1}{4}\right)}(x)}, \quad n \in \mathbb{N}, \; n\sin\vartheta \geq c, \qquad (10)$$

where $c > 0$ is arbitrary, but fixed, $N = n + \frac{\alpha+\beta+1}{2}$ and $\gamma = -\left(\alpha + \frac{1}{2}\right)\frac{\pi}{2}$. As it is well-known, the remainder term $(n\sin\vartheta)^{-1}$ is the best possible in order. The symbol "O" depends on c, α and β.

This is a very important and useful relation. It can be used, for example, to get fairly precise formulae for the roots $x_{kn}^{(\alpha,\beta)} = \cos\vartheta_{kn}^{(\alpha,\beta)}$ of $p_n^{(\alpha,\beta)}$, to investigate many questions concerning with Lagrange-, Hermite- and Hermite-Fejér type interpolations (cf. P. Vértesi [7]–[9], say).

Similar formulae but *without* precise error estimate were obtained by S. Bernstein [10] for the ONP based on the weight $r(x) = H(x)p^{(\alpha,\beta)}(x)$ with $h(\vartheta) = H(\cos\vartheta)$ satisfying (ii).

1.6

In 1983, V. M. Badkov [2] obtained some fairly good asymptotic formulae for φ_n and r_n. He used — among others — relations (10) and (6) to get formulae for

$\psi_n^{(\alpha,-1/2)}(z)$ (based on the weight $\psi^{(\alpha,-1/2)}(\vartheta) = (1 - \cos\vartheta)^{\alpha+1/2}$); then by the simple but useful relation

$$\Phi_n(\Phi_{(\eta)}, z) = e^{in\eta}\Phi_n(\Phi, e^{-i\eta}z), \quad n \in \mathbb{Z}_+, \; z \in \mathbb{C} \tag{11}$$

where $\Phi_{(\eta)} := \Phi(\vartheta - \eta)$, $\eta \in \mathbb{R}$ is fixed (shifted weight), he obtained asymptotics for the ONP based on $\psi^{(\alpha,-1/2)}(\vartheta - \eta)$ (notice that the *only singularity* of the weight $\psi_{(\eta)}^{(\alpha,-1/2)}(\vartheta) \equiv \psi^{(\alpha,-1/2)}(\vartheta - \eta)$ is at $\vartheta = \eta$). Finally, by his statement ([2, §4]), which roughly speaking says that the behaviour of $\varphi_n(e^{i\vartheta})$ in a closed subinterval of the unit cicle line which contains the singularity η only of the weight $\varphi(\vartheta)$, can be described by $\Phi_n(\psi_{(\eta)}^{(\alpha,-1/2)}, e^{i\vartheta})$, he obtained the desired formulae of $\varphi_n(z)$, $z = e^{i\vartheta}$. Finally, using again (6), one gets the formulae for $r_n(x)$, too.

1.7

Here we introduce some further notations. The weight $\Phi \in \mathcal{TW}$ is from the *Szegő-class* S iff $\log\Phi(\vartheta) \in L_1$. Let $\Phi \in S$. The function

$$D(\Phi, z) := \exp\left\{\frac{1}{4\pi}\int_0^{2\pi}\frac{e^{i\tau} + z}{e^{i\tau} - z}\log\Phi(\tau)d\tau\right\}, \quad |z| < 1, \tag{12}$$

nowadays called *Szegő function*, is an analytic function for $|z| < 1$; if $z \in \partial\mathcal{U}$ (unit circle line), let

$$D(\Phi, e^{i\vartheta}) := \lim_{r\to 1-0}D(\Phi, re^{i\vartheta}). \tag{13}$$

This limit exists a.e. (almost everywhere) in ϑ, moreover

$$|D(\Phi, e^{i\vartheta})|^2 = \Phi(\vartheta) \quad \text{a.e. on } \partial\mathcal{U}; \tag{14}$$

if $\Phi(\vartheta)$ satisfies (ii*) then (14) uniformly holds (in ϑ) and $D(\Phi, e^{i\vartheta})$ is continuous (cf. [1, Ch. X] and [2, Part 1] for other details).

By definition, $TGJ \subset S$. Now, for later purposes, we investigate certain $D(\Phi, z)$ for special $\Phi \in TGJ$.

Let, as before,

$$\Phi_{(\eta)}(\vartheta) = |e^{i\vartheta} - e^{i\eta}| = 2\sin\frac{|\vartheta - \eta|}{2}, \quad \eta \in \mathbb{R}, \text{ fixed.} \tag{15}$$

Then, with $\chi = \Phi_{(\eta)}^2$

$$D(\chi, z) = 1 - ze^{-i\eta}, \quad |z| < 1,$$

whence, with $\gamma = \vartheta - \eta$ (cf. (12)–(14) and [1, (10.2.13)], [5, Ch. V, Lemma 3.1]),

$$D(\chi, e^{i\vartheta}) = 1 - e^{i\gamma} = e^{\frac{i\gamma}{2}}\left(e^{-\frac{i\gamma}{2}} - e^{\frac{i\gamma}{2}}\right) = -2ie^{\frac{i\gamma}{2}}\sin\frac{\gamma}{2} = 2\sin\frac{|\gamma|}{2}e^{\frac{i(\gamma - \pi\text{sign}\gamma)}{2}}.$$

So using properties

$$\left.\begin{array}{l} D(\Phi\Psi, z) = D(\Phi, z)D(\Psi, z), \\ D(\Phi^\nu, z) = \{D(\Phi, z)\}^\nu, \quad \nu \in \mathbb{C}, \ \nu \neq 0, \end{array}\right\} \tag{16}$$

([1, (10.2.11)]) we arrive at relations

$$\left.\begin{array}{l} |D(\Phi_{(\eta)}, e^{i\vartheta})| = \left(2\sin\frac{|\vartheta-\eta|}{2}\right)^{1/2}, \\ \arg\left\{D\left(\Phi_{(\eta)}, e^{i\vartheta}\right)\right\} = \frac{1}{4}\{\vartheta - \eta - \pi\mathrm{sign}(\vartheta-\eta)\}, \quad 0 < |\vartheta - \eta| < 2\pi \end{array}\right\} \tag{17}$$

(cf. [2, (6.8)]). By (16) and (17)

$$\left.\begin{array}{ll} D\left(\psi^{(\alpha,\beta)}, e^{i\vartheta}\right) = \left(\psi^{(\alpha,\beta)}(\vartheta)\right)^{1/2} e^{i\left\{\frac{\alpha+\beta+1}{2}\vartheta - (\alpha+1/2)\frac{\pi}{2}\right\}}, & 0 < \vartheta < \pi, \\ D(\Phi_{(\eta)}\Phi_{(-\eta)}, e^{i\vartheta}) = |\cos\vartheta - \cos\eta|^{1/2} e^{i\frac{\vartheta}{2}}, & 0 \le \vartheta < \eta < \pi, \\ D(\Phi_{(\eta)}\Phi_{(-\eta)}, e^{i\vartheta}) = |\cos\vartheta - \cos\eta|^{1/2} e^{i\frac{\vartheta-\pi}{2}}, & 0 < \eta < \vartheta \le \pi. \end{array}\right\} \tag{18}$$

2 New results

2.1

Let the weight function be of the form

$$\varphi^{(\alpha,\beta)}(h, \vartheta) = h(\vartheta)\psi^{(\alpha,\beta)}(\vartheta), \quad \alpha, \beta > -1 \tag{19}$$

with h satisfying (ii*) and let $\varphi_n^{(\alpha,\beta)}(h, z)$ denote the ONP $\Phi_n(\varphi^{(\alpha,\beta)}(h), z)$.

Theorem 2.1 *Let $s \in Z_+$ be fixed. If for the sequence $\{\lambda_n\}$,*

$$2 \le n\lambda_n \le n \quad if \quad n \ge 2 \quad and \quad \lambda_n \searrow 0 \quad when \quad n \to \infty, \tag{20}$$

then, as $n \to \infty$, we have

$$\frac{d^s}{d\vartheta^s}\left\{\varphi_n^{(\alpha,\beta)}(h, e^{i\vartheta})\right\} = \frac{\frac{d^s}{d\vartheta^s}\left\{\psi_n^{(\alpha,\beta)}(e^{i\vartheta})\right\}}{D(h, e^{i\vartheta})}\{1 + O(\nu_{ns}(h))\} \tag{21}$$

uniformly in $\vartheta \in \mathbb{R}$. Here

$$\nu_{n0}(h) = \omega\left(h, \frac{1}{n}\right)\log n + \int_0^{\lambda_n} \frac{\omega(h, \tau)}{\tau}d\tau$$

$$+ \left(\frac{1}{\sqrt{n}} + \omega\left(h, \frac{1}{n}\right)\right)\left(\int_{\lambda_n}^\pi \frac{\omega^2(h, \tau)}{\tau^2}d\tau\right)^{1/2};$$

while for $s \ge 1$

$$\nu_{ns}(h) = \frac{1}{\sqrt{n}},$$

supposing that

(iii*) $\|h^{(s)}\| \le c.$

Let us remark that by (ii*) the sequence λ_n can always be chosen so that it satisfies (20) and $\nu_{n0}(h) \to 0$ (cf. [2, Theorem 1 and (1.32)]).

2.2

Let η_k $(k = 1, 2, \ldots, \mu)$ be arbitrary fixed numbers satisfying

$$-\pi < \eta_1 < \tau_1 < \eta_2 < \ldots < \eta_\mu < \tau_\mu \leq \pi < \eta_{\mu+1} := \eta_1 + 2\pi < \tau_{\mu+1} := \tau_1 + 2\pi,$$

otherwise arbitrary (for τ_k, see (i*) at Point 1.5). Using the notations of Point 1.5, we state a generalization of [2, (6.5)].
Let

$$\chi_{n0}(h) = \omega\left(h, \frac{1}{n}\right) \log n + \int_0^{\lambda_n} \frac{\omega(h, \tau)}{\tau} d\tau$$
$$+ \left(\frac{1}{\sqrt{n}} + \omega\left(h, \frac{1}{n}\right)\right)\left(1 + \int_{\lambda_n}^{\pi} \frac{\omega^2(h, \tau)}{\tau^2} d\tau\right),$$

whence $\chi_{n0}(h) \to 0$ if $\{\lambda_n\}$ is suitably chosen (cf. [2, (1.37)]). Further let

$$\chi_{ns}(h) = \frac{1}{\sqrt{n}} \quad \text{if } s \geq 1.$$

Theorem 2.2 *Let* $\varphi(\vartheta) = \varphi(h, \Delta, \tau; \vartheta)$ *and* $\{\lambda_n\}$ *be defined by* (20), (9), (i*), (ii*), *moreover by* (iii*), *too, if* $s \geq 1$. *Then with* $\varphi_n = \varphi_n(\varphi)$, *we get for* $s \geq 0$, $1 \leq \ell \leq \mu$

$$\frac{d^s \varphi_n(e^{i\theta})}{d\vartheta^s} = (in)^s \frac{e^{in\vartheta}}{D(\varphi, e^{i\vartheta})} \left\{1 + O\left(\chi_{ns}(h) + \frac{1}{n|\vartheta - \tau_\ell|}\right)\right\}, \quad (22)$$

uniformly in $n \in \mathbb{N}$, ℓ *and* $\vartheta \in T_{\ell n}(c)$. *Here* $T_{\ell n}(c) = \left[\eta_\ell, \tau_\ell - \frac{c}{n}\right] \cup \left[\tau_\ell + \frac{c}{n}, \eta_{\ell+1}\right]$, $c > 0$ *is arbitrary fixed.* $(1 \leq \ell \leq \mu, n \in \mathbb{N}.)$

Using this statement one can obtain

Theorem 2.3 *By the notations and conditions of Theorem 2.2, we get for* $s \geq 0$, $1 \leq \ell \leq \mu$

$$\frac{d^s \varphi_n(e^{i\vartheta})}{d\vartheta^s} = (in)^s \varphi_n(e^{i\vartheta}) \left\{1 + O\left(\chi_{ns}(h) + \frac{1}{n|\vartheta - \tau_\ell|}\right)\right\} \quad (23)$$

uniformly in $n \in \mathbb{N}$, μ *and* $\vartheta \in T_{\ell n}(c)$.

2.3

Here we consider the uniform asymptotic representation of the polynomials $r_n^{(s)}(x)$ defined by using (7), (i), (ii).

Let $r(x) \in GJ$ be defined by (7), (i), and (ii). Then

$$\varphi(\vartheta) = r(\cos\vartheta)|\sin\vartheta|, \quad x = \cos\vartheta, \ 0 \le \vartheta \le \pi,$$

is an even TGJ with the singularities $\tau_0 = 0$, $\tau_{m+1} = \pi$, $\pm\tau_k$ $(1 \le k \le m)$; $t_k = \cos\tau_k$, $0 \le \tau_k \le \pi$ and the exponents $2\Gamma_0 + 1$, $2\Gamma_{m+1} + 1$, Γ_k $(1 \le k \le m)$, respectively. Now all the coefficients of $\varphi_n(\varphi)$ are real and

$$r_n(r, x) = \alpha_n Re\left\{e^{-in\vartheta}\varphi_{2n}(\varphi, e^{i\vartheta})\right\}, \tag{24}$$

where $\alpha_n = \left(\frac{2}{\pi}\right)^{1/2}\left(1 + O\left(\frac{1}{\sqrt{n}}\right)\right)$ (cf. (5), (6), [1, (11.5.2)] and [2, (6.19)]). Using Theorem 2.2 and Parts 1.6–1.7 one can get

Theorem 2.4 *By the previous notations and conditions for any fixed $s \ge 0$*

$$\frac{d^s r_n(r, \cos\vartheta)}{d\vartheta^s} = \sqrt{\frac{2}{\pi}}n^s \frac{\cos\left(n\vartheta + \varepsilon_\ell(\vartheta) + s\frac{\pi}{2}\right) + O\left(\chi_{n0}(h) + \frac{1}{n|\vartheta - \tau_\ell|}\right)}{(\varphi(\vartheta))^{1/2}}, \tag{25}$$

uniformly in $n \in \mathbb{N}$, $0 \le \ell \le m + 1$, and $\vartheta \in T_{\ell n}(c) \cap [0, \pi]$. Here $h(\vartheta) = H(\cos\vartheta)$ and with $a(\vartheta) := \arg\{D(h, e^{i\vartheta})\}$

$$\varepsilon_\ell(\vartheta) = a(\vartheta) + \left(1 + \sum_{k=0}^{m+1}\Gamma_k\right)\frac{\vartheta}{2} - \left(\frac{1}{2} + \sum_{k=0}^{L}\Gamma_k\right)\frac{\pi}{2}, \tag{26}$$

$$L = \begin{cases} \ell & \text{if } \tau_\ell < \vartheta \le \eta_{\ell+1}, \\ \ell - 1 & \text{if } \eta_\ell \le \vartheta < \tau_\ell. \end{cases} \tag{27}$$

Moreover, if we apply derivatives with respect to $x = \cos\vartheta$ on the left-hand side $(0 \le \vartheta \le \pi)$, (25) can be replaced by

$$r_n^{(s)}(r, x) = \sqrt{\frac{2}{\pi}}\left(\frac{-1}{\sin\vartheta}\right)^s$$

$$\times n^s \frac{\cos\left(n\vartheta + \varepsilon_\ell(\vartheta) + s\frac{\pi}{2}\right) + O\left(\chi_{n0}(h) + \frac{1}{n|\vartheta - \tau_\ell|}\right)}{(\varphi(\vartheta))^{1/2}}. \tag{28}$$

Remark 2.5 *By definition, if $E_{\ell n}(c) := \left[\tau_\ell + \frac{c}{n}, \tau_{\ell+1} - \frac{c}{n}\right]$ $(0 \le \ell \le m)$, formulae (25)–(28) are valid for $\vartheta \in E_{\ell n}(c)$ writing*

a) $L = \ell$ in $\varepsilon_\ell(\vartheta)$,

b) $(n|\vartheta - \tau_\ell|\,|\vartheta - \tau_{\ell+1}|)^{-1}$ instead of $(n|\vartheta - \tau_\ell|)^{-1}$.

Remark 2.6 *As it is well known if h satisfies* (ii*) *then*

$$\arg\{D(h, e^{i\vartheta}\} = \frac{1}{4\pi} \int_{-\pi}^{\pi} (\log h(\tau) - \log h(\vartheta)) \cot \frac{\vartheta - \tau}{2} d\tau, \quad \vartheta \in \mathbb{R}$$

(cf. [1, §10.3]).

2.4

Many other results in Badkov [2] can also be generalized using the method of the present paper (cf. [2, Theorem 2, (3.56), (3.58), (5.16), (6.3), (6.16), (6.17), (6.22)–(6.29), (6.32)–(6.34), (6.36)]). They may be considered by the interested reader.

3   Applications

3.1

First we give some fairly precise formulae on the roots $x_{kn}(r) = \cos \vartheta_{kn}(r)$ of $r_n(r)$ (cf. Vértesi [8]). Let

$$N(r) = n + \frac{1 + \sum_{k=0}^{m+1} \Gamma_k}{2} \tag{29}$$

$$\gamma(r, \ell, \vartheta) = \left(-\frac{1}{2} - \sum_{k=0}^{\ell} \Gamma_k \right) \frac{\pi}{2} + a(\vartheta), \qquad 0 \le \ell \le m. \tag{30}$$

Then, using the previous notations and conditions, we have

Theorem 3.1 *Let $\vartheta_{kn}(r) \in E_{\ell n}(c_0)$ $(0 \le \ell \le m, n \ge 1)$. Then with $K = k + A(\ell, n)$, $|A(\ell, n)| \le c$,*

$$\vartheta_{kn}(r) = \Theta_{Kn}(r, \ell) + \delta_{n0}, \tag{31}$$

where

$$\Theta_{kn}(r, \ell) = \frac{2k - 1}{2N(r)} \pi - \frac{\gamma(r, \ell, \vartheta)}{N(r)} \tag{32}$$

and for $\delta_{ns} = \delta_{ns}(r, h, \ell)$ we have

$$|\delta_{ns}| = O\left(\frac{\chi_{n0}(h)}{n} + \frac{1}{n^2 |\Theta_{kn} - \tau_\ell||\Theta_{kn} - \tau_{\ell+1}|} \right), \qquad s \ge 0.$$

Let $M = M(s) = n - s$ and denote by $y_{kM}(r) = \cos \xi_{kM}(r)$ the roots of $r_n^{(s)}(x)$ $(s \ge 0)$, by the previous notations and conditions we get for every fixed $s \ge 0$ a generalization of Theorem 3.1. Namely

Theorem 3.2 *For the fixed $s \geq 1$, let $\xi_{kM}(r) \in E_{\ell n}(c_0)$ $(0 \leq \ell \leq m,\ M \geq 1)$ and let us suppose (iii*) for $h^{(s)}$. Then*

$$\xi_{kM}(r) = \Theta_{Kn}(r, \ell) + s\frac{\pi}{2N(r)} + \delta_{ns}. \qquad (33)$$

Remark 3.3 *The formulae (30)–(32) show that the estimation*

$$\xi_{k+1,M} - \xi_{kM} = \frac{\pi}{N(r)}(1 + o(1))$$

holds for $n(1 - o(1))$ roots of $r_n^{(s)}$ $(s \geq 0,$ fixed). For the remaining ones we can use the relation

$$\xi_{k+1,M} - \xi_{kM} \sim \frac{1}{n}, \qquad 0 \leq k \leq M$$

(cf. G. Mastroianni, P. Vértesi [13, Theorem 3.2 and its proof]).

3.2

Finally we generalize a result in P. Vértesi [7]. Let $CBV := \{f;\ f$ is continuous and of bounded variation on $[-1, 1]\}$. If $L_n(f, r, x)$ denotes the n-th Lagrange interpolatory polynomials based on the n roots of $r_n(r)$, we state

Theorem 3.4 *Let $f \in CBV$. Then*

$$\lim_{n \to \infty} \|L_n(f, r) - f\| = 0 \qquad (34)$$

whenever

$$-1 < \Gamma_0, \Gamma_{m+1} < 1/2 \quad and \quad 0 \leq \Gamma_k < 2, \qquad 1 \leq k \leq m. \qquad (35)$$

Remark 3.5 *As it is well known, for a proper continuous f, $\overline{\lim_{n \to \infty}}\|L_n(f, r)\| = \infty$ (Faber theorem).*

4   On the proofs

The detailed proofs of Theorems 2.1–2.4 are rather long (about 25 pages) and use many technical details. The main ingredients are some proper generalizations of the ideas and formulae in subsections 1.5–1.7. They appear soon in [14].

The verification of Theorems 3.1–3.2 uses (25)–(28) and many ideas from [8]. Finally, Theorem 3.4 can be obtained by Theorems 2.4, 3.1 and using the methods in [7]. The detailed proofs will appear soon in our forthcoming paper with B. Della Vecchia and G. Mastroianni.

References

[1] G. Szegő, *Orthogonal Polynomials,* AMS Coll. Publ., Vol. 23, Providence, RI, 1975 (4th ed).

[2] V. M. Badkov, *Uniform asymptotic representation of orthogonal polynomials,* Proc. Steklov Inst. Math., **2** (1985), 5–41.

[3] G. Szegő, *On bi-orthogonal systems of trigonometric polynomials,* MTA Matematikai Kutató Int. Közleményei, **8** (1963/64), 255–273.

[4] Ya. L. Geronimus, *Orthogonal Polynomials,* Pergamon Press, 1960.

[5] G. Freud, *Orthogonal Polynomials,* Pergamon Press, 1971.

[6] P. G. Nevai, *An asymptotic formula for the derivatives of orthogonal polynomials,* SIAM J. Math. Anal., **10** (3) (1979), 472–477.

[7] P. Vértesi, *One-sided convergence conditions for Lagrange interpolation based on Jacobi nodes,* Acta Sci. Math. (Szeged), **45** (1983), 419–428.

[8] P. Vértesi, *On the zeros of Jacobi polynomials,* Studia Sci. Math. Hungar., **25** (1990), 401–405.

[9] P. Vértesi, *Recent results on Hermite–Fejér interpolations of higher order,* Israel Math. Conf., Proc., **4** (1991), 267–271.

[10] N. Bernstein, *On polynomials orthogonal on a finite interval. No. 51 in his Collected Works,* Vol. 2, pp. 7–106, Izd. AN SSS, 1954 (Russian).

[11] J. Szabados and P. Vértesi, *Interpolation of functions,* World Scientific Ltd., 1990.

[12] V. M. Badkov, *On the asymptotic and extremal properties of orthogonal polynomials...,* Proc. Steklov Inst. Math., 1994 (1), 37–82.

[13] G. Mastroianni and P. Vértesi, *Some applications of Generalized Jacobi weights,* Acta Math. Hungar. **77** (1997), 323–357.

[14] P. Vértesi, *Uniform asymptotics of derivatives of orthogonal polynomials based on generalized Jacobi weights,* Acta Math. Hungar. **85** (1–2), 1999.

Mathematical Institute
of the Hungarian Academy of Sciences
H–1364 Budapest, P.O.B. 127
Hungary
Email address: veter@math-inst.hu

List of Participants

Altomare, Francesco, University of Bari, Department of Mathematics, Campus Universitario, Via E. Orabona 4, 70125 Bari, Italy;
e-mail: altomare@pascal.dm.uniba.it

Attalienti, Antonio, Istituto di Matematica Finanziaria, Facolta di Econom, Via camillo Rosalba 53, 70124 Bari, Italy;
e-mail: albano@vm.tno.it

Bastian-Walther, Marion, Technische Universität Dresden, Institut für Numerische Mathematik, 01062 Dresden, Germany;
e-mail: walther@math.tu-dresden.de

Bejancu, Aurelian, D.A.M.T.P., Silver Street, Cambridge, CB3 9EW, England;
e-mail: ab223@damtp.cam.ac.uk

Braess, Dietrich, Ruhr-Universität Bochum, Fakultät für Mathematik, 44780 Bochum, Germany;
e-mail: braess@num.ruhr-uni-bochum.de

Buhmann, Martin D., Universität Dortmund; Lehrstuhl VIII für Mathematik, Vogelpothsweg 87, 44221 Dortmund, Germany;
e-mail: mdb@math.uni-dortmund.de

Butzer, Paul L., RWTH Aachen, Lehrstuhl A für Mathematik, Templergraben 55, 52056 Aachen, Germany;
e-mail: ly010bu@dacth11.bitnet

Campiti, Michele, University of Bari, Department of Mathematics, Campus Universitario, Via E. Orabona 4, 70125 Bari, Italy;
e-mail: campiti@dm.uniba.it

Davydov, Oleg, Universität Dortmund, Lehrstuhl VIII für Mathematik, Vogelpothsweg 87, 44221 Dortmund, Germany;
e-mail: davydov@math.uni-dortmund.de

Delvos, Franz J., Universität - Gesamthochschule - Siegen, Fachbereich Mathematik, Hölderlinstr. 3, 57068 Siegen, Germany

Dyn, Nira, Tel Aviv University, School of Mathematical Sciences, 69978 Tel Aviv, Israel;
e-mail: niradyn@math.tau.ac.il

Felten, Michael, Universität Dortmund, Lehrstuhl VIII für Mathematik, Vogelpothsweg 87, 44221 Dortmund, Germany;
e-mail: felten@math.uni-dortmund.de

Fredebeul, Christoph, Universität Dortmund, Lehrstuhl VIII für Mathematik,
Vogelpothsweg 87, 44221 Dortmund, Germany;
e-mail: christoph.fredebeul@math.uni-dortmund.de

Golitschek, Manfred v., Institut für Angewandte Mathematik und Statistik,
97074 Würzburg, Germany;
e-mail: goli@mathematik.uni-wuerzburg.de

Gonska, Heinz H., Gerhard-Mercator-Universität - GHS - Duisburg,
Fachbereich 11 / Informatik I, Lotharstr. 65, 47057 Duisburg, Germany;
e-mail: gonska@informatik.uni-duisburg.de

Grothmann, René., Katholische Universität Eichstätt, Ostenstraße, 85072
Eichstätt, Germany;
e-mail: grothm@ku-eichstaett.de

Haussmann, Werner, Gerhard-Mercator-Universität - GHS - Duisburg,
Fachbereich Mathematik, Lotharstr. 65, 47057 Duisburg, Germany;
e-mail: haussmann@math.uni-duisburg.de

Heilmann, Margaretha, Universität - Gesamthochschule - Wuppertal,
Gaußstr. 20, 42097 Wuppertal;
e-mail: heilmann@math.uni-wuppertal.de

Knoop, Hans - Bernd, Gerhard-Mercator-Universität - GHS - Duisburg,
Fachbereich Mathematik, Lotharstr. 65, 47057 Duisburg, Germany;
e-mail: knoop@math.uni-duisburg.de

Kubach, Peter, FernUniversität - GH - Hagen, Fachbereich Mathematik,
Lützowstr. 125, 58084 Hagen, Germany;
e-mail: peter.kubach@fernuni-hagen.de

Kunze, Andrea, Universität Dortmund, Lehrstuhl VIII für Mathematik,
Vogelpothsweg 87, 44221 Dortmund, Germany;

LeMéhauté, Alain, Department de Mathématiques, Université de Nantes, 2
rue de la Houssinière, 44072 Nantes Cedex, France;
e-mail: alm@math.univ-nantes.fr

Lenze, Burkhard, Fachhochschule Dortmund, Fachbereich Informatik, Postfach
105018, 44047 Dortmund, Germany;
e-mail: lenze@fh-dortmund.de

Leviatan, Dany, Tel Aviv University, School of Mathematical Sciences, 69978
Tel Aviv, Israel;
e-mail: leviatan@math.tau.ac.il

Light, Will, University of Leicester, Mathematics Departments, Leicester,
United Kingdom;
e-mail: pwl@mcs.le.ac.uk

Lubinsky, Doron S., Witwatersrand University Johannesburg, Mathematics
Department, Wits 2050, South Africa;
e-mail: 036dsl@cosmos.wits.ac.za (and *lubinsky@iweb.co.za*)

Lupaş, Alexandru, Universitatea - Lucian Blaga - Din Sibiu, Facultatea de
Stiinte, Catedra de Matematica, 2400 Sibiu, Romania;
e-mail: lupas@cs.sibiu.ro

Mache, Detlef H., Universität Dortmund, Lehrstuhl VIII für Mathematik, Vogelpothsweg 87, 44221 Dortmund, Germany;
e-mail: mache@math.uni-dortmund.de
z.Zt. Ludwig-Maximilians-Universität München, Lehrstuhl für Angewandte Mathematik, Numerische Analysis, Theresienstraße 39, 80333 München, Germany;
e-mail: mache@rz.mathematik.uni-muenchen.de

Mache, Petra, Technische Fachhochschule - Georg Agricola - für Rohstoff, Energie und Umwelt, Fachbereich Elektrotechnik, Herner Straße 45, 44787 Bochum, Germany;
e-mail: mache@math.uni-dortmund.de

Maier, Ulrike, Universität Dortmund, Lehrstuhl III für Mathematik, Vogelpothsweg 87, 44221 Dortmund, Germany;
e-mail: ulrike.maier@math.uni-dortmund.de

Maier, Volker, Universität Dortmund, Lehrstuhl VIII für Mathematik, Vogelpothsweg 87, 44221 Dortmund, Germany;
e-mail: volker.maier@math.uni-dortmund.de

Mastroianni, Guiseppe, Universita delle Basilicata, Dipartimento di Matematica, Via N. Sauro 85, 85100 Potenza, Italy;
e-mail: mastroianni@unibas.it

Mazure, Marie L., LMC - IMAG, Postbox 53, 38041 Grenoble Cedex 9, France;
e-mail: mazure@imag.fr

Müller, Manfred W., Universität Dortmund, Lehrstuhl VIII für Mathematik, Vogelpothsweg 87, 44221 Dortmund, Germany;
e-mail: mueller@math.uni-dortmund.de

Nürnberger, Günther, Universität Mannheim, Fakultät für Mathematik und Informatik, Lehrstuhl für Mathematik IV, 68131 Mannheim, Germany;
e-mail: nuern@math.uni-mannheim.de

Opfer, Gerhard, Universität Hamburg, Institut für Angewandte Mathematik, Bundesstr. 55, 20146 Hamburg, Germany;
e-mail: opfer@math.uni-hamburg.de

Petras, Knut, Universität (TH) Karlsruhe, Fakultät für Mathematik, Englerstr. 2, 76131 Karlsruhe, Germany;
e-mail: k.petras@tu-bs.de

Powell, M.J.D., DAMTP, University of Cambridge, Silver Street, Cambridge CB3 9Ew, England;
e-mail: m.j.d.powell@damtp.cam.ac.uk

Reimer, Manfred, Universität Dortmund, Lehrstuhl III für Mathematik, Vogelpothsweg 87, 44221 Dortmund, Germany;
e-mail: reimer@math.uni-dortmund.de

Sablonnière, Paul, Laboratoire LANS, IANS de Rennes, 20 Avenue des Buttes de Coesmes, 35043 Rennes cédex, France;
e-mail: sablonni@perceval.univ-rennes1.fr

Schaback, Robert, Universität Göttingen, Institut für Numerische und
Angewandte Mathematik, Lotzestr. 16-18, 37083 Göttingen, Germany;
e-mail: schaback@math.uni-goettingen.de

Scherer, Karl, Universität Bonn, Institut für Angewandte Mathematik,
Wegelerstr. 6, 53115 Bonn, Germany;
e-mail: scherer@iam.uni-bonn.de

Schmidt, Jochen W., Technische Universität Dresden, Institut für Numerische
Mathematik, 01062 Dresden, Germany;
e-mail: jschmidt@math.tu-dresden.de

Sendov, Bl., Academy of Sciences, Sofia 1126, Bulgaria;
e-mail: bsendov@argo.bas.bg

Skrzipek, Michael R., FernUniversität - GH - Hagen, Fachbereich
Mathematik, Lützowstr. 125, 58084 Hagen, Germany;
e-mail: michael.skrzipek@fernuni-hagen.de

Stöckler, Joachim, Universität Hohenheim, Institut für Angewandte
Mathematik und Statistik, 70599 Stuttgart (Hohenheim), Germany;
e-mail: stockler@uni-hohenheim.de

Szabados, Joszef, Mathematical Institute of the Hungarian Academy of
Sciences, Reáltanoda u. 13-15, 1053 Budapest, Hungary;
e-mail: szabados@math-inst.hu

Vértesi, Peter, Mathematical Institute of the Hungarian Academy of Sciences,
Reáltanoda u. 13-15, 1053 Budapest, Hungary;
e-mail: veter@math-inst.hu

Wenz, Jörg, Gerhard-Mercator-Universität - GHS -Duisburg, Fachbereich 11 /
Informatik I, Lotharstr. 65, 47057 Duisburg, Germany;
e-mail: wenz@informatik.uni-duisburg.de

Winckler, Marc, Universität Dortmund, Lehrstuhl VIII für Mathematik,
Vogelpothsweg 87, 44221 Dortmund, Germany;

Zeilfelder, Frank, Universität Mannheim, Fakultät für Mathematik und
Informatik, Lehrstuhl für Mathemaitk IV, 68131 Mannheim, Germany;
e-mail: zeilfelder@fourier.math.uni-mannheim.de

Zhou, Xinlong, Gerhard-Mercator-Universität - GHS - Duisburg, Fachbereich
Mathematik, Lotharstr. 65, 47057 Duisburg, Germany;
e-mail: xzhou@informatik.uni-duisburg.de